PROBABILITY AND STOCHASTIC PROCESSES

Work Examples

PROBABILITY AND STOCHASTIC PROCESSES

Work Examples

Odile Pons

French National Institute for Agricultural Research (INRA), France

NEW JERSEY • LONDON • SINGAPORE • BEIJING • SHANGHAI • HONG KONG • TAIPEI • CHENNAI • TOKYO

Published by

World Scientific Publishing Co. Pte. Ltd.

5 Toh Tuck Link, Singapore 596224

USA office: 27 Warren Street, Suite 401-402, Hackensack, NJ 07601

UK office: 57 Shelton Street, Covent Garden, London WC2H 9HE

British Library Cataloguing-in-Publication Data
A catalogue record for this book is available from the British Library.

PROBABILITY AND STOCHASTIC PROCESSES
Work Examples

ISBN 978-981-121-352-6 (hardcover)
ISBN 978-981-121-446-2 (paperback)
ISBN 978-981-121-353-3 (ebook for institutions)
ISBN 978-981-121-354-0 (ebook for individuals)

For any available supplementary material, please visit
https://www.worldscientific.com/worldscibooks/10.1142/11646#t=suppl

Desk Editor: Liu Yumeng

Preface

The book is intended for undergraduate students, it presents 330 exercises and problems with solutions covering the main subjects of a first course of probability including theories and applications. Each section begins with a short summary of basic notions of probability and more advanced results for processes used in the exercises.

Chapters 1–5 are related to the introduction of the probabilities for real discrete or continuous random variables. The questions are solved using simple mathematical proofs, Laplace and Fourier transforms are the classical methods to prove the convergence for sequences of random variables and to determine the distribution of transformed variables. Chapter 6 presents exercises about the weak convergence of variables and processes. Chapters 7 and 8 contain applications for discrete and time-continuous martingales and the last chapters concern other discrete and continuous processes.

Odile M.-T. Pons
November 2019

Contents

Chapter 1

Probability measures and spaces

1.1 Measure theory

A probability P on a measurable space $(\Omega, \mathcal{A})$ is an additive measure with values in $[0,1]$ such that for all measurable sets A and B

$$P(A \cap B) \leq P(A) + P(B)$$

and for every countable collection of nonintersecting sets $(A_i)_{i \geq 1}$ in $\mathcal{A}$

$$P(\cup_{i \geq 1} A_i) = \sum_{i \geq 1} P(A_i).$$

Let $(\Omega, \mathcal{A}, P)$ be a probability space and let $(A_n)_{n \geq 0}$ be a sequence of measurable sets, their superior and inferior limits are defined as

$$\limsup_n A_n = \cap_{n \to \infty} \cup_{m \geq n} A_m, \quad \liminf_n A_n = \cup_{n \to \infty} \cap_{m \geq n} A_m,$$

they satisfy the inequalities

$$P(\liminf_n A_n) \leq \liminf_n P(A_n) \leq \limsup_n P(A_n) \leq P(\limsup_n A_n)$$

and they converge if and only if $\limsup_n A_n = \liminf_n A_n$.

Lemma 1.1 (Borel–Cantelli's lemma). *A sequence of measurable sets $(A_n)_{n \geq 0}$ on $(\Omega, \mathcal{A})$ satisfies $\limsup_{n \to \infty} A_n = \emptyset$ a.s. if and only if $\sum_{n \leq 0} P(A_n)$ is finite. If the sets A_n are independent and $\sum_{n \leq 0} P(A_n)$ is infinite, then $\limsup_{n \to \infty} A_n = \Omega$ a.s.*

Exercise 1.1.1. Let $(a_n)_{n \geq 1}$ and $(b_n)_{n \geq 1}$ be real sequences and let $A_n =]a_n, b_n[$, determine $\liminf_n A_n$ and $\limsup_n A_n$.

Answer. For every integer n, we have

$$\cup_{m\geq n} A_m =]\inf_{m\geq n} a_n, \sup_{m\geq n} b_n[,$$
$$\cap_{m\geq n} A_m =]\sup_{m\geq n} a_n, \inf_{m\geq n} b_n[$$

then

$$\limsup_n A_n =]\liminf_n a_n, \limsup_{m\geq n} b_n[,$$
$$\liminf_n A_n =]\limsup_n a_n, \liminf_{m\geq n} b_n[.$$

Exercise 1.1.2. For measurable sequences $(A_n)_{n\geq 1}$ and $(B_n)_{n\geq 1}$, prove

$$\limsup_n (A_n \cup B_n) = \limsup_n A_n \cup \limsup_n B_n,$$
$$\liminf_n (A_n \cap B_n) = \liminf_n A_n \cap \liminf_n B_n.$$

Answer. For every integer n, the complementary sets of $\cup_{m\geq n} A_m$ and $\cap_{m\geq n} A_m$ are

$$(\cup_{m\geq n} A_m)^c = \cap_{m\geq n} A_m^c, \quad (\cap_{m\geq n} A_m)^c = \cup_{m\geq n} A_m^c$$

and $\cup_{m\geq n}(A_m \cup B_m) = (\cup_{m\geq n} A_m) \cup (\cup_{m\geq n} B_m)$ the first equality follows by limit as n tends to infinity, the second equality is deduced from the complementary sets.

Exercise 1.1.3. Prove that $P(\limsup_n A_n) \leq \lim_n \sum_{m\geq n} P(A_m)$ and $P(\liminf_n A_n^c) \leq \lim_n \sum_{m\geq n} P(A_n^c)$.

Answer. For every integer n, the increasing sequence $B_n = \cup_{m\geq n} A_m$ has the probability $P(\cup_{m\geq n} A_m) \leq \sum_{m\geq n} P(A_m)$ and the probability of $\lim_n B_n = \limsup_n A_n$ is

$$P(\limsup_n A_n) \leq \lim_n \sum_{m\geq n} P(A_m).$$

The complementary sets $B_n^c = \cap_{m\geq n} A_m^c$ are decreasing with probability $P(B_n^c) \leq P(A_n^c)$, and $B_n^c = \cup_{m\geq n}(B_m^c \setminus B_{m+1}^c)$ has the probability

$$P(B_n^c) = \sum_{m\geq n} P(B_m^c \setminus B_{m+1}^c) = \sum_{m\geq n} \{P(B_m^c) - P(B_{m+1}^c)\}$$

therefore

$$P(\liminf_n A_n^c) = \lim_n P(B_n^c) \leq \lim_n \sum_{m\geq n} P(A_m^c).$$

1.2 Integrability

A random variable X on a probability space (Ω, P) is defined as a measurable map from $(\Omega, \mathcal{A}, P)$ to a metric space $(S, \mathcal{B})$ endowed with the Borel sigma-algebra $\mathcal{B}$. On a probability space $(\Omega, \mathcal{A}, P)$, a variable X is integrable if

$$E(|X|) = \int_\Omega |X(\omega)|\, dP(\omega)$$

is finite. A sequence of variables $(X_n)_{n\geq 1}$ is equi-integrable if, as a tends to infinity

$$\max_{n\geq 1} E(|X_n| 1_{\{|X_n|>a\}}) \to 0.$$

For integrable variables X_n, $n \geq 1$, the following properties are equivalent
1. $(X_n)_{n\geq 1}$ is equi-integrable and X_n converges in probability to X as n tends to infinity,
2. X is integrable and $E|X_n - X|$ converges to zero as n tends to infinity.

Every sequence $(X_n)_{n\geq 1}$ bounded by an integrable variable X satisfies

$$\max_{n\geq 1} E(|X_n| 1_{\{|X_n|>a\}}) \leq \mathrm{E}(|X| 1_{\{|X|>a\}})$$

and it converges to zero as a tends to infinity, so $(X_n)_{n\geq 1}$ is equi-integrable. The equi-integrability of a sequence $(X_n)_{n\geq 1}$ implies that $\max_{n\geq 1} E(|X_n|)$ is finite.

Exercise 1.2.1. Prove that a sequence $(X_n)_{n\geq 1}$ such that $\max_{n\geq 1} E(|X_n|^2)$ is finite is equi-integrable.

Answer. For every $a > 0$

$$\begin{aligned}\max_{n\geq 1} E(|X_n| 1_{\{|X_n|>a\}}) &< a^{-1} \max_{n\geq 1} E(|X_n|^2 1_{\{|X_n|>a\}}) \\ &\leq a^{-1} \max_{n\geq 1} E(|X_n|^2)\end{aligned}$$

and it converges to zero as a tends to infinity.

Exercise 1.2.2. Let $(X_n)_{n\geq 1}$ and $(Y_n)_{n\geq 1}$ be equi-integrable sequences of variables, prove that $(X_n + Y_n)_{n\geq 1}$ is equi-integrable.

Answer. Let $X_n = X_n^+ - X_n^-$ and $Y_n = Y_n^+ - Y_n^-$ with $X_n^+ > 0$ and $Y_n^+ > 0$, $X_n^- < 0$ and $Y_n^- < O$, for all $a > 0$ and $b < 0$, the inequalities $|X_n| > a$ and $|Y_n| > b$ are equivalent to

$$X_n^+ > a,\ Y_n^+ > b,\ X_n^- < -a,\ Y_n^- < -b,$$

they imply $X_n^+ + Y_n^+ > a + b$, $X_n^- + Y_n^- < -(a+b)$ which is equivalent to $|X_n + Y_n| > a + b$ therefore

$$1_{\{|X_n+Y_n|>a+b\}} \leq 1_{\{|X_n|>a,|Y_n|>b\}}.$$

This implies

$$E(|X_n + Y_n|1_{\{|X_n+Y_n|>a+b\}}) \leq E(|X_n|1_{\{|X_n|>a\}}) + E(|Y_n|1_{\{|Y_n|>b\}}$$

and $(X_n + Y_n)_{n\geq 1}$ is equi-integrable.

Exercise 1.2.3. Let X and Y be integrable real variables and let $U = X \vee Y$ and $Z = X \wedge Y$, prove

$$E(U1_{\{U>a\}}) \leq E(X1_{\{X>a\}}) + E(Y1_{\{Y>a\}}),\ a \geq 0,$$
$$E(Z1_{\{Z<a\}}) \leq E(X1_{\{X<a\}}) + E(Y1_{\{Y<a\}}),\ a \leq 0,$$

deduce that if $(X_n)_{n\geq 1}$ is an equi-integrable sequence of variables then $n^{-1}E\max_{1\leq k\leq n}|X_n|$ converges to zero as a tends to infinity.

Answer. For every $a \geq 0$, $P(X > a \vee Y) \leq P(X > a)$ and the set $\{X > a \vee Y\} = \{U = X > a\}$ is included in $\{X > a\}$, in the same way the set $\{X < a \wedge Y\} = \{Z = X < a\}$ is included in $\{X < a\}$, it follows

$$\begin{aligned} E(U1_{\{U>a\}}) &= E(X1_{\{U=X>a\}}) + E(Y1_{\{U=Y>a\}}), \\ &\leq E(X1_{\{X>a\}}) + E(Y1_{\{Y>a\}}) \end{aligned}$$

and

$$\begin{aligned} E(Z1_{\{Z<a\}}) &= E(X1_{\{Z=X<a\}}) + E(Y1_{\{Z=Y<a\}}), \\ &\leq E(X1_{\{X<a\}}) + E(Y1_{\{Y<a\}}). \end{aligned}$$

Let $U_n = \max_{1\leq k\leq n}|X_n|$, then for every $a \geq 0$ the first inequality implies

$$\begin{aligned} E(n^{-1}U_n1_{\{U_n>na\}}) &\leq n^{-1}\sum_{1\leq k\leq n} E(|X_k|1_{\{|X_k|>na\}}) \\ &\leq \max_{1\leq k\leq n} E(|X_k|1_{\{|X_k|>na\}}) \end{aligned}$$

and by assumption this bound converges to zero as n tends to infinity.

For variables X_k with negative values, let

$$Z_n = \min_{1\le k\le n} X_k = -\max_{1\le k\le n}(-X_k),$$

by the second inequality for $a \le 0$, we have

$$\begin{aligned} E(n^{-1}Z_n 1_{\{Z_n<na\}}) &\le n^{-1}\sum_{1\le k\le n} E(X_k 1_{\{X_k<na\}}) \\ &\le \min_{1\le k\le n} E(X_k 1_{\{X_k<na\}}) \end{aligned}$$

and it converges to zero as n tends to infinity.

Exercise 1.2.4. Let $X = \sum_{i=1}^{\infty} x_i 1_{A_i}$, give conditions for a finite expectation of X.

Answer. The expectation of $X_n = \sum_{i=1}^{n} x_i 1_{A_i}$ is $E(X_n) = \sum_{i=1}^{n} x_i P(A_i)$ and a necessary condition for its convergence to $E(X) = \sum_{i=1}^{\infty} x_i P(A_i)$ is the convergence to zero of

$$E(X) - E(X_n) = \sum_{i=n+1}^{\infty} x_i P(A_i).$$

A sufficient condition is the convergence of $\sum_{i=1}^{n} |x_i| P(A_i)$ to a finite limit. Cauchy's necessary and sufficient condition is the convergence to zero of $E(X_n) - E(X_m) = \sum_{i=m+1}^{n} x_i P(A_i)$, as m and n tend to infinity.

1.3 Convergences of random variables

A sequence of variables $(X_n)_{n\ge1}$ converges almost surely (a.s.) if $\limsup_n X_n = \liminf_n X_n$ a.s., then $X = \lim_{\text{a.s.}} X_n$ is defined in the equivalence class of variables such that $P(X_1 \ne X_2) = 0$.

Cauchy criterion: A sequence of a.s. finite variables $(X_n)_{n\ge1}$ converges a.s. to an a.s. finite variable X, as n tends to infinity, if and only if the sequence $(X_m - X_n)_{m,n\ge1}$ converges a.s. to zero as m and n tend to infinity. If there exists a sequence $(\varepsilon_n)_{n\ge1}$ such that $\sum_{n\ge1} \varepsilon_n$ is finite and

$$\sum_{n\ge1} P(|X_n - X| > \varepsilon) < \infty$$

then X_n converges a.s. to X as n tend to infinity.

The upper and lower limits of a sequence of random variables $(X_n)_{n\ge1}$ are defined as

$$\limsup_n X_n = \lim_n \sup_{m\ge n} X_m, \quad \liminf_n X_n = \lim_n \inf_{m\ge n} X_m,$$

X_n converges a.s. if $\limsup_n X_n = \liminf_n X_n$ a.s.

Lemma 1.2 (Fatou–Lebesgue's lemma). *If there exists Y in L^1 such that $Y \le X_n$ a.s. for every n then $E(\liminf_n X_n) \le \liminf_n E(X_n)$. If there exists Z in L^1 such that $X_n \le Z$ a.s. for every n then $E(\limsup_n X_n) \ge \limsup_n E(X_n)$.*

A sequence of a.s. finite variables $(X_n)_{n\ge 1}$ converges in probability to an a.s. finite variable X if for every $\varepsilon > 0$

$$P(|X_n - X| > \varepsilon) \to 0$$

as n tends to infinity. The a.s. convergence of $|X_n - X|$ to zero implies the convergence of $E|X_n - X|$ to zero which implies the convergence in probability of X_n to X by the inequality $P(|X_n - X| > \varepsilon) \le \varepsilon^{-1} E|X_n - X|$ for every $\varepsilon > 0$.

Proposition 1.1 (Strong law of large numbers). *Let $X_1, \dots, X_n$ be independent and identically distributed variables, then $n^{-1}\sum_{i=1}^n X_i$ converges a.s. to $E(X_1)$ if and only if $E|X_1|$ is finite.*

Exercise 1.3.1. Let X_i, $i \ge 1$, be independent and identically distributed variables with values in $\mathbb{R}_+$, prove that $\limsup_{n\to\infty} \max_{i=1,\dots,n} X_i$ is a.s. infinite.

Answer. Let F the distribution function of the variables on $\mathbb{R}_+$ and let $M_n = \max_{i=1,\dots,n} X_i$, for every finite $a > 0$ we have $F(a) < 1$, then for every integer n

$$P(\liminf_n M_n \le a) \le \lim_n P(M_n \le a) = \lim_n F^n(a) = 0,$$
$$P(\limsup_n M_n \le a) \ge \lim_n P(M_n \le a) = \lim_n F^n(a) = 0,$$

then $\liminf_n M_n = \limsup_n M_n$ a.s. and

$$P(\lim_n M_n < \infty) \le \lim_{n,a\to\infty} P(M_n \le a) = 0.$$

Exercise 1.3.2. For an integrable variables, let $d(X,Y) = E(|X - Y|)$ and

$$\rho(X) = E\frac{|X|}{1+|X|},$$

prove that $\rho(X+Y) \le \rho(X) + \rho(Y)$ and the equivalence of the convergence to zero for $\rho(X_n - Y_n)$ and $d(X_n, Y_n)$.

Answer. For the inequality $\rho(X) = E\frac{|X|}{1+|X|}$, it is sufficient to prove that

$$(1+|X|)(1+|Y|)|X+Y| \leq (1+|X+Y|)\{|X|)(1+|Y|) + |Y|)(1+|X|)\},$$
$$|X+Y| \leq |X|\,|Y|\,|X+Y| + |X| + |Y| + 2|X|\,|Y|,$$

and this is true by the triangular inequality $|X+Y| \leq |X| + |Y|$. There is equivalence of $\rho(X_n - Y_n) = 0$ and $d(X_n, Y_n) = 0$. For every positive $a < 1$, the inequality $\rho(X_n - Y_n) \leq a$ is equivalent to $d(X_n, Y_n) \leq a(1-a)$, then $\rho(X_n - Y_n)$ converges to zero as a tends to zero which implies the convergence to zero of $d(X_n, Y_n)$, and the reverse is true.

1.4 Independence and conditional expectation

On a probability space $(\Omega, \mathcal{A}, P)$, measurable sets B_1 and B_2 are independent if

$$P(B_1 \cap B_2) = P(B_1)P(B_2)$$

and the probability of B_1 conditionally on B_2 is

$$P(B_1 \mid B_2) = \frac{P(B_1 \cap B_2)}{P(B_2)},$$

and the independence of B_1 and B_2 is equivalent to $P(B_1 \mid B_2) = P(B_1)$. Integrable variables X and Y are independent if

$$E(XY) = E(X)\,E(Y).$$

Subsigma algebras $\mathcal{B}_1$ and $\mathcal{B}_2$ of $\mathcal{A}$ are independent if for all B_1 in $\mathcal{B}_1$ and B_2 in $\mathcal{B}_2$, $P(B_1 \cap B_2) = P(B_1)P(B_2)$, equivalently all integrable variables X $\mathcal{B}_1$-measurable and Y $\mathcal{B}_2$-measurable are independent.

The expectation of a variable X conditionally on a discrete variable Y with values in $(y_i)_{i \in I}$, where I is a countable set, is

$$E(X \mid Y) = \sum_{i \in I} \frac{E(X1_{\{Y=y_i\}})}{P(y_i)} 1_{\{Y=y_i\}}.$$

For variables X and Y with values in $\mathbb{R}^p$, the expectation of X conditionally on Y is

$$E(X \mid Y) = \frac{\int_{\mathbb{R}^p} x\, P_{XY}(dx, Y)}{\int_{\mathbb{R}^p} P_{XY}(dx, Y)}$$

and it reduces to $E(X)$ if and only X and Y are independent.

The conditional expectation of X with respect to a sigma algebra $\mathcal{B}$ is a random variable $\mathcal{B}$-measurable such that for every B in $\mathcal{B}$

$$\int_B X\,dP = \frac{\int_B E(X \mid \mathcal{B})\,dP}{P(B)}$$

equivalently $E(XY) = E\{E(X \mid \mathcal{B})Y\}$ for every $\mathcal{B}$-measurable variables Y. The expectation $E(X \mid Y)$ of a variable X conditionally on Y satisfies

$$E[\{X - E(X \mid Y)\}Y] = 0.$$

The expectation $E(X \mid \mathcal{B})$ of a variable X conditionally on a sigma-algebra $\mathcal{B}$ generated by a countable partition $(B_i)_{i\in I}$ of Ω is

$$E^{\mathcal{B}}(X) = \sum_{i\in I} \frac{1}{P(B_i)} E(X1_{B_i})1_{B_i}$$

it satisfies $E[\{X - E^{\mathcal{B}}(X)\}1_B] = 0$ for every $\mathcal{B}$-measurable set B. If $\mathcal{B}_1$ is included in $\mathcal{B}_2$ then

$$E\{E(X \mid \mathcal{B}_2) \mid \mathcal{B}_1\} = E(X \mid \mathcal{B}_1)$$

and a $\mathcal{B}$-measurable variable Y is such that

$$E(XY \mid \mathcal{B}) = YE(X \mid \mathcal{B}).$$

The variance of a variable X dependent of a variable Y satisfies

$$\mathrm{Var}(X) = E\{\mathrm{Var}(X \mid Y)\} + \mathrm{Var}\{E(X \mid Y)\}. \tag{1.1}$$

The conditional expectation is linear, for all variables X_1 and X_2 and scalars a and b

$$E(aX_1 + bX_2 \mid Y) = aE(X_1 \mid Y) + bE(X_2 \mid Y).$$

The expectation of a variable Y conditionally on a vector $X = (X_1, X_2)$ of independent variables is the sum of the expectations conditionally on the marginal variables

$$E(Y \mid X_1, X_2) = E(Y \mid X_1) + E(Y \mid X_2)$$

and the sigma-algebra $\mathcal{B}_X$ generated by the variable X is the direct sum of the sigma-algebras generated by the marginal variables $\mathcal{B}_X = \mathcal{B}_{X_1} \oplus \mathcal{B}_{X_2}$.

Let $(X_n)_{n\geq 0}$ be a sequence of real random variables on $(\Omega, \mathcal{A}, P)$, the sigma-algebra generated by X_n is generated by the sets $\{X_n \leq x\}$, x in $\bar{\mathbb{R}}$. The variables are independent if and only if for every finite subset $\mathcal{K}$ of $\mathbb{N}$ and every sequence $(x_k)_{k\in\mathcal{K}}$ of $\bar{\mathbb{R}}$

$$P(\cap_{k\in\mathcal{K}}\{X_k \leq x_k\}) = \prod_{k\in\mathcal{K}} P(X_k \leq x_k).$$

The independence of the sequence $(X_n)_{n\geq 0}$ is equivalent to the equality $E\prod_{k\in\mathcal{K}} f_k(X_k) = \prod_{k\in\mathcal{K}} Ef_k(X_k)$, for every sequence of measurable bounded functions $f_k : \mathbb{R} \mapsto \mathbb{R}$ and for every finite subset $\mathcal{K}$ of $\mathbb{N}$.

Let P_n and Q_n be probability measures on a measurable product space $(\Omega^{\otimes n}, \mathcal{A}_n)$, with marginal probabilities P_{nj} and respectively Q_{nj}, and let

$$L_n = \frac{dP_n}{dQ_n} = \prod_{i=1}^{n} \frac{dP_{nj}}{dQ_{nj}}.$$

The sequence $(Q_n)_n$ is contiguous to $(P_n)_n$ if for every A_n in $\mathcal{A}_n$, the convergence of $Q_n(A_n)$ to zero implies the convergence of $P_n(A_n)$ to zero. Equivalently, every sequence of variables $(X_n)_n$ on $(\Omega^{\otimes n}, \mathcal{A}_n)$ which converges in probability to zero under $(Q_n)_n$, converges in probability to zero under $(P_n)_n$.

Exercise 1.4.1. Extend Fatou–Lebesgue's Lemma to conditional expectations of variables X_n with respect to a subsigma-algebra $\mathcal{B}$ of $\mathcal{A}$.

Answer. If there exists Y in L^1 such that $Y \leq X_n$ a.s. for every n then the equality $E(X_n 1_B) = E\{1_B E^{\mathcal{B}}(X_n)\}$ for every B in $\mathcal{B}$ implies

$$\begin{aligned} E\{E^{\mathcal{B}}(\liminf_n X_n)1_B\} &= E(\liminf_n X_n 1_B) \\ &\leq \liminf_n E(X_n 1_B) \\ &= \liminf_n E\{1_B E^{\mathcal{B}}(X_n)\} \end{aligned}$$

and $E^{\mathcal{B}}(\liminf_n X_n) \leq \liminf_n E^{\mathcal{B}}(X_n)$. If there exists Z in L^1 such that $X_n \leq Z$ a.s. for every n then for every B in $\mathcal{B}$

$$\begin{aligned} E\{E^{\mathcal{B}}(\limsup_n X_n)1_B\} &= E(\limsup_n X_n 1_B) \\ &\geq \limsup_n E(X_n 1_B) \\ &= \limsup_n E\{1_B E^{\mathcal{B}}(X_n)\} \end{aligned}$$

and $E^{\mathcal{B}}(\limsup_n X_n) \geq \limsup_n E^{\mathcal{B}}(X_n)$.

Exercise 1.4.2. Let $(\mathcal{B}_n)_n$ be an increasing sequence of subsigma-algebras of $\mathcal{A}$ converging to $\mathcal{B}$ in $\mathcal{A}$ and let $P^{\mathcal{B}}$ be the restriction of the probability P to $\mathcal{B}$, prove that for every variable X of $L^1(P^{\mathcal{B}})$, $E^{\mathcal{B}_n}(X)$ converges to $E^{\mathcal{B}}(X)$.

Answer. As $(\mathcal{B}_n)_n$ is generated by the finite unions of finite intersections of subsets of $\mathcal{A}$, for every increasing sequence $(B_n)_n$ in $\mathcal{B}_n$ converging to $B = \cup_n B_n$ in $\mathcal{B}$, the equality $\lim_n E(X1_{B_n}) = E(X1_B)$ is equivalent to

$$\lim_n E\{E^{\mathcal{B}_n}(X)1_{B_n}\} = E\{E^{\mathcal{B}}(X)1_B\}$$

and by monotonicity

$$E\{\lim_n E^{\mathcal{B}_n}(X)1_{B_n}\} = \mathrm{E}\{E^{\mathcal{B}}(X)1_B\}$$

which proves the convergence of $E^{\mathcal{B}_n}(X)$ to $E^{\mathcal{B}}(X)$.

Exercise 1.4.3. Let $X_1, \ldots, X_n$ be independent and identically distributed centered variables and let $S_n = \sum_{i=1}^n X_i$, calculate the conditional expectations $E(S_{n+1} \mid S_n)$ and $E(S_{n+1} \mid S_1, \ldots, S_n)$.

Answer. The sigma-algebras generated by $X_1, \ldots, X_n$ and by $S_1, \ldots, S_n$ are identical. The independence of the variables implies that X_{n+1} is independent of $S_1, \ldots, S_n$ then

$$E(X_{n+1} \mid S_1, \ldots, S_n) = E(X_{n+1}) = 0$$

and $E(S_{n+1} \mid S_1, \ldots, S_n) = E(S_n + X_{n+1} \mid S_n) = S_n$.

Exercise 1.4.4. Let X and Y be dependent square integrable variables, calculate the conditional expectation $E(X \mid Y)$ and $\inf_{U \in H^Y} \|X - U\|^2_{H^Y}$, for the Hilbert space $H^Y = L^2(Y)$ generated by a vector Y.

Answer. The conditional expectation $E(X \mid Y)$ belongs to the sigma-algebra generated by Y and $E(X \mid Y) = a^T Y + b$, then

$$E(X) = E\{E(X \mid Y)\} = a^T E(Y) + b,$$

equivalently $E\{X - E(X) \mid Y - E(Y)\} = a^T\{Y - E(Y)\}$. This entails

$$E[\{X - E(X)\} \otimes \{Y - E(Y)\}] = a^T \Sigma_Y,$$

therefore

$$a = \Sigma_Y^{-1} E[\{Y - E(Y)\} \otimes \{X - E(X)\}]$$

and $b = E(X) - a^T E(Y)$.

The squared norm

$$\|X - U\|^2_{H^Y} = E\{XX^T - 2UX^T + UU^T \mid Y\}$$

is minimum as $E\{X \mid Y\} = E\{U \mid Y\}$ and

$$\inf_{U \in H^Y} \|X - U\|_{H^Y}^2 = \|X - E\{X \mid Y\}\|_{H^Y}^2$$
$$= E\{XX^T \mid Y\} - \{E(X \mid Y)^2\},$$

it is the variance matrix of X conditionally on Y.

Exercise 1.4.5. Let P and $Q << P$ be probabilities on a measurable space $(\Omega, \mathcal{A})$ such that $dQ = X\,dP$, determine the conditional expectation $E_Q^{\mathcal{B}}$ on $(\Omega, \mathcal{A}, Q)$ in terms of the expectation conditionally on $\mathcal{B}$, for $\mathcal{B}$ in $\mathcal{A}$ and give conditions for the equality $E_Q^{\mathcal{B}}(Y) = E_P^{\mathcal{B}}(Y)$ a.s. for every positive $\mathcal{A}$ measurable variable Y.

Answer. For every positive variable Y on $(\Omega, \mathcal{A}, Q)$ and every $\mathcal{B}$-measurable set B, we have

$$E_Q(Y \mid B) = \frac{1_B}{\int_B dQ} \int_B Y\,dQ$$
$$= \frac{1_B}{\int_B X\,dP} \int_B YX\,dP.$$

Let $Y = 1_A$, for a measurable set A such that $A \cap B$ is not empty and $A \neq B$, the equality $E_Q^{\mathcal{B}}(Y) = E_P^{\mathcal{B}}(Y)$ a.s. is equivalent to

$$\frac{1}{\int_B X\,dP} \int_{A \cap B} X\,dP = \frac{P(A \cap B)}{P(B)} \text{ a.s.,}$$
$$\frac{1}{P(A \cap B)} \int_{A \cap B} X\,dP = \frac{\int_B X\,dP}{P(B)} \text{ a.s.}$$

this is true if $X = 1$ a.s. and $P = Q$ a.s. on $\mathcal{B}$, or if $X = 0$ a.s. and $Q = 0$ a.s. on $\mathcal{B}$. The equality is also true for $A = B$ and therefore if Y is $\mathcal{B}$ measurable.

Exercise 1.4.6. Let $\mathcal{B}$ a subsigma-algebra of $\mathcal{A}$, prove that the independence of $(X_n)_{n \geq 0}$ conditionally on $\mathcal{B}$ is equivalent to

$$E^{\mathcal{B}} \prod_{k \in \mathcal{K}} f_k(X_k) = \prod_{k \in \mathcal{K}} E^{\mathcal{B}}\{f_k(X_k)\}$$

for every sequence of measurable functions $f_k : \mathbb{R} \mapsto \mathbb{R}$ and for every finite subset $\mathcal{K}$ of $\mathbb{N}$.

Answer. For every B in $\mathcal{B}$

$$\begin{aligned} E\{E^{\mathcal{B}}\prod_{k\in\mathcal{K}} f_k(X_k)1_B\} &= E\{\prod_{k\in\mathcal{K}} f_k(X_k)1_B\} \\ &= \prod_{k\in\mathcal{K}} E\{f_k(X_k)1_B\} = \prod_{k\in\mathcal{K}} E[\{E^{\mathcal{B}} f_k(X_k)\}1_B] \\ &= E[\prod_{k\in\mathcal{K}}\{E^{\mathcal{B}} f_k(X_k)\}1_B]. \end{aligned}$$

Exercise 1.4.7. Square integrable real variables $X_1,\ldots,X_n$ are linearly dependent if there exist constants $c_1,\ldots,c_n$ such that $\sum_{i=1}^n c_iX_i = 0$ almost surely. Prove that $X_1,\ldots,X_n$ are linearly independent if and only if the determinant of their covariance matrix Σ is strictly positive.

Answer. If the determinant of Σ is zero, there exist constants $a_1,\ldots,a_n$ such that $\sum_{i=1}^n a_i\text{Cov}(X_i,X_j) = 0$. Let $Y = \sum_{i=1}^n a_iX_i$, the covariance of Y and X_j is then $\sum_{i=1}^n a_i\text{Cov}(X_i,X_j) = 0$ and the variance of Y is zero hence $Y = 0$ a.s. Reversely if $X_1,\ldots,X_n$ are linearly dependent, the equality $Y = \sum_{i=1}^n c_iX_i = 0$ a.s. entails $\text{Cov}(X_i,Y) = 0$ for every X_i and the determinant of Σ is zero.

Exercise 1.4.8. For dependent variables X, Y, Z, prove

$$\text{Cov}(X,Y) = E\{\text{Cov}(X,Y \mid Z)\} + \text{Cov}\{E(X \mid Z), E(Y \mid Z)\}.$$

Answer. We have

$$\begin{aligned} E\{\text{Cov}(X,Y \mid Z)\} &= E\{E(XY \mid Z)\} - E\{E(X \mid Z)E(Y \mid Z)\} \\ &= E(XY) - E\{E(X \mid Z)E(Y \mid Z)\} \\ \text{Cov}\{E(X \mid Z), E(Y \mid Z)\} &= E\{E(X \mid Z)E(Y \mid Z)\} - E(X)\,E(Y) \end{aligned}$$

and the result follows.

1.5 L^p spaces

Let $L^p = \{X : \|X\|_p = \{E(|X^p)|\}^{\frac{1}{p}} < \infty\}$ with the norms

$$\begin{aligned} \|X\|_p &= \{|X|^p\}^{\frac{1}{p}},\ 1 \le p < \infty, \\ \|X\|_\infty &= \sup\{x : P(|X| > x) > 0\}, \end{aligned}$$

for all variables X and Y, the norms have the properties

$$\begin{aligned} \|X+Y\|_p &\le \|X\|_p + \|Y\|_p,\ p \ge 1, \\ \|\lambda X\|_p &= \lambda\|X\|_p,\ \lambda \le 0, p > 0 \end{aligned}$$

and for $1 \leq p \leq q \leq \infty$

$$0 \leq \|X\|_1 \leq \|X\|_p \leq \|X\|_q \leq \|X\|_\infty, \tag{1.2}$$

which implies $L^\infty \subset L^q \subset L^p \subset L^1$.

The Cauchy–Schwarz inequality for variables X and Y of L^2 is

$$E(XY) \leq \|X\|_2 \|Y\|_2,$$

with equality if and only if X and Y are a.s. proportional. It generalizes as Schwarz's inequality for variables X of L^p and Y of L^q, $1 \leq p, q \leq \infty$ and $p^{-1} + q^{-1}$,

$$\|X\|_1 \leq \|X\|_p \|X\|_q$$

and Hölder's inequality

$$\|X\|_r \leq \|X\|_p \|X\|_q,$$

for $r^{-1} = p^{-1} + q^{-1}$.

A sequence of variables $(X_n)_{n \geq 1}$ converges in L^p if and only if it is a Cauchy sequence in L^p, i.e., $\|X_n - X_m\|_p$ converges to zero as n and m tend to infinity. The convergence of X_n to X in L^p is equivalent to its convergence in probability and the convergence of the norms $\|X_n\|_p$ to $\|X\|_p$ (Vitali's theorem).

Exercise 1.5.1. Prove the inequality $\|X\|_p \leq \|X\|_q$ for $1 \leq p \leq q < \infty$.

Answer. For all integers $p \leq q$ and by convexity we have

$$E(|X|^q) = E\{(|X|^p)^{\frac{q}{p}}\} \geq \{E(|X|^p)\}^{\frac{q}{p}}$$

hence $\|X\|_p \leq \|X\|_q$.

Exercise 1.5.2. Prove that for all real variables X and Y

$$E\{(X+Y)^p\} \leq 2^{p-1} E\{(X)^p\} + E\{(Y)^p\}.$$

Answer. An expansion of $(X+Y)^p$ yields

$$\begin{aligned}
E\{(X+Y)^p\} &= \sum_{k=0}^{p} \binom{k}{p} E(X^k Y^{p-k}) \\
&\leq E(X^p) + E(Y^p) + \sum_{k=1}^{p-1} \binom{p}{k} \Big\{ \frac{kE(X^p)}{p} + \frac{(p-k)E(Y^p)}{p} \Big\} \\
&= E(X^p)\Big\{1 + \sum_{k=1}^{p-1} \binom{p-1}{k-1}\Big\} + E(Y^p)\Big\{1 + \sum_{k=1}^{p-1} \binom{p-1}{k}\Big\} \\
&= 2^{p-1}\{E(X^p) + E(Y^p)\}.
\end{aligned}$$

Exercise 1.5.3. Let X be a random variable of $L_2(\mathbb{R})$ such that $EX = m$ and $varX > 0$. For every $a > 0$ such that $a^2 \leq varX$ prove that

$$P(|X - m| > a) \geq \frac{a^2}{varX}.$$

Answer. By the Cauchy–Schwarz inequality we have

$$a < E\{|X - m|1_{\{|X-m|>a\}}\} \leq \{(varX)P(|X - m| > a)\}^{\frac{1}{2}}.$$

Exercise 1.5.4. Let $(X_i)_{i\geq 1}$ be a sequence of independent and identically distributed centered variables of L^{2k} with variance $\sigma^2 > 0$ and let $S_n = \sum_{i=1}^n X_i$, prove that for every integer $k \geq 1$

$$E(S_n^{2k}) \leq c_k\{\sum_{i=1}^n E^{\frac{1}{k}}(X_i^{2k})\}^k.$$

Answer. For every integer $k \geq 1$, the inequality is an equality with $c_1 = 1$ for $n = 1$. Assuming that it true for n, it follows from the inequality $(a+b)^2 \leq 2(a^2 + b^2)$ that

$$(a^{\frac{1}{2k}} + b^{\frac{1}{2k}})^{2k} \leq 2^k(a^{\frac{1}{k}} + b^{\frac{1}{k}})^k$$

and for S_{n+1}

$$\begin{aligned}
E(S_{n+1}^{2k}) &\leq (\|S_n\|_{L^{2k}} + \|X_{n+1}\|_{L^{2k}})^{2k} \\
&\leq 2^k\{E^{\frac{1}{k}}(S_n^{2k}) + E^{\frac{1}{k}}(X_{n+1}^{2k}\}^k \\
&\leq 2^k\{c_k \sum_{i=1}^n E^{\frac{1}{k}}(X_i^{2k}) + E^{\frac{1}{k}}(X_{n+1}^{2k})\}^k \\
&= 2^k c_k\{\sum_{i=1}^{n+1} E^{\frac{1}{k}}(X_i^{2k})\}^k,
\end{aligned}$$

the inequality is satisfied with the constant $c_{k+1} = 2^k c_k$.

Exercise 1.5.5. Let X in L^p, $p > 1$, prove that $\|X\|_p = \sup_{\|Y\|_q=1} E(XY)$ where $p^{-1} + q^{-1} = 1$.

Answer. The inequality $\sup_{\|Y\|_q=1} E(XY) \leq \|X\|_p$ is a consequence of Hölder's inequality. Let $U = \|X\|_p^{-\frac{p}{q}}|X|^{p-1}$, its L_q-norm is

$$\|U\|_q^q = \|X\|_p^{-p}E(|X|^{(p-1)q}) = \|X\|_p^{-p}\|X\|_p^p = 1$$

where $(p-1)q = p$, and

$$\|X\|_p = \|X\|_p^{-\frac{p}{q}} E(|X|^p) = E(|XU|) \\ \leq \sup_{\|Y\|_q=1} E(XY).$$

Exercise 1.5.6. Let $(X_n)_{n\geq 1}$ and $(Y_n)_{n\geq 1}$ be sequences of $\mathcal{F}_n$-measurable random variables on probability spaces $\Omega, \mathcal{F}_n, P)$, such that $E\Big(\sum_{n\geq 1} |X_n|^p\Big)^{\frac{1}{p}}$ is finite for $p > 1$, prove that

$$E\|(X_n)_n\|_{l_p} = E\Big\{\Big(\sum_{n\geq 1} |X_n|^p\Big)^{\frac{1}{p}}\Big\} = \sup_{E\|(Y_n)_n\|_{l_q}=1} E\Big\{\sum_{n\geq 1} X_n Y_n\Big\}.$$

Answer. Let $U = \sum_{n\geq 0} X_n 1_{]n,n+1]}$ and $V = \sum_{n\geq 0} Y_n 1_{]n,n+1]}$ in L^p, $p > 1$, the result is a consequence of Exercise 1.5.5.

Exercise 1.5.7. For $p \geq 1$ finite and X in $L^p(\Omega, \mathcal{A}, P)$, prove that the family $\{[E^{\mathcal{B}}(X)]^p : \mathcal{B} \subset \mathcal{A}\}$ is equi-integrable. For every sequence $(\mathcal{B}_n)_n$ of sigma-algebras in $\mathcal{A}$ such that $E^{\mathcal{B}_n}(1_A)$ converges in probability for every A in $\mathcal{A}$, prove that $E^{\mathcal{B}_n}(X)$ converges in $L^p(\Omega, \mathcal{A}, P)$.

Answer. For every X in $L^p(\Omega, \mathcal{A}, P)$ we have $\lim_{a\to\infty} E(|X|^p 1_{\{|X|^p > a\}}) = 0$ and for every sub-sigma-algebra $\mathcal{B}$ of $\mathcal{A}$

$$E(|X|^p) = E\{E^{\mathcal{B}}(|X|^p)\} \geq E\{|E^{\mathcal{B}}(X)|^p\}$$

and the variable $\sup_{\mathcal{B}\subset\mathcal{A}} E^{\mathcal{B}}(X)$ belongs to $L^p(\Omega, \mathcal{A}, P)$. It follows that

$$\lim_{a\to\infty} \sup_{\mathcal{B}\subset\mathcal{A}} E(|E^{\mathcal{B}}(X)|^p 1_{\{|E^{\mathcal{B}}(X)|^p > a\}}) = 0.$$

The convergence of $E^{\mathcal{B}_n}(X)$ in $L^p(\Omega, \mathcal{A}, P)$ to a limit X_∞ is deduced from its equi-integrability and from its convergence in probability to X_∞ which is deduced from the convergence in probability of $E^{\mathcal{B}_n}(1_A)$ for every A in $\mathcal{A}$ and by the approximation of X by a step function $X_n = \sum_{i=1}^n x_i 1_{A_i}$.

1.6 Inequalities

A real function φ is convex if

$$\varphi((1-t)x + ty) \leq (1-t)\varphi(x) + t\varphi(y)$$

for every t in $[0,1]$ and for all real x and y, the property is generalized for every integrable variable X by Jensen's inequality

$$\varphi(E(X)) \leq E\{\varphi(X)\}.$$

The space $L^2(\mathbb{R}^m) = \{X : \Omega \mapsto \mathbb{R}^m, \|X\|_2 < \infty\}$ endowed with the scalar product $< X, Y > = E(X^T Y)$ is a Hilbert space. The Cauchy–Schwarz inequality for random variables X and Y of $L^2(\mathbb{R}^m)$ is

$$< X, Y > \leq \|X\|_2 \|Y\|_2$$

with equality if and only if X and Y are a.s. proportional.

The Cauchy–Schwarz inequality extends to the space of complex variables $L^2(\mathbb{C}^m) = \{X : \Omega \mapsto \mathbb{C}^m, \|X\|_2 < \infty\}$ endowed with the scalar product $< X, Y > = E(X^T \bar{Y})$ as $E(X\bar{Y}) \leq \|X\|_2 \|Y\|_2$.

By the Bienaymé–Chebychev inequality, for every variable X of $L^2(\mathbb{R})$ and for every $a > 0$, we have

$$P(|X| > a) \leq a^{-2} E|X|^2.$$

For a variable X in $L^2(\mathbb{R})$ and for every $a > 0$, then

$$P(|X - E(X)| > a) \leq a^{-2} \mathrm{Var}(X). \tag{1.3}$$

In the L^p spaces, $p > 1$, Hölder's inequality extends the Cauchy–Schwarz inequality for conjugate r, p and q such that $p^{-1} + q^{-1} = r^{-1}$

$$\{E(|X^T \bar{Y}|^r)\}^{\frac{1}{r}} \leq \|X\|_p \|Y\|_q,$$

these inequalities are proved by the convexity inequality.

For $p = 1$ and $p' = \infty$, Hölder's inequality is

$$|E(X^T \bar{Y}| \leq \|X\|_1 \|Y\|_\infty.$$

Exercise 1.6.1. Let X be a variable of $L^p(\mathbb{R})$, $p > 0$, prove

$$E(|X|^{-p})) \geq \{E(|X|^p))\}^{-1}.$$

Answer. The Cauchy–Schwarz inequality implies

$$1 = E(|X|^{-\frac{p}{2}} |X|^{\frac{p}{2}}) \leq \{E(|X|^{-p}))\}^{\frac{1}{2}} \{E(|X|^p))\}^{\frac{1}{2}}$$

and the inequality follows.

Exercise 1.6.2. Prove convexity Jensen's inequality.

Answer. The definition of a convex function is extended recursively to

$$\varphi\Big(\sum_{i=1}^n t_i x_i\Big) \geq \sum_{i=1}^n t_i \varphi(x_i)$$

for every integer n and for all real sequences $(t_i)_{i=1,\dots,n}$ in $]0,1[$ such that $\sum_{i=1}^n t_i = 1$ and $(x_i)_{i=1,\dots,n}$. Let X be an integrable variable on a probability space $(\Omega, \mathcal{A}, P)$, let $(B_i)_{i=1,\dots,n}$ be an increasing measurable partition of Ω such that for every n, $X_n = \sum_{i=1}^n x_i 1_{B_i}$ converges a.s. to X and $\sum_{i=1}^n 1_{B_i} = 1$ a.s. then $E(X_n) = \sum_{i=1}^n x_i P(B_i)$ converges to $E(X)$ and on B_i, $\varphi(X_n) = \varphi(x_i)$ therefore

$$\varphi(X_n) = \sum_{i=1}^n \varphi(x_i) 1_{B_i},$$

$$E\varphi(X_n) = \sum_{i=1}^n \varphi(x_i) P(B_i),$$

it converges to $E\varphi(X)$. Furthermore, by convexity

$$\varphi(EX_n) = \varphi\Big\{\sum_{i=1}^n x_i P(B_i)\Big\} \\ \leq \sum_{i=1}^n \varphi(x_i) P(B_i),$$

the result follows.

Exercise 1.6.3. Let $X \geq 0$ in L^p, prove the inequality

$$\sum_{n\geq 1} n^{p-1} P([X] \geq n) \leq E(X^p) \leq \sum_{n\geq 0} (n+2)^{p-1} P([X] \geq n).$$

Answer. The expectation of an integer variable N is

$$EN = \sum_{n\geq 1} nP(N = n) = \sum_{n\geq 1} n\{P(N > n-1) - P(N > n)\} \\ = \sum_{n\geq 1} P(N \geq n),$$

by the same arguments

$$EN^p = \sum_{n\geq 1} n^p P(N = n) = \sum_{n\geq 1} n^{p-1} P(N \geq n),$$

this equality applies to the integer part of a variable $X \geq 0$ of L_p. As $E([X]^p) \leq E(X^p) \leq E\{([X]+1)^p\}$, we have

$$E(X^p) \leq \sum_{n\geq 1} (n+1)^{p-1} P([X] \geq n-1).$$

Exercise 1.6.4. Prove that for all real variables X and Y

$$\frac{1}{2}\{\mathrm{Var}(X) + \mathrm{Var}(Y)\} \geq \mathrm{Cov}(X, Y)$$

deduce that $\mathrm{Var}(X + Y) \leq 2\{\mathrm{Var}(X) + \mathrm{Var}(Y)\}$.

Answer. The first inequality is a consequence of the expectation in the inequality $x^2 + y^2 \geq 2xy$ with $x = X - E(X)$ and $y = Y - E(Y)$. The second inequality is deduced by the expansion

$$\begin{aligned} E(x+y)^2 &= E(x^2) + E(y^2) + 2E(xy) \\ &= \mathrm{Var}(X) + \mathrm{Var}(Y) + 2\mathrm{Cov}(X, Y). \end{aligned}$$

Exercise 1.6.5. Let $(X_i)_{i=1,\dots,n}$ be a sequence of variables with expectation μ and covariances $E(X_iX_j) = \gamma_{|j-i|}$, and let $S_n^2 = (n-1)^{-1}\sum_{i=1}^n (X_i - \bar{X})^2$ where $\bar{X} = n^{-1}\sum_{i=1}^n X_i$, prove that $E\{\sum_{i=1}^n (X_i - \bar{X})^2\} \leq E\{(\sum_{i=1}^n (X_i - \mu)^2\}$ and calculate $E(S_n^2)$.

Answer. As $\sum_{i=1}^n (X_i - \bar{X}) = 0$, we have

$$E\{\sum_{i=1}^n (X_i - \mu)^2\} = E\Big\{\sum_{i=1}^n (X_i - \bar{X})^2\Big\} + nE\{(\bar{X} - \mu)^2\},$$

$$E(S_n^2) = \frac{1}{n-1}E\Big\{\sum_{i=1}^n (X_i - \mu)^2\Big\} - \frac{n}{n-1}E\{(\bar{X} - \mu)^2\},$$

$$E\Big\{\sum_{i=1}^n (X_i - \mu)^2\Big\} = n\gamma_0 + \sum_{i=1}^n \sum_{j\neq i=1}^n \gamma_{|j-i|},$$

$$E\{(\bar{X} - \mu)^2\} = n^{-2}E\Big\{\sum_{i=1}^n (X_i - \mu)^2\Big\},$$

therefore

$$E(S_n^2) = \frac{1}{n}E\Big\{\sum_{i=1}^n (X_i - \mu)^2\Big\} = \gamma_0 + \frac{1}{n}\sum_{i=1}^n \sum_{j\neq i=1}^n \gamma_{|j-i|}.$$

Exercise 1.6.6. Let $(X_n)_{n\geq 1}$ and $(Y_n)_{n\geq 1}$ be independent sets of symmetric independent variables, determine the distribution of $|X_n|$ and $|Y_n|$ according to the distribution of $|X_n + Y_n|$.

Answer. For symmetric variables X_n and Y_n, $|X_n - Y_n|$ has the distribution

$$\begin{aligned}P(|X_n - Y_n| > a) &= P(X_n > a + Y_n) + P(X_n < Y_n - a)\\ &= P(X_n < -a - Y_n) + P(X_n > a - Y_n)\\ &= P(|X_n + Y_n| > a),\end{aligned}$$

$|X_n - Y_n|$ has therefore the same distribution as $|X_n + Y_n|$. The inequalities $|X_n + Y_n| > a$ and $|X_n - Y_n| > a$ imply $|X_n| > \frac{a}{2}$ and $|Y_n| > \frac{a}{2}$.

Exercise 1.6.7. For a real variable X such that $E(e^{tX})$ is finite for every t, prove that for every $a > 0$

$$\log P(X \geq a) \leq -\inf_{b \geq a} \sup_{t \geq 0} \{bt - E(e^{tX})\}.$$

Answer. For all $a > 0$ and $t \geq 0$, we have

$$\begin{aligned}P(X \geq a) &= P(e^{tX} \geq e^{ta}) \leq e^{-ta} E(e^{tX}),\\ \log P(X \geq a) &\leq \inf_{t \geq 0} \log E\{e^{t(X-a)}\}\\ &= -\sup_{t \geq 0} \{-\log E(e^{t(X-a)})\},\end{aligned}$$

by convexity of the exponential, the function $h(t) = at - \log E(e^{tX})$ is such that $h(t) \leq at - E(tX)$ and the bound is minimum as $a = E(X)$ where $h(t) \leq 0$, hence

$$\log P(X \geq a) \leq -\inf_{b \geq a} \sup_{t \geq 0} \{bt - E(e^{tX})\}.$$

Exercise 1.6.8. Let X be a variable with exponential distribution μ with parameter 1 and let f be a function of $L^2(\mathbb{R}_+, \mu) \cap C^1(\mathbb{R}_+)$, prove Poincaré's inequality for the variance of $f(X)$

$$\mathrm{Var}_\mu f(X) \leq 4E_\mu\{f'^2(X)\}.$$

Answer. Assuming that the function f has the value zero at zero does not modifies its variance, then an integration by parts and the Cauchy–Schwarz inequality imply

$$\begin{aligned}\int_{\mathbb{R}_+} f^2(x) e^{-x}\, dx &\leq 2 \int_{\mathbb{R}_+} f(x) f'(x) e^{-x}\, dx\\ &\leq 2 \Big\{\int_{\mathbb{R}_+} f^2(x) e^{-x}\, dx\Big\}^{\frac{1}{2}} \Big\{\int_{\mathbb{R}_+} f'^2(x) e^{-x}\, dx\Big\}^{\frac{1}{2}},\end{aligned}$$

which proves the inequality.

1.7 Hellinger distance, Kullback–Leibler information

On a measurable space $(\Omega, \mathcal{A})$, the Hellinger distance $h(P, Q)$ between probability measures P and Q is defined by

$$h^2(P, Q) = \frac{1}{2} \int (\sqrt{dP} - \sqrt{dQ})^2 = 1 - \rho(P, Q),$$

where $\rho(P, Q) = \int \sqrt{dP\, dQ}$. The L^1 norm, the affinity ρ and the Hellinger distance satisfy the inequalities

$$1 - \rho(P, Q) = h^2(P, Q) \le \frac{1}{2} \|P - Q\|_1 \le \{1 - \rho^2(P, Q)\}^{\frac{1}{2}}.$$

The variational distance of probabilities P and Q absolutely continuous with respect to a dominating measure μ On $(\Omega, \mathcal{A})$ is

$$\begin{aligned} d(P, Q) &= 2 \sup\{|P(A) - Q(A)|, A \in \mathcal{A}\} \\ &= 2 \int_\Omega \left\| \frac{dP}{d\mu} - \frac{dQ}{d\mu} \right\| d\mu \\ &= 2\left\{1 - \int_\Omega \left(\frac{dP}{d\mu} \wedge \frac{dQ}{d\mu} \right) d\mu \right\}, \\ h^2(P, Q) &\le d(P, Q) \le h(P, Q)\{1 - h^2(P, Q)\}^{\frac{1}{2}}. \end{aligned}$$

On the product space $(\Omega^{\otimes n}, \mathcal{A}_n)$ with the Borelian sigma-algebras, the products of probability measures $P_n = \prod_{j=1,\ldots,n} P_{nj}$ and $Q_n = \prod_{j=1,\ldots,n} Q_{nj}$ have the affinity

$$\rho(P_n, Q_n) = \prod_{j=1,\ldots,n} \rho(P_{nj}, Q_{nj})$$

and

$$1 - h^2(P_n, Q_n) = \prod_{j=1,\ldots,n} \{1 - h^2(P_{nj}, Q_{nj})\}.$$

Let P and Q be probability measures on a measurable space $(\Omega, \mathcal{A})$ such that P is absolutely continuous with respect to Q, their Kullback–Leibler information is

$$K(P, Q) = \int_\Omega \log\left(\frac{dP}{dQ}\right) dP.$$

Exercise 1.7.1. On a measurable space $(\Omega, \mathcal{A})$, let P and Q be probability measures absolutely continuous with respect to a dominating measure μ and with respective densities $f = \frac{dP}{d\mu}$ and $g = \frac{dQ}{d\mu}$, prove that

$\rho(P,Q) = \int f^{\frac{1}{2}} g^{\frac{1}{2}}\, d\mu$ and $h^2(P,Q) = \frac{1}{2}\int(\sqrt{f} - \sqrt{g})^2\, d\mu$ do not depend on the dominating measure and give conditions for $\rho(P,Q) = 1$.

Answer. For a probability measure λ equivalent to μ we have

$$\begin{aligned}\rho(P,Q) &= \int \Big(\frac{dP}{d\lambda}\frac{d\lambda}{d\mu}\Big)^{\frac{1}{2}} \Big(\frac{dQ}{d\lambda}\frac{d\lambda}{d\mu}\Big)^{\frac{1}{2}} \frac{d\mu}{d\lambda}\, d\lambda \\ &= \int \Big(\frac{dP}{d\lambda}\Big)^{\frac{1}{2}} \Big(\frac{dQ}{d\lambda}\Big)^{\frac{1}{2}}\, d\lambda,\end{aligned}$$

by the Cauchy–Schwarz inequality

$$\rho(P,Q) \le \Big(\int f\, d\mu\Big)^{\frac{1}{2}} \Big(\int g\, d\mu\Big)^{\frac{1}{2}} = 1,$$

this inequality is an equality if and only f and g are a.s. proportional and this implies $P = Q$ a.s.

By a change of dominating measure, the Hellinger distance becomes

$$\begin{aligned}h^2(P,Q) &= \frac{1}{2}\int \Big\{\Big(\frac{dP}{d\lambda}\frac{d\lambda}{d\mu}\Big)^{\frac{1}{2}} - \Big(\frac{dQ}{d\lambda}\frac{d\lambda}{d\mu}\Big)^{\frac{1}{2}}\Big\}^2 d\mu \\ &= \frac{1}{2}\int \Big\{\Big(\frac{dP}{d\lambda}\Big)^{\frac{1}{2}} - \Big(\frac{dQ}{d\lambda}\Big)^{\frac{1}{2}}\Big\}^2 d\lambda.\end{aligned}$$

Exercise 1.7.2. Let P and Q be probability measures on a measurable space $(\Omega, \mathcal{A})$, prove the inequalities

$$\begin{aligned}\rho(P,Q) &\le 1 \wedge \Big\{2\int (dP \wedge dQ)\Big\}, \\ \rho(P,Q) &\le \rho\Big(P, \frac{P+Q}{2}\Big) \le 1,\end{aligned}$$

$$P(dP \le dQ) + Q(dP > dQ) \le \rho(P,Q) \le P(dP > dQ) + Q(dP \le dQ).$$

Answer. By the Cauchy–Schwarz inequality, we have

$$\begin{aligned}\rho(P,Q) &= \int \sqrt{dP\, dQ} = \int \sqrt{dP \wedge dQ}\sqrt{dP \vee dQ} \\ &\le \Big\{\int (dP \wedge dQ)\Big\}^{\frac{1}{2}} \Big\{\int (dP \vee dQ)\Big\}^{\frac{1}{2}}\end{aligned}$$

where $\int (dP \vee dQ) \le \int (dP + dQ) = 2$. Furthermore, by concavity

$$\begin{aligned}\rho\Big(P, \frac{P+Q}{2}\Big) &= \int \Big(\frac{1}{2} + \frac{dQ}{2dP}\Big)^{\frac{1}{2}} dP \\ &\le \Big\{\frac{1}{2} + \int \frac{dQ}{2dP}\, dP\Big\}^{\frac{1}{2}} = 1\end{aligned}$$

for every Q, this implies $\rho(P,Q) \le 1$. The inequalities $\rho(P,Q) \le 1$ and

$$\Big(\frac{a+b}{2}\Big)^{\frac{1}{2}} \ge \frac{1}{2}\Big(a^{\frac{1}{2}} + b^{\frac{1}{2}}\Big)$$

for all positive a and b entail

$$\rho\Big(P, \frac{P+Q}{2}\Big) \ge \frac{1}{2}\{1 + \rho(P,Q)\} \ge \rho(P,Q).$$

For the last inequalities, we have

$$\begin{aligned}\rho(P,Q) &= \int 1_{\{\frac{dP}{dQ}\le 1\}} \frac{\sqrt{dP}}{\sqrt{dQ}}\, dQ + \int 1_{\{\frac{dP}{dQ}\ge 1\}} \frac{dP}{dQ}\, dQ \\ &\le \int 1_{\{\frac{dP}{dQ}\le 1\}}\, dQ + \int 1_{\{\frac{dP}{dQ}>1\}}\, dP,\end{aligned}$$

and by the same arguments

$$\begin{aligned}\rho(P,Q) &\ge \int 1_{\{\frac{dP}{dQ}\le 1\}}\, dP + \int 1_{\{\frac{dP}{dQ}>1\}} \frac{\sqrt{dP}}{\sqrt{dQ}}\, dQ \\ &\ge \int 1_{\{\frac{dP}{dQ}\le 1\}}\, dP + \int 1_{\{\frac{dP}{dQ}>1\}}\, dQ.\end{aligned}$$

Exercise 1.7.3. Give a first order expansion of $\rho(P_\varepsilon, Q)$ for $P_\varepsilon = (1-\varepsilon)P + \varepsilon Q$, $\varepsilon > 0$, and give the limit of $\rho(P_\varepsilon, Q)$ as ε tends to zero, prove $\rho(P_\varepsilon, Q) \le \rho(P,Q)$.

Answer. Let $P_\varepsilon = P - \varepsilon(P-Q)$, an asymptotic expansion of the square root of

$$\frac{dP_\varepsilon}{dQ} = \frac{dP}{dQ}\Big\{1 - \varepsilon\Big(1 - \frac{dQ}{dP}\Big)\Big\}$$

near zero implies

$$\frac{(dP_\varepsilon)^{\frac{1}{2}}}{(dQ)^{\frac{1}{2}}} = \frac{(dP)^{\frac{1}{2}}}{(dQ)^{\frac{1}{2}}} \Big\{1 - \frac{\varepsilon}{2}\Big(1 - \frac{dQ}{dP}\Big) - \frac{\varepsilon^2}{4}\Big(1 - \frac{dQ}{dP}\Big)^2 + o_p(\varepsilon^2)\Big\}$$

and

$$\begin{aligned}\rho(P_\varepsilon, Q) = (1-\varepsilon)^{\frac{1}{2}} \rho(P,Q) &+ \frac{\varepsilon(1-\varepsilon)}{2} \int \frac{(dQ)^{\frac{1}{2}}}{(dP)^{\frac{1}{2}}}\, dQ \\ &+ \frac{\varepsilon^2}{4} \int \frac{(dQ)^{\frac{3}{2}}}{(dP)^{\frac{3}{2}}}\, dQ + o(\varepsilon^2)\Big\},\end{aligned}$$

as $\varepsilon > 0$ tends to zero, the expansion converges to $\rho(P,Q)$ and

$$\rho(P_\varepsilon, Q) \le (1-\varepsilon)^{\frac{1}{2}} \rho(P,Q) + o(\varepsilon)$$

the inequality $\rho(P_\varepsilon, Q) \le \rho(P,Q)$ follows.

Exercise 1.7.4. Let P_1, P_2, Q_1 and Q_2 be probabilities measures on a measurable space $(\Omega, \mathcal{A})$ and let $P_\lambda = \lambda P_1 + (1-\lambda) Q_1$ and $Q_\lambda = \lambda P_2 + (1-\lambda) Q_2$, for λ in $]0,1[$, prove the inequalities

$$\begin{aligned}
h^2(P_1, Q_\lambda) &\le \lambda h^2(P_1, P_2) + (1-\lambda) h^2(P_1, Q_2),\\
h^2(P_\lambda, Q_\lambda) &\le \lambda^2 h^2(P_1, P_2) + \lambda(1-\lambda)\{h^2(P_1, Q_2) + h^2(P_2, Q_1)\}\\
&\quad + (1-\lambda)^2 h^2(Q_1, Q_2).
\end{aligned}$$

Answer. By concavity of the square root, we have $Q_\lambda^{\frac{1}{2}} \ge \lambda P_2^{\frac{1}{2}} + (1-\lambda) Q_2^{\frac{1}{2}}$

$$\begin{aligned}
\rho(P_1, Q_\lambda) &\ge \lambda \rho(P_1, P_2) + (1-\lambda)\rho(P_1, Q_2),\\
h^2(P_1, Q_\lambda) &\le \lambda\{1 - \rho(P_1, P_2)\} + (1-\lambda)\{1 - \rho(P_1, Q_2)\},
\end{aligned}$$

and the Cauchy–Schwarz inequality implies

$$h^2(P_1, P_2) - h^2(P_1, Q_2) = \int P_1^{\frac{1}{2}} (P_2^{\frac{1}{2}} - Q_2^{\frac{1}{2}}) \le 2h^2(P_2, Q_2).$$

Replacing P_1 by P_λ in the inequality for $\rho(P_1, Q_\lambda)$ yields

$$\begin{aligned}
\rho(P_\lambda, Q_\lambda) &\ge \lambda^2 \rho(P_1, P_2) + \lambda(1-\lambda)\{\rho(P_1, Q_2) + \rho(P_2, Q_1)\}\\
&\quad + (1-\lambda)^2 \rho(Q_1, Q_2),
\end{aligned}$$

where $\lambda^2 + 2\lambda(1-\lambda) + (1-\lambda)^2 = 1$ therefore

$$\begin{aligned}
h^2(P_\lambda, Q_\lambda) &\le \lambda^2 h^2(P_1, P_2) + \lambda(1-\lambda)\{h^2(P_1, Q_2) + h^2(P_2, Q_1)\}\\
&\quad + (1-\lambda)^2 h^2(Q_1, Q_2).
\end{aligned}$$

Exercise 1.7.5. Let P and Q be probability measures on a measurable space $(\Omega, \mathcal{A})$ such that P is absolutely continuous with respect to Q, for their Kullback–Leibler information $K(P,Q)$ prove the inequalities $K(P,Q) \ge 0$

$$\begin{aligned}
K(P,Q) &\ge 2h^2(P,Q),\\
K(P_1, \lambda P_2 + (1-\lambda) Q_2) &\le \lambda K(P_1, P_2) + (1-\lambda) K(P_1, Q_2),
\end{aligned}$$

calculate an approximation of $K((1-\varepsilon)P + \varepsilon Q, Q)$ as ε tends to zero.

Answer. Under the condition $P >> Q$, the inequalities $0 \le \frac{dQ}{dP} \le 1$ and $\log(1-x) \le -x$ for x in $]0,1[$ entail

$$\begin{aligned}
\int \log\Big(\frac{dP}{dQ}\Big)\, dP &= -\int \log\Big(\frac{dQ}{dP}\Big)\, dP\\
&= -\int \log\Big\{1 - \Big(1 - \frac{dQ}{dP}\Big)\Big\}\, dP\\
&\ge 1 - \int \frac{dQ}{dP}\, dP = 0,
\end{aligned}$$

and $K(P,Q)=0$ if and only if $P=Q$. For the second inequality, writing

$$K(P,Q)=-2\int_\Omega \log\Big(\frac{dQ}{dP}\Big)^{\frac{1}{2}}\,dP$$

implies

$$\frac{1}{2}K(P,Q)-h^2(P,Q)=-\int \log\Big(\frac{dQ}{dP}\Big)^{\frac{1}{2}}\,dP-1+\int\Big(\frac{dQ}{dP}\Big)^{\frac{1}{2}}\,dP,$$

for $x>0$, the function $k(x)=-\log x-1+x$ is decreasing from $+\infty$ to zero on $]0,1]$ and increasing from zero to $+\infty$ for $x>1$, if follows that

$$K(P,Q)-2h^2(P,Q)\geq 0.$$

The concavity of the logarithm entails

$$\log(\lambda P_2+(1-\lambda)Q_2)\geq \lambda\log P_2+(1-\lambda)\log Q_2$$

which implies the inequality for $K(P_1,\lambda P_2+(1-\lambda)Q_2)$.

Let $P_\varepsilon=(1-\varepsilon)P+\varepsilon Q=P-\varepsilon(P-Q)$, a first order expansion of the logarithm gives

$$\begin{aligned}\log\frac{dP_\varepsilon}{dQ}&=\log\Big[\frac{dP}{dQ}\Big\{1+\varepsilon\Big(\frac{dQ}{dP}-1\Big)\Big\}\Big]\\&=\log\frac{dP}{dQ}+\varepsilon\Big(\frac{dQ}{dP}-1\Big)+o(\varepsilon),\end{aligned}$$

it follows that

$$K(P_\varepsilon,Q)=(P,Q)-\varepsilon K(Q,P)\}+o(\varepsilon).$$

Exercise 1.7.6. Let X,Y be dependent real variables and let $P_{X,Y}$ and $P_{X|Y}$ be their joint and conditional probability distributions, calculate $K(P_{X,Y},P_{X|Y})$.

Answer. Let λ be the Lebesgue measure on $\mathbb{R}$

$$\begin{aligned}K(P_{X,Y},P_{X|Y})&=\int\log\frac{dP_{X,Y}}{dP_{X|Y}}\,dP_{X,Y}\\&=\int\log\frac{dP_Y}{d\lambda}\,dP_{X,Y}.\end{aligned}$$

Exercise 1.7.7. Let $P_n=\otimes_{j=1,\ldots,n}P_{nj}$ and $Q_n=\otimes_{j=1,\ldots,n}Q_{nj}$ be probability measures on a product space $(\Omega^{\otimes n},\mathcal{A}_n)$, prove

$$h^2(P_n,Q_n)=1-\prod_{j=1}^n\{1-h^2(P_{nj},Q_{nj})\}\leq\sum_{i=1}^n h^2(P_{nj},Q_{nj}).$$

Answer. The affinity of P_n and Q_n factorizes as

$$\rho(P_n, Q_n) = \int (P_n Q_n)^{\frac{1}{2}} = \prod_{j=1}^{n} \int (P_{nj} Q_{nj})^{\frac{1}{2}} = \prod_{j=1}^{n} \rho(P_{nj}, Q_{nj})$$

it entails the equality

$$h^2(P_n, Q_n) = 1 - \prod_{j=1}^{n} \rho(P_{nj}, Q_{nj}) = 1 - \prod_{j=1}^{n} \Big\{1 - h^2(P_{nj}, Q_{nj})\Big\}.$$

For real y_i in $[0,1]$, the inequality

$$\sum_{i=1}^{n} y_i + \prod_{i=1}^{n} (1 - y_i) \geq 1$$

is obvious for $n = 1, 2$, assuming that it is true for n and denoting $S_n = \sum_{i=1}^n y_i$ and $\Pi_n = \prod_{i=1}^n (1 - y_i)$ in $[0,1]$, it follows

$$\begin{aligned} S_n + y_{n+1} + (1 - y_{n+1})\Pi_n &\leq 1 - \Pi_n + y_{n+1} + (1 - y_{n+1})\Pi_n \\ &= 1 + y_{n+1}(1 - \Pi_n) \leq 1, \end{aligned}$$

the inequality is therefore true for every integer $n \geq 1$. Applying it to $y_i = h^2(P_{nj}, Q_{nj})$ implies the inequality

$$1 - \prod_{i=1}^{n} \{1 - h^2(P_{nj}, Q_{nj})\} \leq \sum_{i=1}^{n} h^2(P_{nj}, Q_{nj}).$$

Exercise 1.7.8. Let $P_n = \otimes_{j=1,\dots,n} P_{nj}$ and $Q_n = \otimes_{j=1,\dots,n} Q_{nj}$ be probability measures on a product space $(\Omega^{\otimes n}, \mathcal{A}_n)$ such that $P_{nj} << Q_{nj}$ for every $j \leq n$, calculate $K(P_n, Q_n)$.

Answer. The logarithm of a product of derivatives yields

$$\log \frac{dP_n}{dQ_n} = \sum_{j=1}^{n} \log \frac{dP_{nj}}{dQ_{nj}},$$

it entails $K(P_n, Q_n) = \sum_{j=1}^n K(P_{nj}, Q_{nj})$.

Exercise 1.7.9. For a real variable X such that $E_P(e^{tX})$ is finite for every t, prove that the function $h(a) = \sup_{t \geq 0}\{at - E_P(e^{tX})\}$ satisfies $h(a) = \inf_{\lambda : E_\lambda(X) = a} K(\lambda, P)$.

Answer. Let λ such that $E_\lambda(X) = a$ and let $Y = \frac{d\lambda}{dP}$, then

$$E_P(Y) = 1 = E_\lambda(Y^{-1}) = E_\lambda(e^{-\log Y})$$

and $K(\lambda, P) = E_\lambda(\log Y)$. By convexity of the exponential, for every $t \geq 0$ we have $E_\lambda(e^{tX}) \geq e^{tE_\lambda(X)}$ and

$$\log E_\lambda(e^{-\log Y}) \geq -E_\lambda(\log Y),$$

it follows that

$$\begin{aligned}-\log E_P(e^{tX}) &= -\log E_\lambda(Y^{-1}e^{tX}) = -\log E_\lambda(e^{tX-\log Y}) \\ &\leq K(\lambda, P) - tE_\lambda(X) = K(\lambda, P) - ta,\end{aligned}$$

hence $h(a) \leq \inf_{\lambda:E_\lambda(X)=a} K(\lambda, P)$. This inequality is an equality if there exists t_0 such that

$$\inf_{\lambda:E_\lambda(X)=a} K(\lambda, P) = at_0 - E_P(e^{t_0X}) = \sup_{t\geq 0}\{at - E_P(e^{tX})\}$$

which is denoted h_0. By convexity, it follows

$$at - E_P(e^{tX}) \leq at - tE_P(X)$$

and the bound is zero as $a = E_P(X)$, it follows that $\sup_{t\geq 0}\{at - E_P(e^{tX})\}$ is finite and there exists t_0 such that

$$at_0 - E_P(e^{t_0X}) = \sup_{t\geq 0}\{at - E_P(e^{tX})\} = h_0,$$

let λ_0 a real measure absolutely continuous with respect to P and with density

$$\frac{d\lambda_0}{dP} = e^{h_0+t_0(X-a)},$$

then $K(\lambda_0, P) = h_0$ hence

$$\inf_{\lambda:E_\lambda(X)=a} K(\lambda, P) \leq \sup_{t\geq 0}\{at - E_P(e^{tX})\}.$$

Chapter 2

Probability distributions

2.1 Distribution functions and densities

The distribution of a random variable X on a measurable metric space $(\mathcal{X}, \mathcal{B})$ endowed with the Borel sigma-algebra $\mathcal{B}$ is defined by a measurable map from a probability space $(\Omega, \mathcal{A}, P)$ such that for every B in $\mathcal{B}$

$$P(X^{-1}(B)) = P(X \in B).$$

For a variable X with values in $\mathbb{R}^p$, the distribution function of X is

$$F(x) = PX^{-1}(x) = P(X \leq x)$$

it is an increasing right-continuous function with left-limits. For every measurable function h

$$Eh(X) = \int_{\mathcal{X}} h(x)\, dF(x) = \int_{\Omega} h \circ X(w)\, dP(w).$$

On $\mathbb{R}$, the Lebesgue measure of an interval I is its length

$$\lambda(I) = |I|,$$

on $\mathbb{R}^d$, the Lebesgue measure of a d dimensional set is its volume.

A distribution function on $\mathbb{R}$ absolutely continuous with respect to the Lebesgue measure has a density defined as its derivative, on $\mathbb{R}^d$ a density is the derivative of order d

$$f(x_1, \ldots, x_d) = \frac{\partial^d F(x_1, \ldots, x_d)}{\prod_{j=1}^d \partial x_j}.$$

The uniform distribution on a bounded set A of $\mathbb{R}^d$ is the Lebesgue measure of A

$$f_A(x) = \frac{1_A(x)}{\lambda(A)}.$$

Theorem 2.1 (Fubini's Theorem). *Let f be a measurable function and integrable on a product space $(\mathcal{X}_1 \otimes \mathcal{X}_2, \mathcal{B}_1 \otimes \mathcal{B}_2, \lambda_1 \otimes \lambda_2)$, then the integrals $\int_{\mathcal{X}_1} f\, d\lambda_1$ is integrable with respect to λ_2, $\int_{\mathcal{X}_2} f\, d\lambda_2$ is integrable with respect to λ_1 and*

$$\int_{\mathcal{X}_2} \Big(\int_{\mathcal{X}_1} f\, d\lambda_1 \Big)\, d\lambda_2 = \int_{\mathcal{X}_1} \Big(\int_{\mathcal{X}_2} f\, d\lambda_2 \Big)\, d\lambda_1.$$

For every random variable X with density f_X on $\mathbb{R}^p$ and for every differentiable function $\phi : \mathbb{R}^p \mapsto \mathbb{R}^q$ with inverse function ϕ^{-1}, the density of $Y = \phi(X)$ is

$$f_Y(y) = f_X \circ \phi^{-1}(y) |J(y)|$$

where the Jacobian $|J|(y)$ is the determinant of the matrix $\left(\frac{\partial \phi_i^{-1}(y)}{\partial y_j} \right)_{i \le p, j \le q}$ if it is not singular.

Let F and G be continuous functions on $\mathbb{R}$, the integration by parts formula is

$$F(b)G(b) - F(a)G(a) = \int_a^b F\, dG + \int_a^b G\, dF.$$

Exercise 2.1.1. Let X be a square integrable real variable, find the density of X^2.

Answer. The distribution function of X^2 is $P(X^2 \le y) = F(y^{\frac{1}{2}})$ where F is the distribution function of X, its density is

$$g(y) = 2f(y^{\frac{1}{2}})y^{-\frac{1}{2}}$$

where f is the density f of X.

Exercise 2.1.2. Prove that $\int_{\mathbb{R}} e^{-x^2}\, dx = \sqrt{\pi}$ using polar coordinates.

Answer. Denoting $r = x^2 + y^2$, the square of the integral is

$$\int_{\mathbb{R}^2} e^{-x^2-y^2}\, dx\, dy = \int_{-\pi}^{\pi} \int_{\mathbb{R}} e^{-r^2} r\, dr\, d\theta = \pi.$$

Exercise 2.1.3. Let X and Y be independent uniform variables on $[0, 1]$, find the density of $\varphi = \arctan \frac{X}{Y}$.

Answer. For every bounded function h on $[0, \frac{\pi}{2}]$, the integral $I = \int_0^1 \int_0^1 h(\arctan \frac{x}{y})\, dx\, dy$ is reparametrized by the change of variable

$\varphi_y(x) = \arctan \frac{x}{y}$, its inverse $x = y \tan(\varphi)$ has the derivative $x'_\varphi = y(1 + \tan^2 \varphi)$ and the Jacobian of the reparametrization is $|J_y(\varphi)| = x'_\varphi$ therefore

$$\begin{aligned} I &= \int_0^1 \int_0^{\frac{\pi}{2}} h(\varphi)|J_y(\varphi)|\, d\varphi\, dy \\ &= \int_0^1 y\, dy \int_0^{\frac{\pi}{2}} h(\varphi)(1 + \tan^2 \varphi)\, d\varphi, \end{aligned}$$

the density of the random variable φ on $[0, \frac{\pi}{2}]$ is therefore $f(\varphi) = \frac{1}{2}\{1 + \tan^2(\varphi)\}$.

Exercise 2.1.4. Let $X_k = R\cos\{2\pi(kt + \Phi)\}$ and $Y_k = R\sin\{2\pi(kt + \Phi)\}$ where k and t are integers, R and Φ are independent variables, R has the density $f(r) = re^{-\frac{r^2}{2}}$ on $\mathbb{R}_+$ and Φ has a uniform distribution on $[0, 1]$. Find the distributions of $X_k^2 + Y_k^2$ and (X_k, Y_k).

Answer. The variable $X_k^2 + Y_k^2 = R^2$ has the exponential density with parameter $\frac{1}{2}$. The expectation of a bounded variable h of (X_k, Y_k) is reparametrized as $(u, v) = \varphi(r, \Phi)$ with $u = r\cos\{2\pi(kt + \Phi)\}$ and $v = r\sin\{2\pi(kt + \Phi)\}$ such that $\varphi^{-1}(u, v) = (\sqrt{u^2 + v^2}, \arctan(u^{-1}v))$, $r'_u = (u^2 + v^2)^{-\frac{1}{2}}u$ and $r'_v = (u^2 + v^2)^{-\frac{1}{2}}v$

$$\begin{aligned} \frac{d}{du}\arctan(u^{-1}v) &= -\frac{v}{u^2(1 + u^{-2}v^2} = -\frac{v}{u^2 + v^2}, \\ \frac{d}{dv}\arctan(u^{-1}v) &= \frac{1}{u(1 + u^{-2}v^2)} = \frac{u}{u^2 + v^2}, \end{aligned}$$

the Jacobian of φ is $|J(u, v)| = (u^2 + v^2)^{-\frac{1}{2}}$, then

$$\begin{aligned} Eh(X_k, Y_k) &= \int_0^\infty \int_0^1 h(r\cos\{2\pi(kt + \Phi)\}, r\sin\{2\pi(kt + \Phi)\})re^{-\frac{r^2}{2}}\, d\Phi\, dr \\ &= \frac{1}{2\pi}\int_{-1}^1 \int_{-1}^1 h(u, v)|J(u, v)|\sqrt{u^2 + v^2}e^{-\frac{u^2+v^2}{2}}\, du\, dv \\ &= \frac{1}{2\pi}\int_{-1}^1 \int_{-1}^1 h(u, v)e^{-\frac{u^2+v^2}{2}}\, du\, dv, \end{aligned}$$

the variables X_k and Y_k are therefore independent and normal.

Exercise 2.1.5. Let F be a real and continuous distribution function on $\mathbb{R}$, calculate $\int_{\mathbb{R}} F^n\, d(F^m)$.

Answer. Let X be a variable with distribution function F, the variable $U = F(X)$ has a uniform distribution and

$$\int_{\mathbb{R}} F^n(x)\,d(F^m)(x) = m \int_{\mathbb{R}} u^{n+m-1}\,du = \frac{m}{n+m}.$$

Exercise 2.1.6. Let $X_1, \ldots, X_n$ be independent and identically distributed variables with a continuous distribution function F and let $X_{(1)} < X_{(2)} < \cdots < X_{(n)}$ be the ordered variables, find the distributions of $X_{(k)}$ and $X_{(k+1)} - X_{(k)}$, and the limits of $P(X_{(k)} > n^{-1}x)$ if F has a density in $C_b^1(\mathbb{R}_+)$.

Answer. We have $P(X_{(1)} > t) = \{1 - F(t)\}^n$ and

$$P(X_{(2)} > x) = n\{1 - F(x)\}^{n-1}F(x) + \{1 - F(x)\}^n,$$
$$P(X_{(2)} - X_{(1)} > x) = n \int_{\mathbb{R}} P(X_{(2)} > x + y)\{1 - F(y)\}^{n-1} f(y)\,dy,$$

for every integer k

$$P(X_{(k)} > x) = \sum_{j=k}^{n} \binom{n}{j} \{1 - F(x)\}^{n-j} F^j(x),$$
$$P(X_{(k+1)} - X_{(k)} > x) = \sum_{j=k}^{n} \binom{n}{j} \int_{\mathbb{R}} P(X_{(k+1)} > x + y)\{1 - F(y)\}^{n-j-1} \times F^{j-1}(y)\}[(n-j)F(y) + j\{1 - F(y)\}]f(y)\,dy.$$

As n tends to infinity, $F(n^{-1}x)$ has the expansion

$$F(n^{-1}x) = \frac{x}{n} f(0) + \frac{x^2}{2n^2} f'(0) + o(n^{-2}),$$

let $f_0 = f(0)$, it implies

$$P(X_{(1)} > n^{-1}x) = \{1 - n^{-1}xf_0 + o(1)\}^n = e^{-xf_0} + o(1),$$
$$P(X_{(2)} > n^{-1}x) = e^{-xf_0}(1 + xf_0) + o(1),$$
$$P(X_{(k)} > n^{-1}x) = \sum_{j=k}^{n} e^{-xf_0}\Big\{1 + \frac{x^j f_0^j}{j!}\Big\} + o(1).$$

Exercise 2.1.7. Let X_1 and X_2 be independent variables with densities

$$f(x) = \frac{2}{\pi(1 + x^2)}, \; x > 0,$$

write the densities of X_1X_2 and $X_1^{-1}X_2$.

Answer. The expectation of the variables is infinite, their distribution function is the arc-tangent function $F(x) = \frac{2}{\pi}\arctan(x)$. The distribution function of X_1X_2 is

$$\begin{aligned}P(X_1X_2 \le z) &= \int_{\mathbb{R}_+} P(X_1 \le x^{-1}z) f(x)\,dx\\ &= \frac{2}{\pi}\int_{\mathbb{R}_+} \arctan(x^{-1}z) f(x)\,dx\end{aligned}$$

and its derivative is

$$\begin{aligned}g(z) &= \frac{2}{\pi}\int_{\mathbb{R}_+} \frac{1}{x(1+x^{-2}z^2)} f(x)\,dx\\ &= \frac{2}{\pi}\int_{\mathbb{R}_+} \frac{x}{x^2+z^2} f(x)\,dx,\end{aligned}$$

where the integral is

$$\frac{2}{\pi}\int_{\mathbb{R}_+} \frac{x\,dx}{(z^2+x^2)(1+x^2)} = \frac{2}{\pi}\frac{1}{z^2-1}\Big(\int_{\mathbb{R}_+} \frac{x\,dx}{1+x^2} - \int_{\mathbb{R}_+} \frac{x\,dx}{z^2+x^2}\Big),$$

and

$$\int_{\mathbb{R}_+} \frac{x\,dx}{z^2+x^2} = \int_{\mathbb{R}_+} \frac{t\,dt}{1+t^2}$$

by the change of variable $x = zt$, therefore the integral is degenerated. The distribution function of the ratio $X_1^{-1}X_2$ is

$$P(X_1^{-1}X_2 \le z) = \frac{2}{\pi}\int_{\mathbb{R}_+} \arctan(xz) f(x)\,dx$$

and its density is

$$g(z) = \frac{4}{\pi^2}\int_{\mathbb{R}_+} \frac{x\,dx}{(1+x^2z^2)(1+x^2)},$$

the integrand is proportional to a difference of ratios

$$\frac{x}{(1+x^2z^2)(1+x^2)} = \frac{x}{z^2-1}\Big(\frac{z^2}{1+x^2z^2} - \frac{1}{1+x^2}\Big),$$

$$g(z) = \frac{4}{\pi^2(z^2-1)}\int_{\mathbb{R}_+} x\Big(\frac{z^2}{1+x^2z^2} - \frac{1}{1+x^2}\Big)\,dx,$$

where the integrals satisfy

$$\int_{\mathbb{R}_+} \frac{xz^2\,dx}{1+x^2z^2} = \int_{\mathbb{R}_+} \frac{x\,dx}{1+x^2},$$

they are infinite and the density is degenerated.

Exercise 2.1.8. Let X be a variable with a continuous distribution function F on $\mathbb{R}_+$, calculate $E(x \wedge X)$.

Answer. Let $\bar{F} = 1 - F$, on the set $\{X > x\}$, the expectation is $x\bar{F}(x)$ and, with an integration by parts we obtain

$$E\{(x \wedge X)1_{\{X \le x\}}\} = \int_0^x y\, dF(y) = -\int_0^x y\, d\bar{F}(y)$$
$$= -x\bar{F}(x) + \int_0^x \bar{F}(y)\, dy,$$

therefore $E(x \wedge X) = \int_0^x \bar{F}(y)\, dy$.

Exercise 2.1.9. Let $X = YU$ be a variable on $\mathbb{R}_+$ such that Y is an unobserved positive variable independent of an uniform sampling variable U on $[0, 1]$. Calculate the expectation $E(X1_{\{X<y\}})$, the distribution function of Y according to $E(X1_{\{X<y\}})$, and the distribution function F_X of X.

Answer. The distribution function of X is $F_X(x) = E(xY^{-1} \wedge 1)$ on $\mathbb{R}_+$ and the expectation $E(X1_{\{X<y\}})$ is calculated with integrations by parts

$$E(X1_{\{X<y\}}) = \int_0^y x\, dF_X(x)$$
$$= yF_X(y) - \int_0^y F_X(x)\, dx$$

and the distribution function of Y is obtained by a change of variables $x = uy$

$$F_Y(y) = \int_0^1 \Pr(X \le yu)\, du = \frac{1}{y}\int_0^y F_X(x)\, dx,$$

it implies

$$F_Y(y) = F_X(y) - \frac{E(X1_{\{X<y\}})}{y},$$

and the expression of F_Y gives

$$E(X1_{X<y}) = y\{F_X(y) - F_Y(y)\}.$$

Exercise 2.1.10. Let F_U denote its distribution function of U on $[0, 1]$, find the distribution functions of X and Y under the conditions of Exercise 2.1.9.

Answer. The distribution function of X is

$$F_X(x) = \int_0^\infty F_U(xy^{-1})\,dF_Y(y)$$

for $x > 0$, the distribution function of Y is

$$\begin{aligned} F_Y(y) &= \int_0^1 F_X(uy)\,dF_U(u) = \int_0^y F_X(x)\,dF_U(\frac{x}{y}) \\ &= F_X(y) - \int_0^y F_U(\frac{x}{y})\,dF_X(x) \\ &= F_X(y) - E_X\Big\{F_U\Big(\frac{X}{y}\Big)\Big\}, \end{aligned}$$

with $F_U(0) = 0$ and $F_U(1) = 1$.

Exercise 2.1.11. Let Z be a variable with a finite mean μ, with distribution function G and density g, and let the density of Y at y be proportional to $yg(y)$, find the distribution functions of X and Y.

Answer. The density and the distribution function of Y are

$$f_Y(y) = \frac{yg(y)}{\mu},\ F_Y(y) = \frac{E(Z1_{\{Z<y\}})}{\mu}$$

and for X, the equality $yF_Y(y) = \int_0^y F_X(x)\,dx$ implies

$$\begin{aligned} F_X(x) = F_Y(x) + xf_Y(x) &= \frac{1}{\mu}\Big\{E(Z1_{\{Z<x\}}) + x^2 g(x)\Big\} \\ &= \frac{1}{\mu}\Big\{xG(x) - \int_0^x G(z)\,dz + x^2 g(x)\Big\}. \end{aligned}$$

Exercise 2.1.12. Find distribution functions F and G of a variable X on $\mathbb{R}_+$ such that $E_F h(X)$ is proportional to $E_G h'(X)$ for every function h on $\mathbb{R}_+$, in $\mathcal{H} = \{h : h(0) = 0, h' \text{ bounded}\}$.

Answer. Let X be an exponential variable with parameter λ, integrating by parts $E_\lambda h(X)$ yields

$$\lambda E_\lambda h(X) = Eh'(X).$$

Let X be a variable with distribution function F on $\mathbb{R}_+$, let $\bar{F} = 1 - F$ and μ be the measure on $\mathbb{R}_+$ such that $d\mu(x) = \bar{F}(x)\,dx$, by integrations by parts we obtain

$$E_F h(X) = \int_0^\infty h'(x)\bar{F}(x)\,dx,$$
$$\int_0^\infty d\mu(x) = \int_0^\infty x\,dF(x) = E_F X,$$

let G be the distribution function on $\mathbb{R}_+$ such that $d\mu(x) = E_F X\,dG(x)$, then

$$E_F h(X) = \int_0^\infty h'(x)\,d\mu(x) = E_F X \int_0^\infty h'(x)\,dG(x)$$

therefore $E_F h(X) = E_F X\, E_G h'(X)$ for every h in $\mathcal{H}$.

2.2 Conditional densities

On a probability space $(\Omega, \mathcal{A}, P)$ let X and Y be dependent variables with a joint density f on a product space $\mathcal{X} \times \mathcal{Y}$, their marginal densities are $f_X(x) = \int_{\mathcal{Y}} f(x,y)\,dy$ and $f_Y(y) = \int_{\mathcal{X}} f(x,y)\,dx$, their conditional densities are

$$f_{X|Y=y}(x) = f_Y^{-1}(y) f(x,y), \quad f_{Y|X=x}(y) = f_X^{-1}(x) f(x,y).$$

Exercise 2.2.1. Let X_1 and X_2 be independent exponential variables with distinct parameters, find the probability of $\{X_1 < X_2\}$, the densities of X_1 conditionally on $X_2 1_{\{X_1 > X_2\}}$ and of X_2 conditionally on $X_1 1_{\{X_1 < X_2\}}$, and the densities of X_k conditionally on $X_1 < X_2$, for $k = 1, 2$.

Answer. For every $x > 0$, we have

$$P(X_2 > X_1 > x) = \lambda_2 \int_0^\infty (e^{-\lambda_1 x} - e^{-\lambda_1 y}) e^{-\lambda_2 y}\,dy$$
$$= e^{-\lambda_1 x} - \frac{\lambda_2}{\lambda_1 + \lambda_2},$$

the probability of $\{X_1 < X_2\}$ is then

$$P(X_2 > X_1 \geq 0) = 1 - \frac{\lambda_2}{\lambda_1 + \lambda_2} = \frac{\lambda_1}{\lambda_1 + \lambda_2}$$

and $P(X_2 < X_1) = \lambda_2(\lambda_1 + \lambda_2)^{-1}$. The densities of $X_1 1_{\{X_1 < X_2\}}$ and $X_2 1_{\{X_1 < X_2\}}$ are

$$f_{X_1 1_{\{X_1 < X_2\}}}(x) = \lambda_1 \lambda_2 e^{-\lambda_1 x} \int_x^\infty e^{-\lambda_2 y}\,dy = \lambda_1 e^{-(\lambda_1 + \lambda_2)x},$$
$$f_{X_2 1_{\{X_1 < X_2\}}}(y) = \lambda_1 \lambda_2 e^{-\lambda_2 y} \int_0^y e^{-\lambda_1 x}\,dx = \lambda_2 e^{-\lambda_2 y}\{1 - e^{-\lambda_1 y}\},$$

and the conditional densities of X_1 are

$$\begin{aligned}
f_{X_1|X_1<X_2}(x) &= \frac{f_{X_1 1_{\{X_1<X_2\}}}(x)}{P(X_1 < X_2)} \\
&= (\lambda_1+\lambda_2)e^{-(\lambda_1+\lambda_2)x}\}, \\
f_{X_1|X_1>X_2}(x) &= \frac{f_{X_1 1_{\{X_1>X_2\}}}(x)}{P(X_2 < X_1)} \\
&= (\lambda_1+\lambda_2)e^{-\lambda_1 x}\{1-e^{-\lambda_2 x}\}, \\
f_{X_1|X_2 1_{\{X_1<X_2\}}}(x) &= \frac{\lambda_1 e^{-\lambda_1 x}}{1-e^{-\lambda_2 X_2}} 1_{\{x<X_2\}}, \\
f_{X_1|X_2 1_{\{X_1>X_2\}}}(x) &= \lambda_1 e^{-\lambda_1(x-X_2)} 1_{\{x>X_2\}},
\end{aligned}$$

$f_{X_1|X_1<X_2}$ is an exponential density on $\mathbb{R}_+$ with parameter $\lambda_1+\lambda_2$ and $f_{X_1|X_2 1_{\{X_1>X_2\}}}$ is an exponential density with parameter λ_1 on $[X_2,\infty[$. The conditional densities of X_2 are

$$\begin{aligned}
f_{X_2|X_1<X_2}(y) &= (\lambda_1+\lambda_2)e^{-\lambda_2 y}\{1-e^{-\lambda_1 y}\}, \\
f_{X_2|X_1 1_{\{X_1<X_2\}}}(y) &= \lambda_2 e^{-\lambda_2(y-X_1)} 1_{\{X_1<y\}},
\end{aligned}$$

$f_{X_2|X_1 1_{\{X_1<X_2\}}}$ is an exponential density with parameter λ_2 on $[X_1,\infty[$.

Exercise 2.2.2. Let (X,Y) a random variable (X,Y) having a distribution function F with density f on $\mathbb{R}^2$ and let μ a distribution function on $\mathbb{R}$, calculate the expectation and the variance of a variable Z having the mixture distribution function $G(x) = \int_{\mathbb{R}} F(x,y)\,d\mu(y)$ on $\mathbb{R}$.

Answer. The expectation and the variance of the variables $X1_{\{Y=y\}}$ and $X1_{\{Y\le y\}}$ are

$$\begin{aligned}
m(y) &= E(X1_{\{Y=y\}}) = \int_{\mathbb{R}} xf(x,y)\,dx, \\
\sigma^2(y) &= \int_{\mathbb{R}} x^2 f(x,y)\,dx - m^2(y), \\
M(y) &= E(X1_{\{Y\le y\}}) = \int_{\mathbb{R}} xF(dx,y) \\
&= \int 1_{\{u\le y\}} m(u)\,du, \\
S(y) &= \mathrm{Var}(X1_{\{Y\le y\}}) = \int_{\mathbb{R}} x^2 F(dx,y) - M^2(y) \\
&= \int 1_{\{u\le y\}} \sigma^2(u)\,du.
\end{aligned}$$

The variable Z has the expectation

$$E(Z) = \int_{\mathbb{R}} x\, dG(x) = \int_{\mathbb{R}} xF(dx, y)\, d\mu(y) = \int_{\mathbb{R}} M(y)\, d\mu(y),$$

and the variance of Z is

$$\begin{aligned} E(Z^2) - E^2(Z) &= \int_{\mathbb{R}} x^2 g(x)\, dx - E^2(Z) \\ &= \int_{\mathbb{R}} x^2 f(x, y)\, dx\, d\mu(y) - E^2(Z) \\ &= \int_{\mathbb{R}} S(y)\, d\mu(y) + \int_{\mathbb{R}} M^2(y)\, d\mu(y) - E^2(Z). \end{aligned}$$

Exercise 2.2.3. Let U_1 and U_2 be independent uniform variables on $[0, 1]$ and let $Z_i = (U_1 + U_2)^{-1} U_i$, for $i = 1, 2$, find the density of Z_1 conditionally on Z_2.

Answer. The distribution function of Z_1 is

$$\begin{aligned} F_1(x) &= P\Big(U_1 \le \frac{x}{1-x} U_2\Big) \\ &= \int_0^1 P\Big\{U_1 \le \Big(\frac{x}{1-x} \wedge 1\Big) v\Big\}\, dv \\ &= \frac{x}{2(1-x)} 1_{[0, \frac{1}{2}]}(x). \end{aligned}$$

As the sum of Z_1 and Z_2 is 1, the distribution function of Z_1 at x conditionally on $Z_2 \le y$, hence $Z_1 \ge 1 - y$, is

$$\begin{aligned} F(x \mid y) &= \frac{P(1 - y \le Z_1 \le x)}{F_2(y)} 1_{\{1-y \le x\}} \\ &= \frac{F_1(x) - F_1(1-y)}{F_2(y)} 1_{\{1-y \le x\}} 1_{[0, \frac{1}{2}]}(y). \end{aligned}$$

Problem 2.2.1. (a). Let X_1 and X_2 be independent and identically distributed real variables and let $S = X_1 + X_2$, calculate the conditional expectations $E(X_1 \mid S)$. For exponential variables, find the distribution of the variable X_1 conditionally on S and the conditional expectation $E(X_1 \mid S)$.

(b). Generalize to independent and identically distributed exponential variables and their sum $S_n = X_1 + \ldots + X_n$.

Answer. (a). Let F the distribution function of (X_1, X_2), with density f, for every measurable function h of $C_b(\mathbb{R})$ we have

$$Eh(S, X_1) = \int_{\mathbb{R}^2} h(x+y, x)\, F(dx, dy)$$
$$= \int_{\mathbb{R}^2} h(s, x) 1_{\{s>x\}} f(x, s-x)\, dx\, ds,$$

the density of (X_1, S) is therefore $g(x, s) = 1_{\{s>x\}} f(x, s-x)$ and the marginal density of S is $g_S(s) = \int_0^s f(x, s-x)\, dx$. The conditional expectation $E(X_1 \mid S)$ is

$$E(X_1 \mid S) = \frac{\int_{\mathbb{R}^2} x 1_{\{S>x\}} f(x, S-x)\, dx}{\int_0^S f(x, S-x)\, dx}.$$

If X_1 and X_2 have the same exponential distribution $\mathcal{E}_\lambda$, the joint density of (X_k, S) and the marginal density of S are

$$g(x_1, s) = \lambda^2 1_{\{s>x_1\}} e^{-\lambda s},$$
$$g_S(s) = \lambda^2 s e^{-\lambda s}$$

and $g_{X_1|S=s}(x) = s^{-1} 1_{\{s>x\}}$, the conditional expectation of X_1 is

$$E(X_1 \mid S) = S^{-1} \int_0^S x\, dx = \frac{S}{2}.$$

If $X_1 \sim \mathcal{E}_{\lambda_1}$ and $X_2 \sim \mathcal{E}_{\lambda_2}$, with $\lambda_1 > \lambda_2$, the joint density of (X_1, S) and the density of S are determined by

$$g(x, s) = 1_{\{s>x\}} f(x, s-x)$$
$$= \lambda_1 \lambda_2 1_{\{s>x\}} e^{-(\lambda_1-\lambda_2)x} e^{-\lambda_2 s},$$
$$g_S(s) = \frac{\lambda_1 \lambda_1}{\lambda_1 - \lambda_2} e^{-\lambda_2 s} \{1 - e^{-(\lambda_1-\lambda_2)s}\},$$

the conditional density of X_1 is deduced

$$g_{X_1|S=s}(x) = \frac{g(x, s)}{g_S(s)} = (\lambda_1 - \lambda_2) \frac{e^{-(\lambda_1-\lambda_2)x}}{1 - e^{-(\lambda_1-\lambda_2)s}} 1_{\{s>x\}},$$

the conditional expectation $E(X_1 \mid S)$ is therefore

$$E(X_1 \mid S) = \int_0^\infty x g_{X_1|S}(x)\, dx$$
$$= \frac{1 - e^{-(\lambda_1-\lambda_2)S}\{1 + (\lambda_1 - \lambda_2)S\}}{1 - e^{-(\lambda_1-\lambda_2)S}}$$
$$= 1 - \frac{e^{-(\lambda_1-\lambda_2)S}(\lambda_1 - \lambda_2)S}{1 - e^{-(\lambda_1-\lambda_2)S}}$$
$$= 1 - \frac{e^{-\lambda_1 S}(\lambda_1 - \lambda_2)S}{e^{-\lambda_2 S} - e^{-\lambda_1 S}}.$$

(b). For identically distributed variables $X_1, \ldots, X_n \sim \mathcal{E}_\lambda$, if we assume that S_k has the density

$$g_k(s) = \frac{\lambda^k s^{k-1}}{(k-1)!} e^{-\lambda s}, \tag{2.1}$$

then the density of X_1, S_{k+1} is determined for $x < s$ by

$$\begin{aligned} g_{X,k}(x,s) &= \frac{d^2 P(X_1 \le x, S_{k+1} \le s)}{dx\, ds} \\ &= \frac{\lambda^{k+1}(s-x)^{k-1}}{(k-1)!} e^{-\lambda s} 1_{\{s>x\}}, \end{aligned}$$

and the marginal density of S_{k+1} is g_{k+1} therefore (2.1) is true for every integer n. The density and expectation of X_1 conditionally on S_n are

$$\begin{aligned} g_{X|S_n}(x) &= \frac{(n-1)(S_n - x)^{n-2}}{S_n^{n-1}} 1_{\{S_n > x\}}, \\ E(X_1 \mid S_n) &= \frac{n-1}{S_n^{n-1}} \int_0^{S_n} x(S_n - x)^{n-2}\, dx = \frac{S_n}{n}. \end{aligned}$$

Problem 2.2.2. (a). Let X and Y be independent variables with continuous distribution functions F and respectively G, write the distribution functions of $U = \max(X,Y)$ and $V = \min(X,Y) := X \wedge Y$.

(b). Let $X_1, \ldots, X_n$ be n independent and identically distributed variables with a continuous distribution functions F and let $U = \max_{i=1,\ldots,n} X_i$ and $V = \min_{i=1,\ldots,n} X_i$, write the joint and marginal densities of (U,V) and their conditional densities and distributions.

(c). Give the expectations and the variances of U and V and the conditional expectations $E(U \mid V)$ and $E(V \mid U)$ as the X_i are independent variables with uniform or exponential distributions.

(d). For every $k \le n$, write the joint distribution function of X_k and U and the expectation of X_k conditionally on U.

Answer. (a). The distribution function of U is

$$H(t) = P(X \le t, Y \le t) = F(t)G(t)$$

and for V it is $K(t) = P(X \le t \wedge Y) + P(Y \le X \wedge t)$ where

$$\begin{aligned} P(X \le t \wedge Y) &= F(t) \int 1_{\{y>t\}}\, dG(y) + \int F(y) 1_{\{y \le t\}}\, dG(y) \\ &= F(t)\{1 - G(t)\} + \int F(y) 1_{\{y \le t\}}\, dG(y), \end{aligned}$$

$$K(t) = F(t)\{1 - G(t)\} + G(t)\{1 - F(t)\} + F(t)G(t)$$
$$= F(t) + G(t) - F(t)G(t).$$

(b). The distributions of U and V are defined by

$$P(U \le x) = P(X_1, \ldots, X_n \le x) = F^n(x),$$
$$P(V \ge y) = P(X_1, \ldots, X_n \ge x) = \{1 - F(y)\}^n$$

and for all $y < x$

$$P(U \le x, V \ge y) = P(y \le X_1, \ldots, X_n \le x)$$
$$= \{F(x) - F(y)\}^n,$$
$$P(U \le x \mid V \ge y) = \{1 - F(y)\}^{-n}\{F(x) - F(y)\}^n,$$
$$P(V \ge y \mid U \le x) = F^{-n}(x)\{F(x) - F(y)\}^n.$$

The density of U and V are

$$f_U(x) = nf(x)F^{n-1}(x),$$
$$f_V(y) = nf(y)\{1 - F(y)\}^{n-1}$$

and their joint density is the second derivative of $-P(U \le x, V \ge y)$ with respect to x and y, for $y < x$, namely

$$f_{U,V}(x, y) = n(n-1)f(x)f(y)\{F(x) - F(y)\}^{n-2}.$$

They have the conditional densities

$$f_{U|V}(x, y) = (n-1)f(x)\frac{\{F(x) - F(y)\}^{n-2}}{\{1 - F(y)\}^{n-1}}1_{\{y<x\}},$$
$$f_{V|U}(x, y) = (n-1)f(y)\frac{\{F(x) - F(y)\}^{n-2}}{\{F(x)\}^{n-1}}1_{\{y<x\}}$$

and the probabilities

$$P(U \le x, V = y) = n(n-1)f(y)\int_y^x f(t)\{F(t) - F(y)\}^{n-2}\,dt$$
$$= nf(y)\{F(x) - F(y)\}^{n-1},$$
$$P(V \ge y, U = x) = -n(n-1)f(x)\int_y^x f(t)\{F(x) - F(t)\}^{n-2}\,dt$$
$$= nf(x)\{F(x) - F(y)\}^{n-1}$$

which determine the conditional probabilities

$$P(U \le x \mid V = y) = \{1 - F(y)\}^{1-n}\{F(x) - F(y)\}^{n-1},$$
$$P(V \ge y \mid U = x) = F^{1-n}(x)\{F(x) - F(y)\}^{n-1}.$$

(c). For independent exponential variables X_i on $[0,1]$, the distribution functions of U and V are

$$F_U(x) = (1 - e^{-\lambda x})^n,$$
$$F_V(y) = 1 - e^{-n\lambda x},$$

their expectations and their variances are

$$E(U) = n\lambda \int_0^\infty x e^{-\lambda x}(1 - e^{-\lambda x})^{n-1}\,dx = \frac{1}{\lambda}\sum_{k=1}^{n} \frac{\binom{n}{k}(-1)^{k-1}}{k},$$

$$E(V) = n\lambda \int_0^\infty x e^{-n\lambda x}\,dx = \frac{1}{n\lambda},$$

$$\mathrm{Var}(U) = \frac{2}{\lambda^2}\sum_{k=1}^{n} \frac{\binom{n}{k}(-1)^{k-1}}{k^2},$$

$$\mathrm{Var}(V) = n\lambda \int_0^\infty x^2 e^{-n\lambda x}\,dx = \frac{1}{n^2\lambda^2},$$

and their conditional expectations of U and V are

$$E(U \mid V) = \frac{1-V}{\lambda}\sum_{k=1}^{n-1} \frac{\binom{n}{k}(-1)^{k-1}}{k},$$

$$E(V \mid U) = \frac{1-U}{\lambda}\sum_{k=1}^{n-1} \frac{\binom{n}{k}(-1)^{k-1}}{k}.$$

As the X_i has a uniform distribution on $[0,1]$, the conditional expectations and variances of U and V are $E(U) = n(n+1)^{-1}$, $\mathrm{Var}(U) = n(n+2)^{-1}$, and

$$\begin{aligned}
E(V) &= n\int_0^1 y(1-y)^{n-1}\,dy \\
&= n\int_0^1 (1-y)^{n-1}\,dy - n\int_0^1 (1-y)^n\,dy = (n+1)^{-1}, \\
\mathrm{Var}(V) &= n\int_0^1 y^2(1-y)^{n-1}\,dy \\
&= n\int_0^1 (1-y)^{n-1}\,dy - 2n\int_0^1 (1-y)^n\,dy + n\int_0^1 (1-y)^{n+1}\,dy \\
&= n(n+1)^{-1},
\end{aligned}$$

and their conditional expectations are

$$E(U \mid V) = \frac{n-1}{(1-V)^{n-1}}\int_V^1 t(t-V)^{n-2}\,dt = 1 - n^{-1}(1-V),$$

$$E(V \mid U) = \frac{n-1}{X^{n-1}}\int_0^X y(X-y)^{n-2}\,dy = n^{-1}X.$$

(d). The joint distribution function of X_k and U is

$$\begin{aligned} P(X_k \le x, U \le t) &= P(X_k \le x, X_i \le t, \forall j \ne k) \\ &= F(x)\{F(t)\}^{n-1} 1_{\{x<t\}} + F^n(t) 1_{\{x \le t\}}, \end{aligned}$$

their joint density is

$$f_{X_k,U}(x,t) = (n-1) f(x) f(t) \{F(t)\}^{n-2} 1_{\{x \ge t\}}$$

and the expectation of X_k conditionally on U is

$$E(X_k \mid U = t) = \frac{1}{F(t)} \int s 1_{\{s \ge t\}} \, dF(s).$$

2.3 Convolutions

The convolution of two densities f and g is the density of the sum of independent variables X with density f and Y with density g. On $\mathbb{R}^m$ the convolution is commutative

$$f \star g(x) = \int f(x-t) g(t) \, dt = \int f(t) g(x-t) \, dt = g \star f(x),$$

$f \star g$ belongs to $\mathcal{C}^k$, $k > 1$ if f or g belongs to $\mathcal{C}^k$. If the distribution G of the variable Y is discontinuous, let $G(t) = G^c(t) + \sum_{s \le t} \Delta G(s)$ where $\Delta G(s) = G(s) - G(s^-)$, the distribution of $X + Y$ is

$$G \star f(x) = f \star G(x) = f \star G^c(x) + \sum_t f(x-t) \Delta G(t).$$

For functions f and g defined on subsets of $\mathbb{R}^m$, the convolution is not commutative and its support is included in $\operatorname{Supp} f \cup \operatorname{Supp} g$, for example

$$\begin{aligned} \int_{-\frac{\pi}{2}}^{\frac{\pi}{2}} \sin(x-y) \sin(y) \, dy &= \frac{1}{2} \int_{-\frac{\pi}{2}}^{\frac{\pi}{2}} \cos(x - 2y) \, dy - \frac{\pi}{2} \cos x \\ &= \frac{\cos x}{2} (1 - \pi), \end{aligned}$$

the convolution of uniform densities f on $[0,1]$ is

$$f^{\star 2}(x) = \int_{\inf(0,x-1)}^{sup(1,x)} y \, dy = x 1_{[0,1]}(x) + (2-x) 1_{[1,2]}(x).$$

Let f and g be in $L_1(\mu)$ then $f \star g$ belongs to $L_1(\mu)$ and

$$\|f \star g\|_1 \le \|f\|_1 \, \|g\|_1.$$

Let f be in $L_p(\mu)$, $1 \leq p <$, and let g be in $L_{p'}(\mu)$ such that $p^{-1}+p'^{-1} = 1$, then $f \star g$ belongs to $L_1(\mu)$ and

$$\|f \star g\|_1 \leq \|f\|_p \, \|g\|_{p'}.$$

For sequences $(f_n)_{n\geq 1}$ and $(g_n)_{n\geq 1}$ such that $\|f_n - f\|_p$ and $\|g_n - g\|_{p'}$ converge to zero, the convolutions $f_n \star g_n$ converge uniformly to $f \star g$ with the bounds

$$\|f_n \star g_n - f \star g\|_\infty \leq \|f_n - f\|_p \, \|g\|_{p'} + \|f\|_p \, \|g_n - g\|_{p'}.$$

Exercise 2.3.1. Calculate the convolution of two exponential densities, its integral and an approximation near zero.

Answer. The densities defined on $\mathbb{R}_+$ have convolutions of the form

$$f^{\star 2}(x) = \int_0^\infty f_1(x-y)1_{\{x\geq y\}} f_2(y)\, dy,$$

for exponential densities this yields

$$\begin{aligned} f^{\star 2}(x) &= \theta_1\theta_2 \int_0^x e^{-\theta_1(x-y)-\theta_2 y}\, dy \\ &= \theta_1\theta_2 e^{-\theta_1 x} \int_0^x e^{-(\theta_2-\theta_1)y}\, dy \\ &= \frac{\theta_1\theta_2}{\theta_2-\theta_1}(e^{-\theta_1 x} - e^{-\theta_2 x}), \text{ if } \theta_1 \neq \theta_2, \end{aligned}$$

and $f^{\star 2}(x) = \theta_1^2 x e^{-\theta_1 x}$ if $\theta_1 = \theta_2$. If $\theta_1 \neq \theta_2$, their distribution function is $F \star f(t) = \int_0^t f^{\star 2}(s)\, ds$

$$\begin{aligned} F \star f(t) &= \frac{\theta_1\theta_2}{\theta_1-\theta_2} \int_0^t (e^{-\theta_2 s} - e^{-\theta_1 s})\, ds \\ &= \frac{1}{\theta_1-\theta_2}\{\theta_1(1-e^{-\theta_2 t}) - \theta_2(1-e^{-\theta_1 t})\}, \end{aligned}$$

as t tends to zero, it has the approximation $F \star f(t) = \frac{\theta_1\theta_2 t^2}{2} + o(t^2)$. With a common parameter θ, it is

$$F \star f(t) = \theta^2 \int_0^t s e^{-\theta s}\, ds = \int_0^{\theta t} s e^{-s}\, ds,$$

it has the approximation $F \star f(t) = \theta t e^{-\theta t} + o(t) = \theta t + o(t)$.

Exercise 2.3.2. Let $X_1, \ldots, X_k$ be independent normal variables with density $\phi(x) = (2\pi)^{-\frac{1}{2}} e^{-\frac{x^2}{2}}$, write the densities of $S_2 = X_1 + X_2$ and $S_k = X_1 + \cdots + X_k$.

Answer. The density of $X_1 + X_2$ is the convolution

$$\phi^{\star 2}(x) = \frac{1}{2\pi} \exp\Big\{-\frac{x^2}{4}\Big\} \int_{\mathbb{R}} \exp\Big\{-\frac{1}{2}\Big(\sqrt{2}y - \frac{x}{\sqrt{2}}\Big)^2\Big\}\, dy$$
$$= \frac{1}{2\sqrt{\pi}} \exp\Big\{-\frac{x^2}{4}\Big\},$$

it is the density of a centered Gaussian distribution with variance $\sqrt{2}$. Assuming that the density of S_k is

$$\phi^{\star k}(x) = \frac{1}{\sqrt{2k\pi}} \exp\Big\{-\frac{x^2}{2k}\Big\}, \tag{2.2}$$

the $k+1$ convolution is

$$\phi^{\star(k+1)}(x) = \frac{1}{2\pi\sqrt{k}} \int_{\mathbb{R}} e^{-\frac{(x-y)^2}{2k}} e^{-\frac{y^2}{2}},\, dy$$
$$= \frac{1}{2\pi\sqrt{k}} e^{-\frac{x^2}{2(k+1)}} \int_{\mathbb{R}} e^{-\Big(\frac{\sqrt{k+1}}{2\sqrt{k}}x - \frac{\sqrt{k}y}{2\sqrt{k+1}}\Big)^2} dy$$
$$= \frac{1}{\sqrt{2(k+1)\pi}} \exp\Big\{-\frac{x^2}{2(k+1)}\Big\},$$

by induction on k, (2.2) is true for every interger k.

Exercise 2.3.3. Let U_h^- be the uniform distribution function on $[-h, 0]$ and U_h^+ be the uniform distribution function on $[0, h]$, $h > 0$, for a distribution function F on $\mathbb{R}$, calculate the convolutions $F \star U_h^{\pm}$ and their limits as h tends to zero.

Answer. The convolutions are

$$F \star U_h^-(x) = \frac{1}{h} \int_{\mathbb{R}} 1_{[-h,0]}(x - s)\, dF(s)$$
$$= \frac{1}{h} \int_x^{x+h} dF(s) = \frac{F(x+h) - F(x)}{h},$$
$$F \star U_h^+(x) = \frac{1}{h} \int_{\mathbb{R}} 1_{[0,h]}(x - s)\, dF(s)$$
$$= \frac{1}{h} \int_{x-h}^{x} dF(s) = \frac{F(x) - F(x-h)}{h},$$

if F is differentiable by intervals, $F \star U_h^-(x)$ converges to $f(x^+)$ and $F \star U_h^+(x)$ converges to $f(x^-)$ as h tends to zero.

Problem 2.3.1. (a). Let f be a continuous and bounded function defined on $\mathbb{R}$ and let g be the density $g(x) = c\exp\{-(1-t^2)^{-1}\}$ on $]-1,1[$, prove that the function

$$f_a(x) = a\int_{\mathbb{R}} f(x-t)g(at)\,dt = a\int_{\mathbb{R}} g(a(x-t))f(t)\,dt$$

converges uniformly to f as a tends to infinity.

(b). Let g be a symmetric density on a compact $[-a,a]$ of $\mathbb{R}$ and let $g_n(x) = g(h_n^{-1}x)$, prove that the convolution $f_n(x) = f \star g_n(x)$ converges uniformly to f on every compact set, as h_n tends to zero.

Answer. (a). By changes of variables we obtain

$$f_a(x) = \int_{-1}^{1} f(x-a^{-1}t)g(t)\,dt = a\int_{x-a^{-1}}^{x+a^{-1}} f(y)g(a(x-y))\,dy,$$

$$|f_a(x)-f(x)| \leq a \sup_{]x-a^{-1},x+a^{-1}[} |f(y)-f(x)| \int_{x-a^{-1}}^{x+a^{-1}} g(a(x-y))\,dy$$

where $\sup_{]x-a^{-1},x+a^{-1}[} |f(y)-f(x)|$ converges to zero uniformly in x, as a tends to infinity, and the integral

$$\begin{aligned} I_a &= a\int_{x-a^{-1}}^{x+a^{-1}} g(a(x-y))\,dy \\ &= a\int_{-a^{-1}}^{a^{-1}} g(ay)\,dy = \int_{-a^{-2}}^{a^{-2}} g(y)\,dy \end{aligned}$$

converges to $g(0)$, then $\sup_{|x|} |f_a(x)-f(x)|$ converges to zero.

(b). A change of variables implies

$$f_n(x) = \int_{\mathbb{R}} f(y)g_n(x-y)\,dy = h_n\int_{x-ah_n}^{x+ah_n} f(t+h_nx)g(t))\,dt,$$

$$|f_n(x)-f(x)| \leq h_n \sup_{[x-ah_n,x+ah_n]} |f(y)-f(x)| \int_{x-ah_n}^{x+ah_n} g(t)\,dt$$

as n tends to infinity, $\sup_{[x-ah_n,x+ah_n]} |f(y)-f(x)|$ converges to zero and the integral

$$\begin{aligned} I_n &= \int_{x-ah_n}^{x+ah_n} g(t)\,dt = h_n\{G(x+ah_n) - G(x-ah_n)\} \\ &= 2ah_ng(x) + o(h_n) \end{aligned}$$

converges to zero. Then $\sup_{|x|\leq c} |f_n(x)-f(x)|$ converges to zero for every bound c.

2.4 Inequalities

The inequalities of Section 1.5 extend to the class $L^p(I)$ of functions f defined on a subinterval I of $\mathbb{R}$ with a finite norm $\|f\|_p = \left\{\int_I f^p(x)\,dx\right\}^{\frac{1}{p}}$.

Exercise 2.4.1. Let f be a function of L_p for an integer $p \geq 2$ and let q such that $p^{-1} + q^{-1} = 1$, prove

$$\|f\|_p = \sup_{\|h\|_q=1} \int f(x)h(x)\,dx. \tag{2.3}$$

Answer. The inequality $\sup_{\|h\|_q=1} \int f(x)h(x)\,dx \leq \|f\|_p$ is a consequence of Hölder's inequality. Let $g = \|f\|_p^{-\frac{p}{q}} f^{p-1}$, its L_q-norm is

$$\|g\|_q^q = \|f\|_p^{-p} \int f^{(p-1)q}(x)\,dx = \|f\|_p^{-p} \int f^p(x)\,dx = 1$$

where $(p-1)q = q(q-1)^{-1} = p$, and

$$\begin{aligned}\|f\|_p &= \|f\|_p^{-\frac{p}{q}} \int f^p(x)\,dx = \int f(x)g(x)\,dx \\ &\leq \sup_{\|h\|_q=1} \int f(x)h(x)\,dx.\end{aligned}$$

Exercise 2.4.2. On probability spaces $(\Omega, \mathcal{A}, F_n)$, let X_n and Y_n be F_n-measurable random variables, using (2.3) prove that for $p \geq 2$

$$E\|(X_n)_n\|_{l_p} = E\Big\{\Big(\sum_{n\geq 1} X_n^p\Big)^{\frac{1}{p}}\Big\} = \sup_{E\|(Y_n)_n\|_{l_q}=1} E\Big\{\sum_{n\geq 1} X_n Y_n\Big\}.$$

Answer. Applying (2.3) to the step-wise constant random processes $f = \sum_{n\geq 0} X_n 1_{]n,n+1]}$ and $h = \sum_{n\geq 0} Y_n 1_{]n,n+1]}$ gives the result.

Exercise 2.4.3. On $\mathbb{R}_+$, let f be a density of L_p, $p > 1$, and let $F(x) = \int_0^x f(t)\,dt$, prove Hardy's inequality $I_p = \|x^{-1}F\|_p \leq \frac{p}{p-1}\|f\|_p$.

Answer. As $F(0) = 0$, an integration by parts implies

$$\begin{aligned}I_p &= -\frac{1}{p-1}\int_{\mathbb{R}_+} F^p(x)\,d(x^{1-p}) \\ &= -\Big[\frac{x^{1-p}}{p-1}F^p(x)\Big]_0^\infty + \frac{p}{p-1}\int_{\mathbb{R}_+} F^{p-1}(x)x^{1-p}f(x)\,dx \\ &\leq \frac{p}{p-1}\int_{\mathbb{R}_+} F^{p-1}(x)x^{1-p}f(x)\,dx,\end{aligned}$$

and for q such that $p^{-1}+q^{-1}=1$, Hölder's inequality entails

$$\int_{\mathbb{R}_+} F^{p-1}(x)x^{1-p}f(x), dx \leq \|f\|_p \left\{\int_{\mathbb{R}_+} F^p(x)x^{-p}\,dx\right\}^{\frac{1}{q}},$$

where the last term is the integral $I_p^{\frac{1}{q}}$, the result follows from the definition of q. The inequality is strict unless $x^{-p}F^p$ is proportional to f^p.

Exercise 2.4.4. Let $(P_n)_{n\geq 0}$ and μ be probability measures such that $P_n << \mu$ with a real density $f_n = f_{\theta_n}$ on $\mathbb{R}\times\Theta$ where Θ is a complete subset of $\mathbb{R}^d$ and $\theta \mapsto f_\theta(x)$ belongs to $C_b^2(\Theta)$ uniformly on x and θ_n converges to θ_0 as n tends to infinity. Write a second order approximation of $K(P_0,P_n)$ as n tends to infinity.

Answer. By a second order Taylor expansion of the density f_{θ_n} as n tends to infinity, we have

$$\begin{aligned}
\log\frac{dP_n}{dP_0} &= \log\Big\{1+\frac{f_{\theta_n}-f_0}{f_0}\Big\}\\
&= \frac{f_{\theta_n}-f_0}{f_0}-\frac{1}{2}\frac{(f_{\theta_n}-f_0)^2}{f_0^2}+o(\|f_n-f_0\|^2)\\
&= (\theta_n-\theta_0)^T\frac{\dot f_0}{f_0}+\frac{1}{2}(\theta_n-\theta_0)^T\Big\{\frac{\ddot f_0}{f_0}-\frac{\dot f_0^2}{f_0^2}\Big\}(\theta_n-\theta_0)\\
&\quad +o(\|\theta_n-\theta_0\|^2),
\end{aligned}$$

where $f_0=f_{\theta_0}$, $\dot f_0$ and $\ddot f_0$ are the first two derivatives of $\theta\mapsto f_\theta$. The equality $\int f_\theta\,d\mu=1$ entails $\int \dot f_0\,d\mu=\int \ddot f_0\,d\mu=0$, then

$$\begin{aligned}
K(P_0,P_n) &= -\int \log\frac{dP_n}{dP_0}\,dP_0\\
&= \frac{1}{2}(\theta_n-\theta_0)^T\Big\{\int\frac{\dot f_0^2}{f_0}\,d\mu\Big\}(\theta_n-\theta_0)+o(\|\theta_n-\theta_0\|^2).
\end{aligned}$$

Exercise 2.4.5. Under the conditions of Exercise 2.4.4 prove that the Hellinger distance satisfies $h^2(P_n,P_0)-\frac{1}{8}\|f_0^{-\frac{1}{2}}(f_n-f_{\theta_n})\|_{L^2}^2=o(\|\theta_n-\theta_0\|^2)$ as n tends to infinity.

Answer. The density f_{θ_n} has a second order Taylor expansion

$$f_{\theta_n}=f_0+(\theta_n-\theta_0)^T\dot f_0+\frac{1}{2}(\theta_n-\theta_0)^T\ddot f_0(\theta_n-\theta_0)+o(\|\theta_n-\theta_0\|^2)$$

where $f_0 = f_{\theta_0}$, $\dot{f}_0$ and $\ddot{f}_0$ are the first two derivatives of the function $\theta \mapsto f_\theta$, it implies

$$\|f_0^{-\frac{1}{2}}(f_n - f_0)\|_{L^2}^2 = \int f_0^{-1}(f_{\theta_n} - f_0)^2\, d\mu$$
$$= (\theta_n - \theta_0)^T \Big\{ \int f_0^{-1} \dot{f}_0^2\, d\mu \Big\} (\theta_n - \theta_0) + o(\|\theta_n - \theta_0\|^2).$$

By the Taylor expansion $f_n^{\frac{1}{2}} - f_0^{\frac{1}{2}} = (\theta_n - \theta_0)^T \frac{\dot{f}_0}{2 f_0^{\frac{1}{2}}} + o(\|\theta_n - \theta_0\|)$, the Hellinger distance has the expansion

$$h^2(P_n, P_0) = \frac{1}{2} \int \Big(f_n^{\frac{1}{2}} - f_0^{\frac{1}{2}} \Big)^2 dP_0$$
$$= \frac{1}{8} (\theta_n - \theta_0)^T \Big\{ \int f_0^{-1} \dot{f}_0^2\, d\mu \Big\} (\theta_n - \theta_0) + o(\|\theta_n - \theta_0\|^2)$$

the result follows.

Exercise 2.4.6. Let $P_n = \otimes_{j=1,\ldots,n} P_{nj}$ be a product of probability measures on a space $(\Omega^{\otimes n}, \mathcal{A}_n)$, and let μ be a dominating measure of P_n, prove that if $\sum_{j=1}^n \alpha_j = 1$ for $\alpha = (\alpha_j)_{j=1,\ldots,n}$ in $]0,1[^n$, then the function

$$\phi(\alpha) = E_\mu \prod_{j=1}^n \Big(\frac{dP_{nj}}{d\mu} \Big)^{\alpha_j}$$

does not depend on the dominating measure and that it is convex.

Answer. Let ν be a dominating measure of P_n and let $f_{nj} = \frac{dP_{nj}}{d\mu}$ and $g_{nj} = \frac{dP_{nj}}{d\nu}$, then

$$\phi(\alpha) = \int \prod_{j=1}^n g_{nj}^{\alpha_j} \prod_{j=1}^n \Big(\frac{d\nu}{d\mu} \Big)^{\alpha_j} \frac{d\mu}{d\nu}\, d\nu$$
$$= \int \prod_{j=1}^n g_{nj}^{\alpha_j} \Big(\frac{d\nu}{d\mu} \Big)^{\sum_{j=1}^n \alpha_j - 1} d\nu$$
$$= \int \prod_{j=1}^n g_{nj}^{\alpha_j}\, d\nu.$$

To prove the convexity of ϕ, let λ, $\alpha = (\alpha_j)_{j=1,\ldots,n}$ and $\beta = (\beta_j)_{j=1,\ldots,n}$ in

$]0,1[^n$ such that $\sum_{j=1}^n \alpha_j = \sum_{j=1}^n \beta_j = 1$ then

$$\begin{aligned}
\phi(\lambda\alpha + (1-\lambda)\beta) &= \int \prod_{j=1}^n f_{nj}^{\lambda\alpha_j + (1-\lambda)\beta_j}\, d\mu \\
&= \int \prod_{j=1}^n (f_{nj}^{\alpha_j})^{\lambda} + (1-\lambda)(f_{nj}^{\beta_j})^{1-\lambda}\, d\mu \\
&\le \Big(\int \prod_{j=1}^n f_{nj}^{\alpha_j}\Big)^{\lambda} \Big(\int \prod_{j=1}^n f_{nj}^{\beta_j}\Big)^{1-\lambda} \\
&= \phi^{\lambda}(\alpha)\phi^{1-\lambda}(\beta) \\
&= \exp\{\lambda \log\phi(\alpha) + (1-\lambda)\log\phi(\beta)\}
\end{aligned}$$

and by convexity of the exponential

$$\phi(\lambda\alpha + (1-\lambda)\beta) \le \lambda\phi(\alpha) + (1-\lambda)\phi(\beta)$$

which proves the convexity of ϕ.

Exercise 2.4.7. Let $X_1, \ldots, X_n$ be independent variables on a probability space $(\Omega, \mathcal{A}, P)$, with density f with respect to a probability measure P_0, prove

$$P_0\Big(\prod_{i=1}^n f(X_i) \ge e^{-nh^2(P_0,P)}\Big) \le e^{-\frac{n}{2}h^2(P_0,P)}.$$

Answer. For every $a > 0$, by the inequality $P(|X| > a) \le a^{-1}E|X|$ we have

$$\begin{aligned}
P_0\Big(\prod_{i=1}^n f(X_i) \ge e^{-na}\Big) &\le e^{\frac{na}{2}} E_0\Big\{\prod_{i=1}^n f^{\frac{1}{2}}(X_i)\Big\} \\
&= e^{\frac{na}{2}} \exp\Big\{n \log E_0 f^{\frac{1}{2}}(X)\Big\}
\end{aligned}$$

where

$$E_0 f^{\frac{1}{2}}(X) = E_0\Big\{\frac{dP}{dP_0}(X)\Big\}^{\frac{1}{2}} = \int dP^{\frac{1}{2}} dP_0^{\frac{1}{2}} = 1 - h^2(P_0, P).$$

As $\log\{1 - h^2(P_0,P)\} \le -h^2(P_0,P)$, the inequality becomes

$$P_0\Big(\prod_{i=1}^n f(X_i) \ge e^{-na}\Big) \le e^{\frac{na}{2} - nh^2(P_0,P)}$$

and the results follows with $a = h^2(P_0, P)$.

Chapter 3

Generating function and discrete distributions

An integer variable X with probabilities $p_k = P(X = k)$ has the generating function $G_X(s) = \sum_{k \in \mathbb{N}} p_k s^k$ and its derivatives at zero determine the probabilities of X

$$p_k = \frac{G^{(k)}(0)}{k!}.$$

For every discrete variable X, the first two derivatives of G_X at 1 are $G'_X(1) = \sum_{k \geq 1} kp_k = EX$ and

$$G''_X(1) = \sum_{k \geq 2} k(k-1)p_k = E(X^2) - EX,$$

then the variance of X is

$$\mathrm{Var}(X) = G''_X(1) + G'_X(1) - G'^2_X(1).$$

3.1 Generating functions

Exercise 3.1.1. Let Y be a geometric variable with probabilities $p_k = p(1-p)^{k-1}$, give the generating function of Y, its expectation and its variance.

Answer. The expectation is $E(Y) = \sum_{k \geq 0} kp_k = p \sum_{k \geq 0} k(1-p)^{k-1}$ where $\sum_{k \geq 1} k(1-p)^{k-1}$ is the derivative with respect to $q = 1-p$ of $\sum_{k \geq 0} q^k = (1-q)^{-1}$, therefore $E(Y) = p^{-1}$. The generating function of Y is

$$G(s) = ps \sum_{k \geq 1} (qs)^{k-1} = \frac{ps}{1-qs},$$

its first two derivatives are

$$G'(s) = \frac{p}{(1-qs)^2}, \quad G''(s) = \frac{2pq}{(1-qs)^3},$$

$G'(1) = E(Y)$, $G''(1) = 2qp^{-2}$ and $\mathrm{E}(Y^2) = 2qp^{-2} + p^{-1} = (2-p)p^{-2}$. The variance of Y is $\mathrm{Var}(Y) = (1-p)p^{-2}$.

Exercise 3.1.2. Let $X_1, \ldots, X_k, \ldots$ and N be independent geometric variables with probabilities $p_k = p(1-p)^{k-1}$, and respectively $\pi_k = \pi(1-\pi)^{k-1}$, for $k \geq 1$, prove that $S_N = X_1 + \cdots + X_N$ has a geometric distribution with parameter $p\pi$.

Answer. Let $q = 1-p$ and $\rho = 1-\pi$, for all intergers k and $n \geq 1$ the generating function of X_k is $G_X(x) = xp\sum_{j\geq 1} x^{j-1}q^{j-1} = (1-xq)^{-1}xp$ and the generating function of S_n is $G_n(x) = G_X^n(x)$ then the generating function of S_N is

$$\begin{aligned}
G(x) &= \sum_{n,k\geq 1} x^k P(S_n = k)P(N = n) = \pi \sum_{n\geq 1} G_X^n(x)\rho^{n-1} \\
&= \frac{\pi G_X(x)}{1-\rho G_X(x)} = \frac{xp\pi}{1-xq-\rho px} \\
&= \frac{xp\pi}{1-x(1-p\pi)},
\end{aligned}$$

it is the generating function of a geometric distribution with parameter $p\pi$. The first derivatives of G_X are

$$\begin{aligned}
G'(x) &= \frac{\pi G'_X(x)}{\{1-\rho G_X(x)\}^2} = \frac{p\pi}{(1-xq)^2\{1-\rho G_X(x)\}^2}, \\
G''(x) &= \frac{\pi G''_X(x)\{1-\rho G_X(x)\} + 2\rho\pi G'^2_X(x)}{\{1-\rho G_X(x)\}^3}, \\
G''(1) &= \frac{2(1-p\pi)}{p^2\pi^2}.
\end{aligned}$$

The expectation of S_N is $E(S_N) = E(N)E(X) = (p\pi)^{-1} = G'(1)$ and $E(S_N^2) = G''(1) + G'(1) = p^{-2}\pi^{-2}(2-p\pi)$ it follows that the variance of S_N is

$$\mathrm{Var}(S_N) = \frac{1-p\pi}{p^2\pi^2}.$$

Exercise 3.1.3. Calculate the generating function, the expectation and the variance of $S = aX + bY$, for independent binomial variables X and Y, and for independent Poisson variables.

Answer. Let $B(m,p)$ and respectively $B(n,q)$ be the binomial distributions of the variables X and Y. The generating function of S is

$$G(s) = \sum_{j=1}^{m}\sum_{k=1}^{n}\binom{m}{j}\binom{n}{k}s^{aj}s^{bk}p^j(1-p)^{m-j}q^k(1-q)^{n-k}$$
$$= \{1-p(1-s^a)\}^m\{1-q(1-s^b)\}^n,$$

the expectation of S is $E(S) = amp + bnq$ and its variance is $\text{Var}(S) = a^2mp(1-p) + b^2nqp(1-q)$. With Poisson variables X with distribution $\mathcal{P}_\lambda$ and Y with distribution $\mathcal{P}_\mu$, S has the generating function

$$G(s) = e^{-\lambda-\mu}\sum_{j,k\geq 0}\frac{s^{aj}\lambda^j}{j!}\frac{s^{bk}\mu^k}{k!} = e^{-\lambda(1-s^a)}e^{-\mu(1-s^b)},$$

$E(S) = a\lambda + b\mu$ and $\text{Var}(S) = a^2\lambda + b^2\mu$.

Problem 3.1.1. (a). Let X_1 and X_2 be integer variables, with joint probabilities $p_{ij} = P(X_1 = i, X_2 = j)$ and with the bivariate generating function $G(s_1,s_2) = \sum_{i\in\mathbb{N}}\sum_{j\in\mathbb{N}} p_{ij}s_1^i s_2^j$, write the generating function of the sum $X_1 + X_2$.

(b). Prove that the independence of X_1 and X_2 is equivalent to the factorization of their generating function.

Answer. (a). The probabilities of $X_1 + X_2$ are

$$\pi_k = P(X_1 + X_2 = k) = \sum_{i,j\in\mathbb{N}, i+j=k} p_{ij}$$

and its generating function is

$$G_{X_1+X_2}(s) = \sum_{k\in\mathbb{N}}\pi_k s^k = \sum_{k\in\mathbb{N}}\sum_{i,j\in\mathbb{N}} p_{ij}s^i s^j 1_{\{i+j=k\}}$$
$$= \sum_{i,j\in\mathbb{N}} p_{ij}s^i s^j.$$

(b). Independent variables X_1 and X_2 have product probabilities $p_{ij} = p_{1i}p_{2j}$ with the marginal probabilities $p_{1i} = P(X_1 = i) = \sum_{j\in\mathbb{N}} p_{ij}$ and $p_{2j} = P(X_2 = j) = \sum_{i\in\mathbb{N}} p_{ij}$, it follows

$$G_{X_1}(s_1) = G(s_1,1), \quad G_{X_2}(s_2) = G(1,s_2)$$

and

$$G(s_1,s_2) = G_{X_1}(s_1)G_{X_2}(s_2).$$

Reversely, the factorization of the generating functions of $X_1 + X_2$ implies the independence of X_1 and X_2, by the differentiation

$$
\begin{aligned}
p_{ij} &= \frac{\partial^i \partial^j G_{X_1+X_2}(s_1, s_2)}{\partial s_1^i \partial s_2^j} \\
&= \frac{\partial^i G_{X_1}(s_1)}{\partial s_1^i} \frac{\partial^j G_{X_2}(s_2)}{\partial s_2^j} = p_{1i} p_{2j}.
\end{aligned}
$$

3.2 Multinomial variables

A Bernoulli variable X has the values in $\{0, 1\}$ with probabilities $P(X = 1) = p$ in $]0, 1[$ and $P(X = 0) = 1 - p$, a binomial variable $Y \sim \mathcal{B}(n, p)$ is the sum of n Bernoulli variables with probability p, it has the probabilities

$$
\pi_k = P(Y = k) = \binom{n}{k} p^k (1 - p)^{n-k}.
$$

A multinomial variable X with probabilities $p_1, \ldots, p_J$ such that $\sum_{j=1}^J p_j = 1$ has the distribution

$$
p(X) = \prod_{j=1}^J p_j^{\delta_j(X)},
$$

with the indicator variables $\delta_j(X) = 1_{\{X=j\}}$.

A variable N has a negative binomial distribution $\mathcal{N}B(\lambda, a)$ if it has the probabilities

$$
P(N = k) = \frac{\Gamma(k + a^{-1})}{k!\Gamma(a^{-1})} \Big(\frac{a\lambda}{1 + a\lambda}\Big)^k \Big(\frac{1}{1 + a\lambda}\Big)^{a^{-1}}, \quad k \in \mathbb{N}.
$$

Exercise 3.2.1. Let $(X_1, \ldots, X_n)$ be a vector of independent Bernoulli variables with sum S_n, prove that its probability conditionally on S_n does not depend on $p = P(X_1 = 1)$.

Answer. The conditional probability of $A_n(k) = \{X_1 = k_1, \ldots, X_n = k_n\}$ given S_n is

$$
\begin{aligned}
P(A_n(k) \mid S_n) &= 1_{\{S_n = \sum_{i=1}^n k_i\}} \frac{\prod_{i=1}^n p^{k_i} (1 - p)^{1-k_i}}{\binom{n}{S_n} p^{S_n} (1 - p)^{n-S_n}} \\
&= 1_{\{S_n = \sum_{i=1}^n k_i\}} \frac{1}{\binom{n}{S_n}}.
\end{aligned}
$$

Exercise 3.2.2. Let X be Bernoulli variable with a random probability exponentially distributed, find conditions for the unconditional distribution of X.

Answer. Let θ be the parameter of the exponentially distribution, the probability of $\{X = 1\}$ is

$$\begin{aligned} p &= \theta \int_0^\infty x e^{-\theta x}\, dx \\ &= \frac{1}{\theta} \int_0^\infty x e^{-x}\, dx = \frac{1}{\theta}, \end{aligned}$$

it is a probability if $\theta > 1$.

Exercise 3.2.3. Let X be a Bernoulli variable with a random probability, find the probabilities of $\{X = 1\}$ as p follows an arc-sine distribution with density $f(p) = \pi^{-1}(1-p^2)^{-\frac{1}{2}}$, and the density of p conditionally on X.

Answer. The joint density of X and p and the probability of $\{X = 1\}$ are

$$\begin{aligned} P(X = 1, p = y) &= \frac{y}{\pi(1-y^2)^{\frac{1}{2}}}, \\ P(X = 1) &= \frac{1}{\pi} \int_0^1 \frac{y}{(1-y^2)^{\frac{1}{2}}}\, dy \\ &= \frac{1}{2\pi} \int_0^1 \frac{dx}{(1-x)^{\frac{1}{2}}} = \frac{1}{\pi}, \end{aligned}$$

the density of p conditionally on X is

$$P(p = y \mid X) = \frac{y^X (1-y)^{1-X}}{(1-y^2)^{\frac{1}{2}}}.$$

Exercise 3.2.4. Let X be a binomial variable with a random probability p in $[0, 1]$, find the probabilities of X for a uniform probability p.

Answer. For a uniform probability p, the density of (X, p) is proportional to a Beta density (Section 5.2) and the probabilities of X is

$$\begin{aligned} P(X = k) &= \binom{n}{k} E\{p^k (1-p)^{n-k}\} \\ &= \binom{n}{k} B_{k+1, n-k+1} \end{aligned}$$

where

$$B_{k+1,n-k+1} = \frac{\Gamma_{k+1}\Gamma_{n-k+1}}{\Gamma_{n+2}} = \frac{k!(n-k)!}{(n+1)!},$$

hence $P(X = k) = \frac{1}{n+1}$ for every $k \leq n$, and X has a uniform distribution on $\{0, \dots, n\}$.

Exercise 3.2.5. Prove that

$$E\Big(\frac{1}{X+1}\Big) > \frac{1}{E(X+1)}$$

for Bernoulli and binomial variables.

Answer. For a Bernoulli variable, let

$$P(\frac{1}{X+1} = 1) = P(X = 0) = 1 - p,$$
$$P(\frac{1}{X+1} = \frac{1}{2}) = P(X = 1) = p,$$

with $0 < p < 1$, it follows

$$E\Big(\frac{1}{X+1}\Big) = 1 - \frac{p}{2} > \frac{1}{p+1} = \frac{1}{E(X+1)}.$$

A binomial variable $B(n,p)$, $0 < p < 1$, the variable $Y = \frac{1}{X+1}$ has the probabilities

$$P(Y = \frac{1}{k} = P(X = k-1) = \binom{n}{k-1} p^{k-1}(1-p)^{n-k+1},\ 1 \leq k \leq n+1,$$

$$\begin{aligned}
E(Y) &= \sum_{k=0}^{n} \frac{n!}{(k+1)!(n-k)!} p^k (1-p)^{n-k} \\
&= \frac{(1-p)}{(n+1)p} \sum_{k=1}^{n+1} \binom{n}{k} p^k (1-p)^{n-k} \\
&= \frac{(1-p)}{(n+1)p} \{1 - (1-p)^n\} \\
&= \frac{(1-p)}{(n+1)} \{1 + (1-p) + (1-p)^2 + \cdots + (1-p)^{n-1}\} \\
&= \frac{1-(1-p)^{n+1}}{(n+1)p}
\end{aligned}$$

and $E(X+1) = 1 + np$, the result is deduced from the equivalences

$$\begin{aligned}
(n+1)p &< 1 + np - (1+np)(1-p)^{n+1} \\
0 &< (1-p)\{1 - (1+np)(1-p)^n\} \\
&= (1-p)(1-p^2)(1-p)^{n-1}.
\end{aligned}$$

Exercise 3.2.6. Let Y be a geometric variable with probabilities $p_k = p(1-p)^{k-1}$, find the limit of the distribution of Y as p converges to zero.

Answer. The variable Y has the distribution function

$$P(Y \leq n) = p \sum_{k=1}^{n-1} (1-p)^k = 1 - (1-p)^n$$

where

$$(1-p)^n = \exp\{n \log(1-p)\} = \exp\{-np + o(np)\}$$

and it converges to $e^{-\lambda}$ as np converges to λ, then $P(Y \leq n)$ converges to $1 - e^{-\lambda}$.

Exercise 3.2.7. Let $(X_i)_{i=1,\dots,n}$ be independent multinomial variables with probabilities $p_1, \dots, p_J$, find the empirical distribution of $(X_i)_{i=1,\dots,n}$. Calculate the variance and the covariance of the empirical probabilities.

Answer. For $j = 1, \dots, J$, let $N_{nj} = \sum_{i=1}^n 1_{\{X_i = j\}}$ be the number of variables X_i with value j, the empirical probability $p_{nj} = n^{-1} N_{nj}$ converges a.s. to $p_j = P(X = j)$, as n tends to infinity. The empirical distribution of the observations is

$$\pi_n = \prod_{j=1}^J p_{nj}^{N_{nj}}.$$

The variance of p_{nj} is

$$\begin{aligned} v_{nj} &= n^{-2} E[\sum_{i=1}^n 1_{\{X_i = j\}} + \sum_{i_1 \neq i_2 = 1}^n 1_{\{X_{i_1} = j\}} 1_{\{X_{i_2} = j\}}] - p_j^2 \\ &= n^{-1} p_j (1 - p_j) \end{aligned}$$

and for $j \neq j'$, the covariance of p_{nj} and $p_{nj'}$ is

$$\begin{aligned} v_{njj'} &= n^{-2} E[\{\sum_{i_1=1}^n 1_{\{X_{i_1} i = j\}}\}\{\sum_{i_2=1}^n 1_{\{X_{i_2} = j'\}}] - p_j p_{j'} \\ &= -n^{-1} p_j p_{j'}. \end{aligned}$$

Exercise 3.2.8. Calculate the moments of orders 3 and 4 of the empirical probabilities.

Answer. For distinct j, j', j'', we obtain

$$E(p_{nj}^3) = n^{-2}\{(n-1)(n-2)p_j^3 + (n-1)p_j^2 + p_j\},$$

$$\begin{aligned}
E(p_{nj}^2 p_{nj'}) &= n^{-2}(n-1)p_j p_{j'}\{(n-2)p_j + 1\},\\
E(p_{nj}^4) &= n^{-3}\{(n-1)(n-2)(n-3)p_j^3 + 6(n-1)(n-2)p_j^3\\
&\quad +7(n-1)p_j^2 + p_j\},\\
E(p_{nj}^2 p_{nj'}^2) &= n^{-3}(n-1)p_j p_{j'}\{(n-2)(n-3)p_j p_{j'}\\
&\quad +(n-2)(p_j + p_{j'}) + 1\},\\
E(p_{nj}^3 p_{nj'}) &= n^{-3}p_j p_{j'}\{(n-2)(n-3)p_j^2 + 3(n-2)p_j + 1\},\\
E(p_{nj}^2 p_{nj'} p_{nj''}) &= n^{-3}(n-1)(n-2)(n-3)p_j p_{j'} p_{j''}\{(n-3)p_j + 1\},
\end{aligned}$$

the centered moments are deduced

$$\begin{aligned}
E\{(p_{nj} - p_j)^3\} &= n^{-2}\{(3n^2 - 6n + 5)p_j^3 - 3np_j^2 + p_j\},\\
E\{(p_{nj} - p_j)^4\} &= n^{-3}[3(n-2)p_j^4 + 2\{(n-1)(n-6) - 3n^2\}p_j^3\\
&\quad +(3n-7)p_j^2 + p_j],
\end{aligned}$$

$$\begin{aligned}
E\{(p_{nj} - p_j)^2(p_{nj'} - p_{j'})^2\} &= n^{-3}(n^3 - 5n^2 + 11n - 6)p_j^2 p_{j'}^2\\
&\quad +n^{-3}(n-1)p_j p_{j'}\{1 - 2(p_j + p_{j'})\}.
\end{aligned}$$

Furthermore, $E\{(p_{nj} - p_j)^2(p_{nj'} - p_{j'})\}$, $E\{(p_{nj} - p_j)^3(p_{nj'} - p_{j'})\}$ and $E\{(p_{nj} - p_j)^2(p_{nj'} - p_{j'})(p_{nj''}p_{j''})\}$ are zero.

3.3 Poisson variables

A Poisson variable X with parameter $\lambda > 0$ is an integer variable with probabilities

$$P(X = k) = e^{-\lambda}\frac{\lambda^k}{k!}.$$

The difference of two independent Poisson variables X and Y with parameter $\lambda > 0$ has a distribution

$$P(X - Y = k) = e^{-2\lambda} I_{|k|}(2\lambda)$$

where I_n is the Bessel function

$$I_n(x) = \sum_{k\geq 0} \frac{1}{k!(k+n)!}\left(\frac{x}{2}\right)^{2k+n}, \; n \in \mathcal{Z}.$$

Exercise 3.3.1. Determine recursively the distribution of the sum S_n of n independent and identically distributed Poisson variables.

Answer. The distribution of two Poisson variables X_k with intensity λ is the convolution of the distribution of the marginal distributions

$$\begin{aligned} P(X_1 + X_2 = k) &= \sum_{j\geq 0} P(X_1 = j)P(X_2 = k - j) \\ &= e^{-2\lambda} \sum_{j\geq 0} \frac{\lambda^k}{j!(k-j)!} \\ &= e^{-2\lambda} \frac{\lambda^k}{k!} \sum_{j\geq 0} \binom{k}{j} = e^{-2\lambda} \frac{2^k \lambda^k}{k!}, \end{aligned}$$

this is the distribution of a Poisson variable with intensity 2λ.

Let $S_n = X_1 + \cdots + X_n$ for n i.i.d. Poisson variables X_k with intensity λ, assuming that S_{n-1} has a Poisson distribution with intensity $(n-1)\lambda$ the distribution of S_n is given by

$$\begin{aligned} P(S_n = k) &= \sum_{j\geq 0} P(X_1 = j)P(S_n = k - j) \\ &= e^{-n\lambda} \sum_{j\geq 0} (n-1)^k \frac{\lambda^k}{j!(k-j)!} \\ &= e^{-n\lambda} \frac{\lambda^k}{k!} (n-1)^{k-j} \sum_{j\geq 0} \binom{k}{j} = e^{-n\lambda} \frac{n^k \lambda^k}{k!}, \end{aligned}$$

S_n is Poisson variable with intensity $n\lambda$, this is therefore true for every integer n.

Exercise 3.3.2. Let X be a Poisson variable, give a condition on the parameter for $P(X = k+1) > P(X = k)$.

Answer. The ratio of the probabilities for a Poisson variable with parameter λ is larger than 1 if

$$\frac{\lambda k!}{(k+1)!} = \frac{\lambda}{k+1} > 1,$$

i.e., $\lambda > k + 1$.

Exercise 3.3.3. Let P_1 and P_2 be the probability distributions of Poisson variables with parameters λ_1 and λ_2, find the Hellinger distance $h(P_1, P_2)$

and the Kullback–Leibler information $K(P_1, P_2)$.

Answer. The affinity of P_1 and P_2 is

$$\begin{aligned}\rho(P_1, P_2) &= e^{-\frac{1}{2}(\lambda_1+\lambda_2)} \sum_{k\geq 0} \frac{(\lambda_1\lambda_2)^{\frac{k}{2}}}{k!} \\ &= \exp\{-\frac{1}{2}(\lambda_1 + \lambda_2) + (\lambda_1\lambda_2)^{\frac{1}{2}}\} \\ &= \exp\{-\frac{1}{2}(\lambda_1^{\frac{1}{2}} - \lambda_2^{\frac{1}{2}})^2\}\end{aligned}$$

and their Hellinger distance is deduced.
The information $K(P_1, P_2)$ is defined as

$$\begin{aligned}K(P_1, P_2) &= e^{-\lambda_1} \sum_{k\geq 0} \frac{\lambda_1^k}{k!} \log\{e^{-(\lambda_1-\lambda_2)} \frac{\lambda_1^k}{\lambda_2^k}\} \\ &= \lambda_2 - \lambda_1 + \lambda_1 \log \frac{\lambda_1}{\lambda_2}.\end{aligned}$$

Exercise 3.3.4. Let X be a Poisson variable with an uniform intensity λ on $[0, a]$, $a > 0$, find the distribution of X and the density of λ conditionally on X.

Answer. The conditional probabilities of X are $\pi_k = \frac{1}{k!} E(e^{-\lambda}\lambda^k)$ and, using integrations by parts, we have

$$\begin{aligned}\pi_k &= \frac{1}{a\,k!} \int_0^a e^{-x} x^k \, dx \\ &= \frac{1}{a}\Big\{1 - e^{-a}(1 + a + \frac{a^2}{2!} + \ldots + \frac{a^k}{k!}\Big\},\end{aligned}$$

it follows that $\pi_k = a^{-1} P_a(X > k)$ where P_a is the distribution of a Poisson variable with intensity a. The conditional density of λ is

$$f(\lambda \mid X) = \frac{e^{-\lambda}\lambda^k}{k! P_a(X > k)}.$$

Exercise 3.3.5. Calculate the moments $E(X^k)$ and $\mu_k = E\{(X - EX)^k\}$ of a Poisson variable X, for $k = 3, 4$.

Answer. The generating function of a Poisson variable X with intensity λ is $G_\lambda(s) = e^{\lambda(s-1)}$ and its derivatives at 1 are $G_\lambda^{(k)}(1) = \lambda^k$. The expectation of X is λ, $E(X^2) = G_\lambda^{(2)}(1) + G'_\lambda(1) = \lambda^2 + \lambda$ and the variance of X is λ. The third derivative of G_λ yields

$$G_\lambda^{(3)}(1) = E(X^3) - 3E(X^2) + 2E(X),$$

equivalently

$$E(X^3) = \lambda^3 + 3(\lambda + \lambda^2) - 2\lambda = \lambda^3 + \lambda + 3\lambda^2$$

and $\mu_3 = E\{(X - \lambda)^3\} = \lambda$.

For $k = 4$, we have $G_\lambda^{(k)}(1) = E(X^4) - 6E(X^3) + 11E(X^2) - 6E(X)$ which yields

$$E(X^4) = \lambda^4 + 6\lambda^3 + 7\lambda^2 + \lambda$$

and $\mu_4 = 3\lambda^2 + \lambda$.

Exercise 3.3.6. Let $(X_i)_{i=1,\dots,n}$ be independent Poisson variables with intensity λ, determine the empirical distribution of $(X_i)_{i=1,\dots,n}$.

Answer. For $k \leq n$, let n_k be the number of variables X_i with value k, the empirical probability of the event $\{X_i = k\}$ is

$$\pi_{nk} = n^{-1} n_k = n^{-1} \sum_{i=1}^{n} 1_{\{X_i = k\}},$$

it converges a.s. to $\pi_k = P(X = k)$, as n tends to infinity. The empirical expectation of the variables X_i is

$$\mu_n = \sum_{i=1}^{n} k\pi_{nk}$$

and the empirical distribution of the observations $(X_i)_{i=1,\dots,n}$ is

$$\prod_{k \leq n} e^{-\mu_n} \frac{\mu_n^{n_k}}{n_k!}.$$

Exercise 3.3.7. Let Y be an exponential variable with parameter μ and let X be a Poisson variable with conditional parameter λY, find the distribution of X and the distribution of Y conditionally on X.

Answer. The probabilities of X are

$$\begin{aligned}\pi_k &= P(X = k) = E\{\pi_k(\lambda Y)\}\\ &= \mu \frac{\lambda^k}{k!} \int_0^\infty e^{-(\mu+\lambda)y} y^k \, dy\\ &= \frac{\mu\lambda^k}{(\mu + \lambda)^{k+1}}.\end{aligned}$$

The conditional density of Y is

$$f(y \mid X = k) = \frac{(\mu + \lambda)^{k+1}}{k!} e^{-(\mu+\lambda)y} y^k.$$

Exercise 3.3.8. Let Y be a Poisson variable with parameter λ and let X be an exponential variable with parameter μY conditionally on Y, find the distribution of X.

Answer. The conditional density of X is $f_{X|Y}(x) = \mu Y e^{-\mu Y x}$ and its density is

$$\begin{aligned} f(x) = E f_{X|Y}(x) &= e^{-\lambda} \sum_{k \geq 0} \frac{\lambda^k}{k!} \mu k e^{-\mu k x} \\ &= \lambda \mu e^{-\lambda(1-e^{-\mu x})}. \end{aligned}$$

Exercise 3.3.9. Let N be a variable with a negative binomial distribution $NB(\lambda, a)$. Prove that its distribution is equivalent to a Poisson distribution as a tends to zero and that the L^1 distance between the distribution of N and its limit converge to zero.

Answer. By an expansion of the logarithm, as a tends to zero, $(1+a\lambda)^{-a^{-1}}$ has the approximation

$$\exp\Bigl\{a^{-1} \log(1 - \frac{a\lambda}{1+a\lambda})\Bigr\} = \exp\Bigl\{-\frac{\lambda}{1+a\lambda} + o(1)\Bigr\}$$

and it converges to $e^{-\lambda}$. By Stirling's formula, as a tends to zero

$$\frac{\Gamma(k+a^{-1})}{\Gamma(a^{-1})} \sim (1+ak)^{\frac{1}{2}} a^{-k} \sim a^{-k},$$

the probabilities $\pi_k = P(N = k)$ have the approximation

$$\pi_k = \frac{\Gamma(k+a^{-1})}{k!\Gamma(a^{-1})} \Bigl(\frac{a\lambda}{1+a\lambda}\Bigr)^k \Bigl(\frac{1}{1+a\lambda}\Bigr)^{a^{-1}} \sim e^{-\lambda} \frac{\lambda^k}{k!}.$$

The L^1 distance between the distributions is

$$\begin{aligned} d &= \sum_{k \geq 0} \Bigl| P(N = k) - e^{-\lambda} \frac{\lambda^k}{k!} \Bigr| \\ &\leq \sum_{k \geq 0} \frac{\lambda^k}{k!} \Bigl| \exp\Bigl\{-\frac{\lambda}{1+a\lambda}\Bigr\} \Bigl(1 + \frac{ak}{2}\Bigr) \frac{1}{(1+a\lambda)^k} - e^{-\lambda} \Bigr| \\ &\leq \sum_{k \geq 1} ake^{-\lambda} \frac{\lambda^k}{k!} |\lambda - \frac{1}{2}| = a\lambda |\lambda - \frac{1}{2}| \end{aligned}$$

and it converges to zero with a.

Exercise 3.3.10. Let $Y = \sqrt{N+c}$ for a Poisson variable N with a parameter λ and $c > 0$, let μ be the expectation of Y, find a bound for μ^2 as λ tends to infinity.

Answer. By the equality $E(Y) \leq \{E(Y^2)\}^{\frac{1}{2}}$, the expectation μ of Y is such that $\mu^2 \leq \lambda + c$. A Taylor expansion of the square root of $\lambda + c$, as λ tends to infinity, yields

$$\mu \leq (\lambda + c)^{\frac{1}{2}} = \lambda^{\frac{1}{2}} + \frac{c}{2\lambda^{\frac{1}{2}}} - \frac{c^2}{8\lambda^{\frac{3}{2}}} + o(\lambda^{-\frac{3}{2}}),$$

$$\mu^2 \leq \lambda + c - \frac{c^3}{8\lambda^2} + o(\lambda^2).$$

Problem 3.3.1. (a). Let X_1 and X_2 be independent Poisson variables with parameters λ_1 and λ_2, and let $S = X_1 + X_2$, find the probabilities of X_k conditionally on S and the conditional expectations $E(X_k \mid S)$.

(b). Generalize to n independent Poisson variables and their sum S_n.

Answer. (a). The variable S is Poisson variable with parameter $\lambda_1+\lambda_2$. For all integers $k > i$, the probabilities of (X_1, S) are $p_{i,k} = P(X_1 = i, S = k)$

$$p_{i,k} = P(X_1 = i, X_2 = k - i) = e^{-\lambda_1-\lambda_2} \frac{\lambda_1^i \lambda_2^{k-i}}{i!(k-i)!},$$

$$p_k = P(S = k) = e^{-(\lambda_1+\lambda_2)} \frac{(\lambda_1 + \lambda_2)^k}{n!},$$

and the conditional probability of X_1 given $S = k$ is

$$p_{i|k} = P(X_1 = i \mid S = k) = \binom{k}{i} \frac{\lambda_1^i}{(\lambda_1 + \lambda_2)^i} \frac{\lambda_2^{k-i}}{(\lambda_1 + \lambda_2)^{k-i}},$$

the distribution of X_i conditionally on S is binomial with parameters S and $p_i = \lambda_i(\lambda_1 + \lambda_2)^{-1}$, its conditional expectation is Sp_i.

(b). For n independent Poisson variables and integers $k > i$, the variables S_n and $S_n - X_1$ are Poisson variables with parameter $\mu_n = \lambda_1 + \cdots + \lambda_n$ and $\mu_n - \lambda_1$, the probabilities of (X_1, S_n) are $p_{i,k} = P(X_1 = i, S_n = k)$

$$\begin{aligned} p_{i,k} &= P(X_1 = i, S_n - X_1 = k - i) \\ &= e^{-\mu_n} \frac{\lambda_1^i (\mu_n - \lambda_1)^{k-i}}{i!(k-i)!}, \end{aligned}$$

and the distribution of X_i conditionally on S_n is

$$p_{i|k} = P(X_1 = i \mid S = k) = \binom{k}{i} \frac{\lambda_1^i}{\mu_n^i} \frac{(\mu_n - \lambda_1)^{k-i}}{\mu_n^{k-i}},$$

it is a binomial distribution with parameters S_n and $p_{in} = \lambda_i \mu_n^{-1}$, its conditional expectation is $S_n p_{in}$.

Problem 3.3.2. (a). Write the generating function of a Poisson variable X_λ with parameter λ, and the generating functions of the sum and the empirical mean of Poisson variables.

(b). Write the generating function of a binomial variable X with distribution $B(n,p)$ with the probabilities $p_k = \binom{n}{k} p^k (1-p)^{n-k}$ and prove that it converges to the generating function of a Poisson variable with parameter λ if and only if $np(n)$ converges to λ as n tends to infinity. Show that $p_{Xk} = p_{\lambda k}\{1 + n^{-1}k^2\lambda + o(n^{-1})\}$.

(c). Write the generating function of the sum of n independent binomial variables and prove its convergence to the generating function of a sum of independent Poisson variables.

Answer. (a). A Poisson variable X with parameter λ has the generating function

$$G_X(s) = \sum_{k \in \mathbb{N}} e^{-\lambda} \frac{\lambda^k s^k}{k!} = e^{-\lambda(1-s)},$$

its expectation and its variance are λ. The sum $X_1 + X_2$ of independent Poisson variables X_1 and X_2, with parameters λ_k for $k = 1, 2$, has the generating function

$$G_{X_1+X_2}(s) = e^{-\lambda_1+\lambda_2} \sum_{i,j \in \mathbb{N}} \frac{\lambda_1^i s^i}{i!} \frac{\lambda_2^j s^j}{j!} = e^{-(\lambda_1+\lambda_2)(1-s)},$$

this is the generating function of a Poisson variable with parameter $\lambda_1 + \lambda_2$. The expectation of the empirical mean of n independent Poisson variables with parameter λ is λ and its variance is $n^{-1}\mathrm{Var}(X) = n^{-1}\lambda$.

(b). A binomial variable X with distribution $B(n,p)$ has the probabilities $p_k = \binom{n}{k} p^k (1-p)^{n-k}$ and the generating function

$$G_X(s) = \sum_{k=0}^{n} p_k s^k = \{sp + 1 - p\}^n,$$

its expectation is $EX = np$ and its variance is $np(1-p)$.

The Poisson approximation of a binomial variable is proved by an expansion of its generating function. Let X be binomial variable $B(n, p_n)$ such that np_n converges to λ, then the expectation and the variance of X converge to λ. The logarithm of $G_X(s)$ has the expansion

$$\begin{aligned}\log G_X(s) &= n \log\{(s-1)p_n + 1\}\\ &= (s-1)np_n + \frac{(s-1)^2 np_n^2}{2}\{1 + o(1)\}\end{aligned}$$

where np_n^2 converges to zero as np_n converges to λ and $G_X(s)$ converges to the generating function $G_\lambda(s) = e^{-\lambda(1-s)}$ of a Poisson variable X_λ.

The second order expansion of G_X is

$$G_X(s) = G_\lambda(s) \exp\Big\{\frac{(s-1)^2 np_n^2}{2}\Big\}\{1 + o(1)\}$$

denoted $G_X = G_\lambda G_n\{1 + o(1)\}$ with

$$G_n(s) = \exp\Big\{\frac{(s-1)^2 np_n^2}{2}\Big\}.$$

The probabilities of the binomial variable are determined by the derivatives of G_X for every integer $k \geq 1$

$$\begin{aligned}p_{Xk} &= G_X^{(k)}(1) = \Big\{\sum_{j=0}^{k} \binom{k}{j} G_\lambda^{(j)}(1) G_n^{(k-j)}(1)\Big\}\{1 + o(1)\},\\ p_{\lambda j} &= G_\lambda^{(j)}(1),\\ G_n^{(k-j)}(1) &= n^{-(k-j)} \lambda^{2(k-j)}\{1 + o(1)\},\end{aligned}$$

and

$$\begin{aligned}p_{Xk} &= e^{-\lambda}\Big\{\frac{\lambda^k}{k!} + kn^{-1}\lambda^2 \frac{\lambda^{k-1}}{(k-1)!} + o(n^{-1})\Big\}\{1 + o(1)\}\\ &= p_{\lambda k}\{1 + n^{-1}\lambda k^2 + o(n^{-1})\}.\end{aligned}$$

(c). The sum $X_1 + X_2$ of two independent binomial variables X_k with distributions $B(n_k, p)$ has the generating function

$$G_{X_1+X_2}(s) = G_{X_1}(s) G_{X_2}(s) = \{sp + 1 - p\}^{n_1+n_2}.$$

The sum of N independent binomial variables $B(n, p)$ has the generating function $G_n(s) = \{sp + 1 - p\}^{n_1+\cdots+n_N}$.

The expectation of the empirical mean $\bar{X}_N$ of N independent binomial variables converges to λ and its variance is asymptotically equivalent to $N^{-1}\lambda$, like the empirical mean of N independent Poisson variables with parameter λ. The generating function $\{sp+1-p\}^{N^2}$ of $X_1+\cdots+X_N$ has the expansion

$$\begin{aligned} G_{X_1+\cdots+X_N} &= \exp[N^2 \log\{(s-1)p_N+1\}] \\ &= \exp\{-N\lambda(1-s)\}\{1+o(1)\}, \end{aligned}$$

and it is asymptotically equivalent to the generating function of a sum of N independent Poisson variables with parameter λ.

Problem 3.3.3. Prove that the L^1 distance between the Poisson distribution and the binomial distribution $B(n,p)$ with the probabilities $p_n(k) = \binom{n}{k} p_n^k (1-p_n)^{n-k}$, for $k = 0, \ldots, n$, converges to zero as np_n converges to λ.

Answer. The L^1 distance between the distributions is

$$\begin{aligned} d_\lambda &= \sum_{k=0}^{\infty} \frac{1}{k!} \left| \frac{n!}{(n-k)!} p_n^k (1-p_n)^{n-k} - e^{-\lambda}\lambda^k \right| \\ &= \sum_{k=0}^{n} \frac{1}{k!} \left| \frac{n!}{(n-k)!} p_n^k (1-p_n)^{n-k} - e^{-\lambda}\lambda^k \right| + \sum_{k=n+1}^{\infty} e^{-\lambda} \frac{\lambda^k}{k!} \end{aligned}$$

where the last term is bounded by $\frac{\lambda^{n+1}}{(n+1)!}$ and $n!$ has the de Moivre–Stirling approximation

$$n! \sim \sqrt{2\pi} n^n e^{-n},$$

then the last term converges to zero as n tends to infinity. For the first term of the distance, we have

$$\begin{aligned} \log \frac{n!}{(n-k)!} &\sim \log(n-k)^k - n\log(1-\frac{k}{n}) \\ &\sim \log(n-k)^k + k + \frac{k^2}{2n}, \\ \frac{n!}{(n-k)!} p_n^k (1-p_n)^{n-k} &\sim e^{\frac{k^2}{2n}} \{(n-k)p_n\}^k (1-p_n)^{n-k} \end{aligned}$$

where $(1-p_n)^n$ converges to $e^{-\lambda}$, $a_{nk} = \{(n-k)p_n\}(1-p_n)^{-1}$ converges to λ for every finite k, and

$$\frac{p_n}{1-p_n} = \frac{np_n}{n-np_n} \sim \frac{\lambda}{n-\lambda}$$

converges to zero. It follows that the first term S_{1n} has the approximation

$$\begin{aligned} S_{1n} &\sim \sum_{k=0}^{n} \frac{e^{-\lambda}}{k!} \left| \lambda^k - e^{\frac{k^2}{2n}} a_{nk}^k \right| \\ &\leq \sum_{k=0}^{n} \frac{e^{-\lambda}}{k!} \left| \lambda^k - a_{nk}^k \left(1 + \frac{k^2}{2n}\right) \right| \\ &\leq \sum_{k=0}^{n} \frac{e^{-\lambda}\lambda^k}{k!} \left\{ \frac{k^2}{2n} + o(1) \right\} \end{aligned}$$

with the bound $e^x \geq 1 + x$ for x converging to zero. Finally, there exists a constant C_λ depending only on λ such that

$$\sum_{k=0}^{n} \frac{k^2 \lambda^k}{k!} \leq C_\lambda e^{\lambda},$$

$$\sum_{k=0}^{n} \frac{e^{-\lambda}\lambda^k}{k!} \frac{k^2}{2n} = O(n^{-1})$$

and d_λ converges to zero as n tends to infinity.

Problem 3.3.4. Prove that the distribution of the sum S_n of n independent Bernoulli variables X_k with respective probabilities p_{nk}, for $k = 0, \ldots, n$, converges to the distribution of a Poisson variable with parameter λ if and only if $\pi_n = \sum_{k=1}^{n} p_{nk}$ converges to λ and $\rho_n = \sum_{k=1}^{n} p_{nk}^2$ converges to zero.

Answer. The probabilities of S_n are

$$P_{nk} = P(S_n = k) = \sum_{A_k \subset \{1,\ldots,n\}} \prod_{i \in A_k} p_{ni} \prod_{i \in A_k^c} (1 - p_{ni})$$

where $|A_k| = k$. The generating function of X_i is $1 - p_{ni} + s p_{ni}$, for S_n it is

$$\begin{aligned} G_n(s) &= \prod_{i=1}^{n} (1 - p_{ni} + s p_{ni}) = \exp\Big\{ \sum_{i=1}^{n} \log(1 - p_{ni}(1-s)) \Big\} \\ &= \exp\Big[-(1-s)\pi_n - \frac{(1-s)^2 \rho_n}{2} \{1 + o(1)\} \Big], \end{aligned}$$

G_n converges if and only if the condition are satisfied and its limit is the generating function $G_\lambda(s) = e^{-(1-s)\lambda}$ of a Poisson variable.

Problem 3.3.5. Let $X_1, \ldots, X_k, \ldots$ be independent and identically distributed integer variables and let N be a Poisson variable with parameter λ independent of $X_1, \ldots, X_k, \ldots$, write the generating function of $S_N = \sum_{k=1}^N X_k$, its expectation and its variance.

Answer. Let $p_k = P(X_i = k)$ and let G be the generating function of the variables X_i, their expectation is $E(X) = G'(1) = \mu_X$ and their variance is $\mathrm{Var}(X) = v_X$. For every integer n, the generating function S_n is $G^n(t)$ and its probabilities are defined by the derivatives of G^n at zero

$$\pi_n(k) = P(S_n = k) = (G^n)^k(0),$$

therefore $\pi_n(1) = np_1$, $\pi_n(2) = 2np_2 + n(n-1)p_1^2$

$$\pi_n(3) = 3!\, np_3 + 3!\, n(n-1)p_2 + n(n-1)(n-2)p_1^3, \ldots$$

The generating function of S_N is

$$\begin{aligned} G_\lambda(t) &= \sum_{k \in \mathbb{N}} s^k P(S_N = k) = \sum_{k,n \in \mathbb{N}} s^k P(S_n = k) P(N = n) \\ &= e^{-\lambda} \sum_{k,n \in \mathbb{N}} s^k \frac{\lambda^n}{n!} \pi_n(k). \end{aligned}$$

The first two derivatives of G_λ are

$$\begin{aligned} G'_\lambda(t) &= e^{-\lambda} \sum_{k,n \in \mathbb{N}} k \pi_n(k) s^{k-1} \frac{\lambda^n}{n!}, \\ G''_\lambda(t) &= e^{-\lambda} \sum_{k,n \in \mathbb{N}} k(k-1) \pi_n(k) s^{k-2} \frac{\lambda^n}{n!}, \end{aligned}$$

the expectation of S_n is deduced as

$$E(S_N) = G'_\lambda(1) = e^{-\lambda} \sum_{k,n \in \mathbb{N}} k \pi_n(k) = \frac{\lambda^n}{n!} = \lambda \mu_X$$

and its variance is

$$\begin{aligned} \mathrm{Var}(S_N) &= G''_\lambda(1) + G'_\lambda(1) - G'^2_\lambda(1) \\ &= e^{-\lambda} \sum_{k,n \in \mathbb{N}} k^2 \pi_n(k) \frac{\lambda^n}{n!} - E^2(S_N), \end{aligned}$$

more simply

$$\begin{aligned} \mathrm{Var}(S_N) &= E\{\mathrm{Var}(S_N) \mid N = n\} + \mathrm{Var}\{E(S_N \mid N = n)\} \\ &= v_X E(N) + \mu_X^2 \mathrm{Var}(N) = \lambda(v_X + \mu_X^2), \end{aligned}$$

hence $\mathrm{Var}(S_N) = \lambda E(X^2)$.

(b). Let X and $Y > 0$ be independent Gamma distributions with pa-
meters a and respectively b, write the Laplace transform of the variable
$= X + Y$ and the Laplace transform of the empirical mean $\bar{X}_n$ of n
dependent and identically distributed Gamma variables, find $E(\bar{X}_n^k)$, for
tegers $k \geq 1$.
(c). Prove the convergence of $L_{\bar{X}_n}$ to e^{-at}.

nswer. (a). The Laplace transform of a Gamma variable X is

$$L_X(t) = \frac{1}{\Gamma(a)} \int_0^\infty x^{a-1} e^{-(1+t)x}\, dx = \frac{1}{(1+t)^a},$$

has the derivatives $L_X^{(k)}(t) = (-1)^k E(X^k e^{-tX})$ for every $k \geq 1$ and they
termine the moments of X

$$E(X^k) = (-1)^k L_X^{(k)}(0),$$

have $E(X) = a$, $E(X^2) = a(a+1)$, and for every integer $k \geq 1$

$$E(X^k) = a(a+1) \cdots (a+k-1).$$

(b). The variable $S = X + Y$ has the Laplace transform

$$L_{X+Y}(t) = (t+1)^{-a-b},$$

has a Gamma distribution with parameter $a + b$. The empirical mean
of n independent and identically distributed variables with density f_a
s the Laplace transform

$$\begin{aligned} L_{\bar{X}_n}(t) &= \prod_{k=1}^n E\left(e^{n^{-1}tX_k}\right) = (n^{-1}t + 1)^{-an} \\ &= L_a^n(n^{-1}t). \end{aligned}$$

e transform $L_{\bar{X}_n}$ has infinitely many derivatives and its moments are
$\bar{X}_n) = a$

$$\begin{aligned} E(\bar{X}_n^2) &= n^{-1}a(a+1) + n^{-1}(n-1)a^2 = a^2 + n^{-1}a, \\ E(\bar{X}_n^k) &= L_{\bar{X}_n}^{(k)}(0) \end{aligned}$$

every integer $k \geq 1$.
(c). As n tends to infinity, $L_{\bar{X}_n}$ has the expansion

$$L_{\bar{X}_n}(t) = \{1 - n^{-1}at + o(n^{-2})\}^n$$

d it converges e^{-at}, this is the Laplace transform of a Dirac measure at
nd $\bar{X}_n$ converges in probability to a.

Chapter 4

Laplace transform and characteristic function

4.1 Laplace transform

The Laplace transform of a positive random variable X is $L_X(t) = E(e^{-tX})$.
It determines the distribution function F_X of X at every point of continuity
of F_X, by Feller's inversion formula

$$F_X(x) = \lim_{t \to \infty} \sum_{k \leq xt} \frac{(-1)^k t^k}{k!} L_X^{(k)}(t).$$

For every real variable X, the derivatives of the Laplace transform of a
variable X satisfy

$$(-1)^k L_X^{(k)}(t) = \int_{\mathbb{R}} x^k e^{-tx}\, dF_X(x)$$

and the k-th moment of X is $\mu_k = (-1)^k L_X^{(k)}(0)$. For every positive variable
X having a distribution function such that $\lim_{x\to\infty} xF(x) = 0$, the derivative of $L_X(t)$ satisfies $L'_X(t) = tLF_X(t)$. By convexity of the exponential
function, L_X is a convex function and $L_X(t) \geq e^{-tEX}$ therefore $L_X(t) \geq 1$
for every variable X such that $EX \leq 0$.

The Laplace transform of convolution $f * g$ of functions f and g is the
product $Lf * g(t) = Lf(t)\, Lg(t)$ and for every integer n, the convolution
$x^n = 1 * x^{n-1}$ has the transform

$$Lx^n(t) = \frac{t^n}{n!}.$$

The Laplace transform of a symmetric real variable X is

$$L_X(t) = \int_{-\infty}^0 e^{-tx} f(x)\, dx + \int_0^\infty e^{-tx} f(x)\, dx = 2E(e^{-t|X|}),$$

for a centered Gaussian variable X with variance σ^2, $L_X(t) = e^{\frac{\sigma^2 t^2}{2}}$.

Exercise 4.1.1. For $a > 0$ and $t > 0$, write the Laplace transform of the density

$$f(x) = \frac{t}{\sqrt{2\pi x^3}} \exp\Big\{-\frac{1}{2}\Big(\frac{t}{\sqrt{x}} - a\sqrt{x}\Big)^2\Big\}, \ x > 0.$$

Answer. The Laplace transform of f is obtained by a change of the centering term in the density

$$\begin{aligned} L(s) &= \frac{t}{\sqrt{1\pi x^3}} \int_0^\infty e^{-sx} \exp\Big\{-\frac{1}{2}\Big(\frac{t}{\sqrt{x}} - a\sqrt{x}\Big)^2\Big\}\, dx \\ &= \frac{t}{\sqrt{1\pi x^3}} \exp\Big\{-(\sqrt{a^2+2s} - a)\Big\} \\ &\quad \cdot \int_0^\infty \exp\Big\{-\frac{1}{2}\Big(\frac{t}{\sqrt{x}} - \sqrt{a^2+2s}\sqrt{x}\Big)^2\Big\}\, dx \\ &= \exp\Big\{-(\sqrt{a^2+2s} - a)\Big\}. \end{aligned}$$

Exercise 4.1.2. Write the Laplace transform of the sum $X_1^2 + \cdots + X_n^2$ for n independent centered Gaussian variables X_i with variance 1.

Answer. The Laplace transform of X^2 is

$$\begin{aligned} L(s) = E(e^{-sX^2}) &= \frac{1}{\sqrt{2\pi}} \int_{\mathbb{R}} e^{-\frac{(2s+1)x^2}{2}}\, dx \\ &= (2s+1)^{-\frac{1}{2}} \end{aligned}$$

and the Laplace transform of $X_1^2 + \cdots + X_n^2$ is $L^n(s) = (2s+1)^{-\frac{n}{2}}$.

Exercise 4.1.3. Calculate the Laplace transform of the functions $\sin(x)$ and $\cos(x)$. Calculate $\int_0^\infty x^{-1}\sin(x)\, dx$ using the Laplace transform of $x^{-1}\sin(x)$.

Answer. The Laplace transform of $e^{ix} = \cos(x) + i\sin(x)$ is

$$L(t) = \int_0^\infty e^{-tx} e^{ix}\, dx = \frac{1}{t-i} = \frac{t+i}{1+t^2},$$

it follows that the Laplace transform of $\sin(x)$ is $\frac{1}{1+t^2}$ and for $\cos(x)$ it is $\frac{t}{1+t^2}$.

The integral $L(t) = \int_0^\infty e^{-tx} x^{-1} \sin(x)\, dx$ has the derivative

$$\begin{aligned} L'(t) &= -\int_0^\infty e^{-tx} \sin(x)\, dx \\ &= -\mathrm{Im} \int_0^\infty e^{-tx} e^{itx}\, dx = -\frac{1}{1+t^2}, \end{aligned}$$

it implies $L(t) = \frac{\pi}{2} - \arctan t$. To prove that $L(t)$ conver to zero, let

$$|I - L(t)| \le \Big|\int_0^a (1 - e^{-tx}) x^{-1} \sin(x)\, dx\Big| + \Big|\int_a^\infty x^{-1}\sin(x)\, dx\Big| + \Big|\int_a^\infty e^{-tx} x^{-1}$$

by the mean value theorem, for all a and $b > 0$, there exis that

$$\Big|\int_a^b e^{-tx} x^{-1} \sin(x)\, dx\Big| \le e^{-ta} a^{-1} \Big|\int_a^c \sin(x)\, dx\Big| \le$$

$$\Big|\int_a^b x^{-1} \sin(x)\, dx\Big| \le a^{-1} \Big|\int_a^c \sin(x)\, dx\Big| \le 2a^{-1}$$

and the bounds converge to zero as a and b tend to i grand of the first term converges to zero as t tends to ze $\lim_{t\to 0} |\int_0^a (1 - e^{-tx}) x^{-1} \sin(x)\, dx| = 0$ for every finite a.

Problem 4.1.1. Let X_1 and X_2 be positive continuous var density f, write their bivariate Laplace transform and prov pendence of X_1 and X_2 is equivalent to its factorization.

Answer. The bivariate Laplace transform is $L(s_1, s_2) = E$ dependent variables with density $f = f_X f_Y$ have the Lapla

$$L(s_1, s_2) = L_{X_1}(s_1) L_{X_2}(s_2).$$

Reversely, the factorization of the Laplace functions of X_1

$$\begin{aligned} E\{X_1^i X_2^j\} &= \frac{\partial^i \partial^j L_{X_1+X_2}(s_1, s_2)}{\partial s_1^i \partial s_2^j} \\ &= \frac{\partial^i L_{X_1}(s_1)}{\partial s_1^i} \frac{\partial^j L_{X_2}(s_2)}{\partial s_2^j} = E\{X_1^i\} E\{ \end{aligned}$$

for all integers i and j and the variables are independent.

Problem 4.1.2. (a). A Gamma variable X has a distributio strictly parameter a and its density is

$$f_a(x) = \frac{1}{\Gamma(a)} x^{a-1} e^{-x}, \ x \ge 0,$$

write its Laplace transform and its moments $E(X^k)$.

Problem 4.1.3. (a). Let X be a centred Gaussian variable, write the Laplace transform of $|X|$ and prove the inequality

$$P(|X| > a) \leq 2e^{-\frac{a^2}{2\sigma^2}}, \ a > 0.$$

(b). For the sum $S_n = X_1 + \cdots + X_n$ of n centred Gaussian variables prove the inequality

$$P(n^{-1}|S_n| > a) \leq e^{-\frac{na^2}{2\sigma^2}}, \ a > 0.$$

Answer. The Laplace transform of $|X|$ is

$$E(e^{-t|X|}) = \frac{2}{\sigma\sqrt{2\pi}} \int_0^\infty e^{-tx-\frac{x^2}{2\sigma^2}} = e^{\frac{\sigma^2 t^2}{2}}.$$

By the Bienaymé–Chebychev inequality, for all $a > 0$ and $t > 0$ the probability

$$P(|X| > a) = P\Big(e^{t|X|} > e^{ta}\Big)$$

has the bound

$$P(|X| > a) \leq \inf_{t>0} e^{-ta} E\Big(e^{t|X|}\Big)$$
$$= \inf_{t>0} e^{-ta+\frac{\sigma^2 t^2}{2}},$$

its minimum is achieved at $t = -2a\sigma^2$ where the inequality becomes

$$P(|X| > a) \leq e^{-\frac{a^2}{2\sigma^2}}.$$

The Laplace transform of $n^{-1}S_n$ is

$$E\Big(e^{-tn^{-1}S_n}\Big) = \Big\{E\Big(e^{-tn^{-1}X}\Big)\Big\}^n = e^{\frac{\sigma^2 t^2}{2n}},$$

it follows that for all $a > 0$ and $t > 0$

$$P(n^{-1}|S_n| > a) = P\Big(e^{tn^{-1}S_n} > e^{ta}\Big)$$
$$\leq \inf_{t>0} e^{-ta+\frac{\sigma^2 t^2}{2n}} = e^{-\frac{na^2}{2\sigma^2}}.$$

Problem 4.1.4. Let $(X_i)_{i\geq 1}$ be independent and identically distributed random variables with values in $\mathbb{R}_+$ and let N be a Poisson variable with parameter λ, independent of $(X_i)_{i\geq 1}$. Write the Laplace transform of $S_N = \sum_{k=1}^N X_k$, its expectation and its variance.

Answer. Let L_X be the Laplace transform of X, the Laplace transform of S_k is L_X^k for every integer k, and S_N has the Laplace transform

$$L_\lambda(t) = E(e^{-tS_N}) = e^{-\lambda} \sum_{k\geq 0} \frac{\lambda^k L_X^k(t)}{k!} = e^{\lambda\{L_X(t)-1\}},$$

its derivatives at zero are

$$L'_\lambda(0) = \lambda L'_X(0), \quad L''_\lambda(0) = \lambda L''_X(0) + \lambda^2 L'^2_X(0)$$

with $L'_\lambda(0) = -E(X)$ and $L''_\lambda(0) = E(X^2)$. The expectation of S_N is

$$E(S_N) = -L'_\lambda(0) = \lambda E(X),$$

the variance of X is $\mathrm{Var}(X) = L''_X(0) - L'^2_X(0)$ and the variance of S_N is

$$\mathrm{Var}(S_N) = L''_\lambda(0) - L'^2_\lambda(0) = \lambda L''_X(0) = \lambda E(X^2).$$

Problem 4.1.5. Let $X_1, \ldots, X_n$ be independent and identically distributed variables in $\mathbb{R}_+$, with expectation μ and variance σ^2. Write the Laplace transforms of their empirical mean $\bar{X}_n$ and of $U_n = n^{\frac{1}{2}}(\bar{X}_n - \mu)$, prove their convergence.

Answer. The Laplace transform of $\bar{X}_n$ is

$$L_{\bar{X}_n}(t) = L^n_X(n^{-1}t) = \exp\{n \log L_X(n^{-1}t)\},$$

as n tends to infinity, $L_X(n^{-1}t)$ converges to 1 and it has an expansion

$$\begin{aligned} L_X(n^{-1}t) &= L_X(0) + \frac{t}{n} L'_X(0) + \frac{t^2}{2n^2} L''_X(0) + o(n^{-2}t^2) \\ &= 1 - \frac{t\mu}{n} + \frac{t^2}{2n^2} E(X^2) + o(n^{-2}t^2) \end{aligned}$$

which entails

$$L_{\bar{X}_n}(t) = \exp\Big\{-t\mu + \frac{\sigma^2 t^2}{2n} + o(n^{-1}t^2)\Big\},$$

it converges to $\exp\{-t\mu\}$ for every $t > 0$, therefore $\bar{X}_n$ converges a.s. to μ.

The Laplace transform of U_n is

$$L_{U_n}(t) = e^{n^{\frac{1}{2}}\mu t} E(e^{-n^{\frac{1}{2}}\bar{X}_n t}) = e^{n^{\frac{1}{2}}\mu t} L^n_X(n^{-\frac{1}{2}}t),$$

as n tends to infinity, $L_X(n^{-\frac{1}{2}}t)$ has the expansion

$$\begin{aligned} L_X(n^{-\frac{1}{2}}t) &= 1 - n^{-\frac{1}{2}} t\mu + \frac{E(X^2)t^2}{2n} + o(n^{-1}t^2), \\ \log L_X(n^{-\frac{1}{2}}t) &= -n^{-\frac{1}{2}} t\mu + \frac{\sigma^2 t^2}{2n} + o(n^{-1}t^2) \end{aligned}$$

which entails

$$L_{U_n}(t) = e^{n^{\frac{1}{2}}\mu t} \exp\{-n^{-\frac{1}{2}} t\mu + \frac{\sigma^2 t^2}{2} + o(t^2)$$

and it converges to $\exp\{\frac{\sigma^2 t^2}{2}\}$ for every $t > 0$, this is the Laplace transform of a centered Gaussian variable U with variance σ^2 therefore U_n converges weakly to U.

4.2 Characteristic function

The characteristic function of a real variable X with distribution function F_X is defined as

$$\varphi_X(t) = E(e^{itX}) = \int_{\mathbb{R}} e^{itx}\, dF_X(x),$$

$\varphi_X(t)$ is a real function if and only if X is symmetric. The Fourier transform of a centered Gaussian variable with variance σ^2 is $\sqrt{2\pi}\sigma^{-1}\phi_{\sigma^{-1}}$, where $\phi_{\sigma^{-1}}$ is the density of a centered Gaussian variable with variance σ^{-2}, reciprocally the inverse Fourier transform of a density ϕ_σ is the density $\phi_{\sigma^{-1}}$

$$\phi_\sigma(x) = \frac{1}{2\pi} \int_{\mathbb{R}} e^{-itx} \phi_{\sigma^{-1}}(t)\, dt.$$

The characteristic function of a sum of independent variables is the product of their characteristic functions. If the characteristic function of a real variable X belongs to L^1, then X has a bounded and continuous density

$$f_X(x) = \frac{1}{2\pi} \int_{\mathbb{R}} e^{-itx} \varphi_X(t)\, dt.$$

If the characteristic function of a real variable X has the value zero at a non zero value a, then X belongs a.s. to $2\pi a^{-1}\mathcal{Z}$ and its distribution function is $\sum_{k\in\mathcal{Z}} p_k \delta_{\{2\pi a^{-1}k\}}$ with the probabilities $p_k = P(X = 2\pi a^{-1}k)$, its characteristic function is

$$\varphi_X(t) = \sum_{k\in\mathcal{Z}} p_k e^{2\pi a^{-1}kit},$$

with

$$p_k = a^{-1} \int_0^a e^{2\pi a^{-1}kit}\, dt,$$

where the set of functions $(u_k)_{k\in\mathcal{Z}}$ defined as $u_k(x) = e^{i\omega kx}$ is an orthonormal basis of $L^2[0,a]$, for $\omega = 2\pi a^{-1}$. The L^2 norm of a periodic function f of $L^2[0,a]$ is

$$\|f\|_2 = \int_0^a f(x)\bar{f}(x)\, dx = \sum_{k\in\mathcal{Z}} \lambda_k^2.$$

Examples of characteristic functions: A binomial variable $B(n,p)$ is the sum of n Bernoulli variables and its characteristic function is

$$\varphi_{n,p}(t) = \ (pe^{it} + 1 - p)^n,$$

for a Poisson variable with parameter λ, $\varphi_\lambda(t) = e^{\lambda(e^{it}-1)}$,
for a uniform variable U on $[-1,1]$, $\varphi_U(t) = t^{-1}\sin t$,
for a Gaussian variable $\mathcal{N}(\mu,\sigma^2)$, $\varphi_{\mu,\sigma}(t) = e^{i\mu t}e^{-\frac{\sigma^2t^2}{2}}$.

Let X and Y be dependent real variables with joint distribution function F and with marginal distribution F_X and F_Y and with conditional distributions $F_{X|Y}$ and $F_{Y|X}$, then their characteristic functions are

$$\varphi_X(t) = \int_{\mathbb{R}^2} e^{itx}\,dF_{X|Y}(x;y)\,dF_Y(y) = \int_{\mathbb{R}} \varphi_{X|Y}(t;y)\,dF_Y(y),$$

$$\varphi_Y(t) = \int_{\mathbb{R}^2} e^{ity}\,dF_{Y|X}(y;x)\,dF_X(x) = \int_{\mathbb{R}} \varphi_{Y|X}(t;x)\,dF_X(x),$$

the same rule applies to random vectors X and Y.

Exercise 4.2.1. Let $H_k(x)$ be the kth Hermite polynomial defined by the derivatives of the normal density ϕ as

$$H_k(x) = (-1)^k\frac{\phi^{(k)}(x)}{\phi(x)},\ k \geq 0. \tag{4.1}$$

Write the derivatives of order $k \geq 1$ of the normal density from the inverse of its characteristic function and determine their Fourier transforms.

Answer. The normal density ϕ has the characteristic function $\varphi(t) = e^{-\frac{t^2}{2}}$ and the derivatives of φ are

$$\varphi^{(k)}(t) = (-1)^kH_k(t)\varphi(t).$$

The inverse of the Fourier transform is the normal density

$$\phi(x) = \frac{1}{2\pi}\int_{\mathbb{R}} e^{-itx}\varphi(t)\,dt = \frac{1}{2\pi}\int_{\mathbb{R}} e^{-itx-\frac{t^2}{2}}\,dt,$$

it has the derivatives

$$\phi^{(k)}(x) = \frac{1}{2\pi}\int_{\mathbb{R}}(-it)^k e^{-itx}\varphi(t)\,dt,$$

integrating by parts, we obtain their Fourier transforms

$$\int_{\mathbb{R}} e^{itx}\phi^{(k)}(x)\,dx = (-it)^k\varphi(t).$$

Exercise 4.2.2. Find bounds for the characteristic function φ of a real variable with density f and for $\varphi_n - \varphi$ where φ_n is the empirical characteristic function defined with respect to the empirical distribution function of a sample $(X_i)_{i=1,\ldots,n}$ of independent variables with density f.

Chapter 4

Laplace transform and characteristic function

4.1 Laplace transform

The Laplace transform of a positive random variable X is $L_X(t) = E(e^{-tX})$. It determines the distribution function F_X of X at every point of continuity of F_X, by Feller's inversion formula

$$F_X(x) = \lim_{t\to\infty} \sum_{k\le xt} \frac{(-1)^k t^k}{k!} L_X^{(k)}(t).$$

For every real variable X, the derivatives of the Laplace transform of a variable X satisfy

$$(-1)^k L_X^{(k)}(t) = \int_{\mathbb{R}} x^k e^{-tx}\, dF_X(x)$$

and the k-th moment of X is $\mu_k = (-1)^k L_X^{(k)}(0)$. For every positive variable X having a distribution function such that $\lim_{x\to\infty} xF(x) = 0$, the derivative of $L_X(t)$ satisfies $L'_X(t) = tLF_X(t)$. By convexity of the exponential function, L_X is a convex function and $L_X(t) \ge e^{-tEX}$ therefore $L_X(t) \ge 1$ for every variable X such that $EX \le 0$.

The Laplace transform of convolution $f * g$ of functions f and g is the product $Lf * g(t) = Lf(t)\, Lg(t)$ and for every integer n, the convolution $x^n = 1 * x^{n-1}$ has the transform

$$Lx^n(t) = \frac{t^n}{n!}.$$

The Laplace transform of a symmetric real variable X is

$$L_X(t) = \int_{-\infty}^{0} e^{-tx} f(x)\, dx + \int_0^{\infty} e^{-tx} f(x)\, dx = 2E(e^{-t|X|}),$$

for a centered Gaussian variable X with variance σ^2, $L_X(t) = e^{\frac{\sigma^2 t^2}{2}}$.

Exercise 4.1.1. For $a > 0$ and $t > 0$, write the Laplace transform of the density

$$f(x) = \frac{t}{\sqrt{2\pi x^3}} \exp\Big\{-\frac{1}{2}\Big(\frac{t}{\sqrt{x}} - a\sqrt{x}\Big)^2\Big\},\ x > 0.$$

Answer. The Laplace transform of f is obtained by a change of the centering term in the density

$$\begin{aligned} L(s) &= \frac{t}{\sqrt{1\pi x^3}} \int_0^\infty e^{-sx} \exp\Big\{-\frac{1}{2}\Big(\frac{t}{\sqrt{x}} - a\sqrt{x}\Big)^2\Big\}\, dx \\ &= \frac{t}{\sqrt{1\pi x^3}} \exp\Big\{-(\sqrt{a^2+2s} - a)\Big\} \\ &\qquad \cdot \int_0^\infty \exp\Big\{-\frac{1}{2}\Big(\frac{t}{\sqrt{x}} - \sqrt{a^2+2s}\sqrt{x}\Big)^2\Big\}\, dx \\ &= \exp\Big\{-(\sqrt{a^2+2s} - a)\Big\}. \end{aligned}$$

Exercise 4.1.2. Write the Laplace transform of the sum $X_1^2 + \cdots + X_n^2$ for n independent centered Gaussian variables X_i with variance 1.

Answer. The Laplace transform of X^2 is

$$\begin{aligned} L(s) = E(e^{-sX^2}) &= \frac{1}{\sqrt{2\pi}} \int_{\mathbb{R}} e^{-\frac{-(2s+1)x^2}{2}}\, dx \\ &= (2s+1)^{-\frac{1}{2}} \end{aligned}$$

and the Laplace transform of $X_1^2 + \cdots + X_n^2$ is $L^n(s) = (2s+1)^{-\frac{n}{2}}$.

Exercise 4.1.3. Calculate the Laplace transform of the functions $\sin(x)$ and $\cos(x)$. Calculate $\int_0^\infty x^{-1}\sin(x)\, dx$ using the Laplace transform of $x^{-1}\sin(x)$.

Answer. The Laplace transform of $e^{ix} = \cos(x) + i\sin(x)$ is

$$L(t) = \int_0^\infty e^{-tx} e^{ix}\, dx = \frac{1}{t-i} = \frac{t+i}{1+t^2},$$

it follows that the Laplace transform of $\sin(x)$ is $\frac{1}{1+t^2}$ and for $\cos(x)$ it is $\frac{t}{1+t^2}$.

The integral $L(t) = \int_0^\infty e^{-tx}x^{-1}\sin(x)\, dx$ has the derivative

$$\begin{aligned} L'(t) &= -\int_0^\infty e^{-tx}\sin(x)\, dx \\ &= -\mathrm{Im}\int_0^\infty e^{-tx}e^{itx}\, dx = -\frac{1}{1+t^2}, \end{aligned}$$

it implies $L(t) = \frac{\pi}{2} - \arctan t$. To prove that $L(t)$ converges to I as t tends to zero, let

$$|I - L(t)| \leq \left|\int_0^a (1 - e^{-tx})x^{-1} \sin(x)\, dx\right| + \left|\int_a^\infty x^{-1} \sin(x)\, dx\right| + \left|\int_a^\infty e^{-tx}x^{-1} \sin(x)\, dx\right|,$$

by the mean value theorem, for all a and $b > 0$, there exists c in $]a, b[$ such that

$$\left|\int_a^b e^{-tx}x^{-1} \sin(x)\, dx\right| \leq e^{-ta}a^{-1}\left|\int_a^c \sin(x)\, dx\right| \leq 2e^{-ta}a^{-1},$$

$$\left|\int_a^b x^{-1} \sin(x)\, dx\right| \leq a^{-1}\left|\int_a^c \sin(x)\, dx\right| \leq 2a^{-1},$$

and the bounds converge to zero as a and b tend to infinity, the integrand of the first term converges to zero as t tends to zero which implies $\lim_{t\to 0} |\int_0^a (1 - e^{-tx})x^{-1} \sin(x)\, dx| = 0$ for every finite a.

Problem 4.1.1. Let X_1 and X_2 be positive continuous variables with joint density f, write their bivariate Laplace transform and prove that the independence of X_1 and X_2 is equivalent to its factorization.

Answer. The bivariate Laplace transform is $L(s_1, s_2) = Ee^{s_1X_1+s_2X_2}$. Independent variables with density $f = f_X f_Y$ have the Laplace transform

$$L(s_1, s_2) = L_{X_1}(s_1)L_{X_2}(s_2).$$

Reversely, the factorization of the Laplace functions of $X_1 + X_2$ implies

$$E\{X_1^i X_2^j\} = \frac{\partial^i \partial^j L_{X_1+X_2}(s_1, s_2)}{\partial s_1^i \partial s_2^j} = \frac{\partial^i L_{X_1}(s_1)}{\partial s_1^i} \frac{\partial^j L_{X_2}(s_2)}{\partial s_2^j} = E\{X_1^i\}E\{X_2^j\}$$

for all integers i and j and the variables are independent.

Problem 4.1.2. (a). A Gamma variable X has a distribution $\Gamma(a)$ with a strictly parameter a and its density is

$$f_a(x) = \frac{1}{\Gamma(a)} x^{a-1}e^{-x},\ x \geq 0,$$

write its Laplace transform and its moments $E(X^k)$.

(b). Let X and $Y > 0$ be independent Gamma distributions with parameters a and respectively b, write the Laplace transform of the variable $S = X + Y$ and the Laplace transform of the empirical mean $\bar{X}_n$ of n independent and identically distributed Gamma variables, find $E(\bar{X}_n^k)$, for integers $k \geq 1$.

(c). Prove the convergence of $L_{\bar{X}_n}$ to e^{-at}.

Answer. (a). The Laplace transform of a Gamma variable X is

$$L_X(t) = \frac{1}{\Gamma(a)} \int_0^\infty x^{a-1} e^{-(1+t)x}\,dx = \frac{1}{(1+t)^a},$$

it has the derivatives $L_X^{(k)}(t) = (-1)^k E(X^k e^{-tX})$ for every $k \geq 1$ and they determine the moments of X

$$E(X^k) = (-1)^k L_X^{(k)}(0),$$

we have $E(X) = a$, $E(X^2) = a(a+1)$, and for every integer $k \geq 1$

$$E(X^k) = a(a+1)\cdots(a+k-1).$$

(b). The variable $S = X + Y$ has the Laplace transform

$$L_{X+Y}(t) = (t+1)^{-a-b},$$

its has a Gamma distribution with parameter $a + b$. The empirical mean $\bar{X}_n$ of n independent and identically distributed variables with density f_a has the Laplace transform

$$\begin{aligned} L_{\bar{X}_n}(t) &= \prod_{k=1}^n E\left(e^{n^{-1}tX_k}\right) = (n^{-1}t+1)^{-an} \\ &= L_a^n(n^{-1}t). \end{aligned}$$

The transform $L_{\bar{X}_n}$ has infinitely many derivatives and its moments are $E(\bar{X}_n) = a$

$$\begin{aligned} E(\bar{X}_n^2) &= n^{-1}a(a+1) + n^{-1}(n-1)a^2 = a^2 + n^{-1}a, \\ E(\bar{X}_n^k) &= L_{\bar{X}_n}^{(k)}(0) \end{aligned}$$

for every integer $k \geq 1$.

(c). As n tends to infinity, $L_{\bar{X}_n}$ has the expansion

$$L_{\bar{X}_n}(t) = \{1 - n^{-1}at + o(n^{-2})\}^n$$

and it converges e^{-at}, this is the Laplace transform of a Dirac measure at a and $\bar{X}_n$ converges in probability to a.

Problem 4.1.3. (a). Let X be a centred Gaussian variable, write the Laplace transform of $|X|$ and prove the inequality

$$P(|X| > a) \leq 2e^{-\frac{a^2}{2\sigma^2}},\ a > 0.$$

(b). For the sum $S_n = X_1 + \cdots + X_n$ of n centred Gaussian variables prove the inequality

$$P(n^{-1}|S_n| > a) \leq e^{-\frac{na^2}{2\sigma^2}},\ a > 0.$$

Answer. The Laplace transform of $|X|$ is

$$E(e^{-t|X|}) = \frac{2}{\sigma\sqrt{2\pi}} \int_0^\infty e^{-tx-\frac{x^2}{2\sigma^2}} = e^{\frac{\sigma^2 t^2}{2}}.$$

By the Bienaymé–Chebychev inequality, for all $a > 0$ and $t > 0$ the probability

$$P(|X| > a) = P\Big(e^{t|X|} > e^{ta}\Big)$$

has the bound

$$\begin{aligned} P(|X| > a) &\leq \inf_{t>0} e^{-ta} E\Big(e^{t|X|}\Big) \\ &= \inf_{t>0} e^{-ta+\frac{\sigma^2 t^2}{2}}, \end{aligned}$$

its minimum is achieved at $t = -2a\sigma^2$ where the inequality becomes

$$P(|X| > a) \leq e^{-\frac{a^2}{2\sigma^2}}.$$

The Laplace transform of $n^{-1}S_n$ is

$$E\Big(e^{-tn^{-1}S_n}\Big) = \Big\{E\Big(e^{-tn^{-1}X}\Big)\Big\}^n = e^{\frac{\sigma^2 t^2}{2n}},$$

it follows that for all $a > 0$ and $t > 0$

$$\begin{aligned} P(n^{-1}|S_n| > a) &= P\Big(e^{tn^{-1}S_n} > e^{ta}\Big) \\ &\leq \inf_{t>0} e^{-ta+\frac{\sigma^2 t^2}{2n}} = e^{-\frac{na^2}{2\sigma^2}}. \end{aligned}$$

Problem 4.1.4. Let $(X_i)_{i\geq 1}$ be independent and identically distributed random variables with values in $\mathbb{R}_+$ and let N be a Poisson variable with parameter λ, independent of $(X_i)_{i\geq 1}$. Write the Laplace transform of $S_N = \sum_{k=1}^N X_k$, its expectation and its variance.

Answer. Let L_X be the Laplace transform of X, the Laplace transform of S_k is L_X^k for every integer k, and S_N has the Laplace transform

$$L_\lambda(t) = E(e^{-tS_N}) = e^{-\lambda} \sum_{k\geq 0} \frac{\lambda^k L_X^k(t)}{k!} = e^{\lambda\{L_X(t)-1\}},$$

its derivatives at zero are

$$L'_\lambda(0) = \lambda L'_X(0), \quad L''_\lambda(0) = \lambda L''_X(0) + \lambda^2 L'^2_X(0)$$

with $L'_\lambda(0) = -E(X)$ and $L''_\lambda(0) = E(X^2)$. The expectation of S_N is

$$E(S_N) = -L'_\lambda(0) = \lambda E(X),$$

the variance of X is $\mathrm{Var}(X) = L''_X(0) - L'^2_X(0)$ and the variance of S_N is

$$\mathrm{Var}(S_N) = L''_\lambda(0) - L'^2_\lambda(0) = \lambda L''_X(0) = \lambda E(X^2).$$

Problem 4.1.5. Let $X_1, \ldots, X_n$ be independent and identically distributed variables in $\mathbb{R}_+$, with expectation μ and variance σ^2. Write the Laplace transforms of their empirical mean $\bar{X}_n$ and of $U_n = n^{\frac{1}{2}}(\bar{X}_n - \mu)$, prove their convergence.

Answer. The Laplace transform of $\bar{X}_n$ is

$$L_{\bar{X}_n}(t) = L^n_X(n^{-1}t) = \exp\{n \log L_X(n^{-1}t)\},$$

as n tends to infinity, $L_X(n^{-1}t)$ converges to 1 and it has an expansion

$$\begin{aligned} L_X(n^{-1}t) &= L_X(0) + \frac{t}{n} L'_X(0) + \frac{t^2}{2n^2} L''_X(0) + o(n^{-2}t^2) \\ &= 1 - \frac{t\mu}{n} + \frac{t^2}{2n^2} E(X^2) + o(n^{-2}t^2) \end{aligned}$$

which entails

$$L_{\bar{X}_n}(t) = \exp\Big\{-t\mu + \frac{\sigma^2 t^2}{2n} + o(n^{-1}t^2)\Big\},$$

it converges to $\exp\{-t\mu\}$ for every $t > 0$, therefore $\bar{X}_n$ converges a.s. to μ.

The Laplace transform of U_n is

$$L_{U_n}(t) = e^{n^{\frac{1}{2}}\mu t} E(e^{-n^{\frac{1}{2}}\bar{X}_n t}) = e^{n^{\frac{1}{2}}\mu t} L^n_X(n^{-\frac{1}{2}}t),$$

as n tends to infinity, $L_X(n^{-\frac{1}{2}}t)$ has the expansion

$$\begin{aligned} L_X(n^{-\frac{1}{2}}t) &= 1 - n^{-\frac{1}{2}}t\mu + \frac{E(X^2)t^2}{2n} + o(n^{-1}t^2), \\ \log L_X(n^{-\frac{1}{2}}t) &= -n^{-\frac{1}{2}}t\mu + \frac{\sigma^2 t^2}{2n} + o(n^{-1}t^2) \end{aligned}$$

which entails

$$L_{U_n}(t) = e^{n^{\frac{1}{2}}\mu t} \exp\{-n^{-\frac{1}{2}}t\mu + \frac{\sigma^2 t^2}{2} + o(t^2)$$

and it converges to $\exp\{\frac{\sigma^2 t^2}{2}\}$ for every $t > 0$, this is the Laplace transform of a centered Gaussian variable U with variance σ^2 therefore U_n converges weakly to U.

4.2 Characteristic function

The characteristic function of a real variable X with distribution function F_X is defined as

$$\varphi_X(t) = E(e^{itX}) = \int_{\mathbb{R}} e^{itx}\, dF_X(x),$$

$\varphi_X(t)$ is a real function if and only if X is symmetric. The Fourier transform of a centered Gaussian variable with variance σ^2 is $\sqrt{2\pi}\sigma^{-1}\phi_{\sigma^{-1}}$, where $\phi_{\sigma^{-1}}$ is the density of a centered Gaussian variable with variance σ^{-2}, reciprocally the inverse Fourier transform of a density ϕ_σ is the density $\phi_{\sigma^{-1}}$

$$\phi_\sigma(x) = \frac{1}{2\pi} \int_{\mathbb{R}} e^{-itx} \phi_{\sigma^{-1}}(t)\, dt.$$

The characteristic function of a sum of independent variables is the product of their characteristic functions. If the characteristic function of a real variable X belongs to L^1, then X has a bounded and continuous density

$$f_X(x) = \frac{1}{2\pi} \int_{\mathbb{R}} e^{-itx} \varphi_X(t)\, dt.$$

If the characteristic function of a real variable X has the value zero at a non zero value a, then X belongs a.s. to $2\pi a^{-1}\mathcal{Z}$ and its distribution function is $\sum_{k\in\mathcal{Z}} p_k \delta_{\{2\pi a^{-1}k\}}$ with the probabilities $p_k = P(X = 2\pi a^{-1}k)$, its characteristic function is

$$\varphi_X(t) = \sum_{k\in\mathcal{Z}} p_k e^{2\pi a^{-1}kit},$$

with

$$p_k = a^{-1} \int_0^a e^{2\pi a^{-1}kit}\, dt,$$

where the set of functions $(u_k)_{k\in\mathcal{Z}}$ defined as $u_k(x) = e^{i\omega kx}$ is an orthonormal basis of $L^2[0,a]$, for $\omega = 2\pi a^{-1}$. The L^2 norm of a periodic function f of $L^2[0,a]$ is

$$\|f\|_2 = \int_0^a f(x)\bar{f}(x)\, dx = \sum_{k\in\mathcal{Z}} \lambda_k^2.$$

Examples of characteristic functions: A binomial variable $B(n,p)$ is the sum of n Bernoulli variables and its characteristic function is

$$\varphi_{n,p}(t) = (pe^{it} + 1 - p)^n,$$

for a Poisson variable with parameter λ, $\varphi_\lambda(t) = e^{\lambda(e^{it}-1)}$,
for a uniform variable U on $[-1,1]$, $\varphi_U(t) = t^{-1}\sin t$,
for a Gaussian variable $\mathcal{N}(\mu,\sigma^2)$, $\varphi_{\mu,\sigma}(t) = e^{i\mu t}e^{-\frac{\sigma^2 t^2}{2}}$.

Let X and Y be dependent real variables with joint distribution function F and with marginal distribution F_X and F_Y and with conditional distributions $F_{X|Y}$ and $F_{Y|X}$, then their characteristic functions are

$$\varphi_X(t) = \int_{\mathbb{R}^2} e^{itx}\, dF_{X|Y}(x;y)\, dF_Y(y) = \int_{\mathbb{R}} \varphi_{X|Y}(t;y)\, dF_Y(y),$$
$$\varphi_Y(t) = \int_{\mathbb{R}^2} e^{ity}\, dF_{Y|X}(y;x)\, dF_X(x) = \int_{\mathbb{R}} \varphi_{Y|X}(t;x)\, dF_X(x),$$

the same rule applies to random vectors X and Y.

Exercise 4.2.1. Let $H_k(x)$ be the kth Hermite polynomial defined by the derivatives of the normal density ϕ as

$$H_k(x) = (-1)^k \frac{\phi^{(k)}(x)}{\phi(x)}, \ k \geq 0. \tag{4.1}$$

Write the derivatives of order $k \geq 1$ of the normal density from the inverse of its characteristic function and determine their Fourier transforms.

Answer. The normal density ϕ has the characteristic function $\varphi(t) = e^{-\frac{t^2}{2}}$ and the derivatives of φ are

$$\varphi^{(k)}(t) = (-1)^k H_k(t)\varphi(t).$$

The inverse of the Fourier transform is the normal density

$$\phi(x) = \frac{1}{2\pi}\int_{\mathbb{R}} e^{-itx}\varphi(t)\, dt = \frac{1}{2\pi}\int_{\mathbb{R}} e^{-itx-\frac{t^2}{2}}\, dt,$$

it has the derivatives

$$\phi^{(k)}(x) = \frac{1}{2\pi}\int_{\mathbb{R}} (-it)^k e^{-itx}\varphi(t)\, dt,$$

integrating by parts, we obtain their Fourier transforms

$$\int_{\mathbb{R}} e^{itx}\phi^{(k)}(x)\, dx = (-it)^k\varphi(t).$$

Exercise 4.2.2. Find bounds for the characteristic function φ of a real variable with density f and for $\varphi_n - \varphi$ where φ_n is the empirical characteristic function defined with respect to the empirical distribution function of a sample $(X_i)_{i=1,\dots,n}$ of independent variables with density f.

Answer. By definition

$$|\varphi(t)| = \left|\int_{\mathbb{R}} e^{itx} f(x)\,dx\right| \le \int_{\mathbb{R}} f(x)\,dx = 1$$

and $\varphi(t) = 1$ is equivalent to $t = 0$, we obtain $|\varphi(t)| < 1$, for $t > 0$. For the empirical characteristic function $\varphi_n(t) = n^{-1}\sum_{i=1}^n e^{-tX_i}$, we have

$$\begin{aligned}|\varphi_n(t) - \varphi(t)| &= \left|\int_{\mathbb{R}} e^{itx}\,\{d\widehat{F}_n(x) - dF(x)\}\right| \\ &\le \int_{\mathbb{R}} |d\widehat{F}_n(x) - dF(x)| \le 2\end{aligned}$$

and $\|\varphi_n(t) - \varphi(t)\|_\infty$ converges to zero as $\|\widehat{F}_n(x) - F\|_\infty$ converges to zero. The inversion formula implies

$$|\widehat{F}_n(x) - F(x)| \le \frac{1}{2\pi}\int_{\mathbb{R}} |\varphi_n(t) - \varphi(t)|\,dt$$

and $\|\widehat{F}_n(x) - F\|_\infty$ converges to zero if $\|\varphi_n(t) - \varphi(t)\|_\infty$ converges to zero.

Exercise 4.2.3. Prove that the norm of a characteristic function $\varphi = u + iv$ has the bound $|\varphi(t)|^2 \le \frac{1}{2}\{1 + |\varphi(2t)|\}$.

Answer. The norm of $\varphi(t) = Ee^{itX} = E\cos(tX) + iE\sin(tX)$ is

$$\begin{aligned}|\varphi(t)|^2 &= \{E\cos(tX)\}^2 + \{E\sin(tX)\}^2 \\ &\le E\{\cos^2(tX)\} + E\{\sin^2(tX)\}\end{aligned}$$

where $u(t) = E\cos(tX)$ and $v(t) = E\sin(tX)$ satisfy

$$\begin{aligned}u^2(t &= E^2\{\cos(tX)\} \le E\{\cos^2(tX)\} \\ &= \frac{1}{2}[1 + E\{\cos(2tX)\}] \\ &= \frac{1}{2}\{1 + u(2t)\}, \\ v^2(t) &= E^2\{\sin(tX)\} \le \frac{1}{2}\{1 - u(2t)\} \\ &\le \frac{1}{2}|v(2t)|\end{aligned}$$

as $|u(2t)| \le 1$ entails $\{1 - u(2t)\}^2 \le v^2(2t) = 1 - u^2(2t)$, therefore

$$|\varphi(t)|^2 \le \frac{1}{2}\{1 + |u(2t)| + |v(2t)|\}.$$

Exercise 4.2.4. Prove the inequality $|\varphi(t) - \varphi(s)|^2 \leq 2\{1 - u(t-s)\}$ and find its limit as $|t - s|$ tends to zero.

Answer. Applying the equalities

$$\cos x - \cos y = 2 \sin \frac{x+y}{2} \sin \frac{x-y}{2},$$
$$\sin x - \sin y = 2 \sin \frac{x-y}{2} \cos \frac{x+y}{2}$$

to $\varphi(t) = Ee^{itX}$ leads to

$$\begin{aligned}|\varphi(t) - \varphi(s)|^2 &= 4E^2 \sin \frac{tX - sX}{2} \\ &\leq 4E \sin^2 \frac{tX - sX}{2} \\ &\leq 2\{1 - u(t-s)\}.\end{aligned}$$

As $|t - s|$ tends to zero, $u(t-s) = E\cos\{(t-s)X\}$ converges to one and $|\varphi(t) - \varphi(s)|$ converges to zero.

Exercise 4.2.5. Prove the inequality $|\varphi(t) + \varphi(s)|^2 \leq 2\{1 + u(t-s)\}$.

Answer. The equalities

$$\cos x + \cos y = 2 \sin \frac{x+y}{2} \cos \frac{x-y}{2},$$
$$\sin x + \sin y = 2 \cos \frac{x+y}{2} \cos \frac{x-y}{2}$$

imply

$$\begin{aligned}|\varphi(t) + \varphi(s)|^2 = 4 &\leq 4E \cos^2 \frac{tX - sX}{2} \\ &\leq 2\{1 + u(t-s)\}.\end{aligned}$$

Exercise 4.2.6. Express the L^2 norm of a density on $\mathbb{R}$ using the L^2 norm of its characteristic function.

Answer. For a density f on $\mathbb{R}$, with characteristic function φ and for a density g with characteristic functions $\widehat{g}$, we have

$$\begin{aligned}\int_{\mathbb{R}^2} \varphi(t-s)g(s)g(t)\,dt\,ds &= \int_{\mathbb{R}} \widehat{g}(x)\bar{\widehat{g}}(x)f(x)\,dx \\ &= \int_{\mathbb{R}} |\widehat{g}(x)|^2 f(x)\,dx,\end{aligned}$$

and by the Cauchy–Schwarz inequality

$$\int_{\mathbb{R}} |\widehat{g}(x)|^2 f(x)\,dx \le \Big\{\int_{\mathbb{R}} |g(t)|\,dt\Big\}^2 \le \int_{\mathbb{R}} |g(t)|^2\,dt$$

in particular $\int_{\mathbb{R}} |\widehat{g}(x)|^2\,dx \le \int_{\mathbb{R}} |g(t)|^2\,dt$. Applying the Cauchy–Schwarz inequality to the inverse of the Fourier transform implies

$$\begin{aligned}\int_{\mathbb{R}} |f(x)|^2 g(x)\,dx &= \frac{1}{4\pi^2}\int_{\mathbb{R}}\Big|\int_{\mathbb{R}} e^{-itx}\varphi(t)\,dt\Big|^2 g(x)\,dx\\ &\le \frac{1}{4\pi^2}\int_{\mathbb{R}} |\varphi(t)|^2\,dt \le \int_{\mathbb{R}} |\varphi(t)|^2\,dt\end{aligned}$$

for every density g hence $\int_{\mathbb{R}} |f(x)|^2\,dx \le \int_{\mathbb{R}} |\varphi(t)|^2\,dt$.

Problem 4.2.1. Let $X_1,\ldots,X_k,\ldots$ be independent and identically distributed real random variables and let N be an integer random variable independent of $X_1,\ldots,X_k,\ldots$ Write the characteristic function of $S_N = \sum_{k=1}^N X_k$, its expectation and its variance if N is a Poisson variable with parameter λ and if N is a geometric variable.

Answer. Let φ be the characteristic function of the variables X_k, their variance is

$$\mathrm{Var}(X) = -\varphi''(0) + \varphi'^2(0)$$

with $\varphi'(0) = iE(X)$ and $\varphi''(0) = - E(X^2)$. For every integer k, the characteristic function S_k is $\varphi^k(t)$ and the characteristic function of S_N, if N is a Poisson variable, is

$$\varphi_\lambda(t) = E(e^{itS_N}) = e^{-\lambda}\sum_{k\ge 0}\frac{\lambda^k\varphi^k(t)}{k!} = e^{-\lambda\{1-\varphi(t)\}}$$

such that $\varphi_\lambda(0) = 1$, $\varphi'_\lambda(t) = \lambda\varphi'(t)\varphi_\lambda(t)$, $\varphi'_\lambda(0) = i\lambda E(X)$ and

$$\begin{aligned}\varphi''_\lambda(t) &= \{\lambda\varphi''(t) + \lambda^2\varphi'^2(t)\}\varphi_\lambda(t),\\ \varphi''_\lambda(0) &= -\lambda E(X^2) - \lambda^2 E^2(X).\end{aligned}$$

The expectation and the variance of S_N are

$$\begin{aligned}E(S_N) &= -i\varphi'_\lambda(0) = \lambda E(X),\\ \mathrm{Var}(S_N) &= -\varphi''_\lambda(0) + \varphi'^2_\lambda(0) = \lambda E(X^2).\end{aligned}$$

The formula (1.1) provides a direct method to calculate the variance of S_N

$$\begin{aligned}\mathrm{Var}(S_N) &= E\{\mathrm{Var}(S_N \mid N)\} + \mathrm{Var}\{E(S_N \mid N)\}\\ &= E\{N\mathrm{Var}(X)\} + E^2(X)\mathrm{Var}(N)\end{aligned}$$

therefore

$$\text{Var}(S_N) = \lambda\text{Var}(X) + \lambda E(X^2) = \lambda E(X^2)$$

with $E(N) = \text{Var}(N) = \lambda$.

A geometric variable N with parameters n and p has the probabilities $P(N = k) = p(1-p)^{k-1}$, has the expectation $EN = p^{-1}$ and $E(S_N) = p^{-1}E(X)$. The characteristic function of S_N is

$$\begin{aligned}\varphi_p(t) &= \sum_{k\geq 1}(1-p)^{k-1}p\varphi^k(t)\\ &= p\varphi(t)\sum_{k\geq 1}\{q\varphi(t)\}^{k-1} = \frac{p\varphi(t)}{1-q\varphi(t)}\end{aligned}$$

with $q = 1-p$ then $\varphi_p(0) = 1$, $\varphi_p'(0) = p^{-1}\varphi'(0)$ and

$$\varphi_p''(0) = p^{-2}\{2q\varphi'^2(0) + p\varphi''(0)\} = -p^{-2}\{2qE^2(X) + pE(X^2)\},$$

the variance of S_N is deduced

$$\text{Var}(S_N) = p^{-1}\text{Var}(X) + p^{-2}(2-p)E^2(X).$$

Problem 4.2.2. Let $X_1, \ldots, X_n$ be independent and identically distributed variables in $\mathbb{R}^p$, with expectation μ and variance matrix Σ. Write the characteristic function of their empirical mean $\bar{X}_n$, prove its convergence. Prove the convergence of the characteristic function of the variable $U_n = n^{-\frac{1}{2}}(\bar{X}_n - \mu)$.

Answer. The characteristic function of $\bar{X}_n - \mu$ is $\varphi_n(t) = \varphi^n(n^{-1}t)$ where $\varphi(s) = E(e^{is^T(X-\mu)})$ is the characteristic function of the vector $X_1 - \mu$. A first order expansion of $\varphi(t_n)$ as t_n converges to zero, with $\varphi'(0) = 0$, implies that $\varphi(t_n) = 1 + o(\|t_n\|)$ converges to one,

$$\log\varphi_n(t) = n\log\varphi(n^{-1}t) = o(1)$$

and $\varphi_n(t)$ converges to one. This convergence is equivalent to the convergence in probability of $\bar{X}_n$ to μ.

The characteristic function of U_n is $\varphi_{U_n}(t) = \varphi^n(n^{-\frac{1}{2}}t)$ and by a second order expansion of $\varphi(t_n)$ as t_n converges to zero, we have

$$\varphi(t_n) = 1 - \frac{1}{2}t_n^T\Sigma t_n + o(\|t_n\|^2),$$

$$\log\varphi_{U_n}(t) = n\log\varphi(n^{-\frac{1}{2}}t) = -\frac{1}{2}t^T\Sigma t + o(1)$$

and $\varphi_{U_n}(t)$ converges to $e^{-\frac{1}{2}t^T\Sigma t}$ which is the characteristic function of a centered Gaussian vector U with variance matrix Σ. It follows that U_n converges in distribution to U.

Problem 4.2.3. Let $X_1, \ldots, X_n$ be independent variables in $\mathbb{R}$, with respective expectations μ_i and variances σ_i^2.

(a). Prove the convergence of their empirical mean $\bar{X}_n = n^{-1}S_n$ as $\bar{\mu}_n = n^{-1}\sum_{i=1}^n \mu_i$ converges to a limit μ.

(b). Let $(b_n)_n$ be a real sequence tending to infiny as n tends to infinity, and let $\bar{\sigma}_n^2 = b_n^{-2}\sum_{i=1}^n \sigma_i^2$ converges to a limit σ^2, prove the convergence of the characteristic function of the variable $U_n = (\sum_{i=1}^n \sigma_i^2)^{-\frac{1}{2}}(S_n - n\mu_n)$.

Answer. (a). The variable $\bar{X}_n - \bar{\mu}_n$ is the empirical mean of the centered variables $Y_i = X_i - \mu_i$, its characteristic function is

$$\varphi_n(t) = \prod_{i=1}^n \varphi_{Y_i}(n^{-1}t)$$

such that $\varphi'_{Y_i}(0) = iE(Y_i) = 0$. As t_n converges to zero, a first order expansion of $\varphi(t_n)$ implies that $\varphi_{Y_i}(t_n) = 1 + o(\|t_n\|)$ converges to one and

$$\log \varphi_n(t) = \sum_{i=1}^n \log \varphi_{Y_i}(n^{-1}t) = o(1),$$

therefore $\varphi_n(t)$ converges to one and $\bar{X}_n$ converges in probability to the limit μ of μ_n.

(b). The characteristic function of U_n is

$$\varphi_n(t) = \prod_{i=1}^n \varphi_{Y_i}(b_n^{-1}\bar{\sigma}_n^{-1}t)$$

and by a second order expansion of $\varphi_{Y_i}(t_n)$ as t_n converges to zero, we have

$$\begin{aligned}
\varphi_{Y_i}(t_n) &= 1 - \frac{1}{2}t_n^2\sigma_i^2 + o(\|t_n\|^2), \\
\log \varphi_{U_n}(t) &= \sum_{i=1}^n \log \varphi_{Y_i}(b_n^{-1}\bar{\sigma}_n^{-1}t) \\
&= -\frac{1}{2b_n^2\bar{\sigma}_n^2}\sum_{i=1}^n \sigma_i^2 t^2 + o(1) \\
&= -\frac{1}{2}t^2 + o(1) \qquad (4.2)
\end{aligned}$$

and $\varphi_{U_n}(t)$ converges to $e^{-\frac{t^2}{2}}$ which is the characteristic function of a centered Gaussian variable U with variance one. It follows that U_n converges in distribution to a normal Gaussian variable.

Problem 4.2.4. Let f_n be the density of the sum S_n of n independent and identically distributed centered variables $(X_i)_{i=1,\ldots,n}$ with variance σ^2, such that $\mu_3 = E|X_i|^3$. Prove the uniform approximation

$$f_n(x) - \phi(x) = \frac{\mu_3}{6\sigma^3\sqrt{n}}(x^3 - 3x)\phi(x) + o\Big(\frac{1}{\sqrt{n}}\Big)$$

where ϕ is the normal density.

Answer. The characteristic function of $(x^3 - 3x)\phi(x)$ is

$$\begin{aligned}\int_{\mathbb{R}}(x^3 - 3x)e^{-itx}e^{-\frac{x^2}{2}}\,dx &= e^{-\frac{t^2}{2}}\int_{\mathbb{R}}(x^3 - 3x)e^{-itx-\frac{(x+it)^2}{2}}\,dx\\ &= e^{-\frac{t^2}{2}}(it)^3,\end{aligned}$$

the characteristic function of

$$g_n(x) = f_n(x) - \phi(x) - \frac{\mu_3}{6\sigma^3\sqrt{n}}(x^3 - 3x)\phi(x)$$

is

$$\widehat{g}_n(x) = \varphi_X^n\Big(\frac{t}{\sigma\sqrt{n}}\Big) - e^{-\frac{t^2}{2}} - \frac{\mu_3}{6\sigma^3\sqrt{n}}(it)^3e^{-\frac{t^2}{2}}.$$

As n tends to infinity, the function $\log\varphi_X^n(\frac{t}{\sigma\sqrt{n}})$ has a third order Taylor expansion

$$n\log\varphi_X(n^{-\frac{1}{2}}\sigma^{-1}t) = -\frac{t^2}{2} + \frac{\mu_3}{6\sigma^3\sqrt{n}}(it)^3 + o(\frac{1}{\sqrt{n}}),$$

it follows that

$$\begin{aligned}\widehat{g}_n(t) &= e^{n\log\varphi_X(t)} - e^{-\frac{t^2}{2}} - \frac{\mu_3}{6\sigma^3\sqrt{n}}(it)^3e^{-\frac{t^2}{2}}\\ &= e^{-\frac{t^2}{2}-\frac{\mu_3}{6\sigma^3\sqrt{n}}(it)^3+o(\frac{1}{\sqrt{n}})} - e^{-\frac{t^2}{2}} + \frac{\mu_3}{6\sigma^3\sqrt{n}}(it)^3e^{-\frac{t^2}{2}},\end{aligned}$$

it has the form $\widehat{g}_n(t) = e^{-a}(e^{b_n}\{1 + o(n^{-\frac{1}{2}})\} - 1 - b_n)$ where $b_n = 0(n^{-\frac{1}{2}})$, hence

$$\widehat{g}_n(t) = e^{-a}\frac{b_n^2}{2} + o(n^{-\frac{1}{2}}) = o(n^{-\frac{1}{2}}).$$

By the inverse of the Fourier transform, the convergence rate of g_n is the same as $\widehat{g}_n$ and $g_n = o(n^{-\frac{1}{2}})$.

Problem 4.2.5. Let $X_1, \ldots, X_n$ be independent variables in $\mathbb{R}$, with respective expectations μ_i and variances σ_i^2 and third moment $\rho_i = E|X_i|^3$, and let $S_n = \sum_{i=1}^n X_i$. Prove the uniform approximation for the density f_n of S_n

$$f_n(x) - \phi(x) = \frac{\mu_3}{6s^3\sqrt{n}}(x^3 - 3x)\phi(x) + o\Big(\frac{1}{\sqrt{n}}\Big)$$

where ϕ is the normal density, $\bar{\mu}_n = n^{-1}\sum_{i=1}^n \mu_i$, $s_n^2 = \sum_{i=1}^n \sigma_i^2$ and $\mu_{3n} = \sum_{i=1}^n \rho_i$ such that $n^{-1}s_n^2$ and respectively $n^{-1}\mu_{3n}$ converge to strictly positive and finite limits s^2 and respectively μ_3.

Answer. The characteristic function of $U_n = n^{-\frac{1}{2}}s_n^{-1}(S_n - n\bar{\mu}_n)$ is

$$\varphi_{U_n}(t) = \prod_{i=1}^n \varphi_i\Big(n^{-\frac{1}{2}}s_n^{-1}t\Big)$$

where φ_i is the characteristic function of $X_i - \mu_i$, φ_{U_n} has a third order expansion

$$\varphi_{U_n}(t) = \prod_{i=1}^n \varphi_i\Big(n^{-\frac{1}{2}}s_n^{-1}t\Big)$$

and by a second order expansion of $\varphi_i(t_n)$ as t_n converges to zero, we have

$$\varphi_i(t_n) = 1 - \frac{1}{2}t_n^2\sigma_i^2 + \frac{(it_n)^3\rho_i}{6\sigma_i^3} + o(\|t_n\|^3),$$

$$\begin{aligned}\log\varphi_{U_n}(t) &= \sum_{i=1}^n \log\varphi_i\Big(n^{-\frac{1}{2}}s_n^{-1}t\Big)\\ &= -\frac{t^2}{2} + \frac{(it)^3\mu_3}{6\sqrt{n}s_n^3} + o\Big(\frac{1}{\sqrt{n}}\Big).\end{aligned} \qquad (4.3)$$

The function

$$g_n(x) = f_n(x) - \phi(x) - \frac{\mu_3}{6\sigma^3\sqrt{n}}(x^3 - 3x)\phi(x)$$

has the characteristic function

$$\begin{aligned}\widehat{g}_n(t) &= e^{n\log\varphi_{U_n}(t)} - e^{-\frac{t^2}{2}} - \frac{\mu_3}{6s^3\sqrt{n}}e^{-\frac{t^2}{2}}\\ &= e^{-\frac{t^2}{2} - \frac{\mu_3(it)^3}{6s^3\sqrt{n}}(it)^3 + o(\frac{1}{\sqrt{n}})} - e^{-\frac{t^2}{2}} + \frac{\mu_3}{6s^3\sqrt{n}}(it)^3e^{-\frac{t^2}{2}},\end{aligned}$$

it has the form $\widehat{g}_n(t) = e^{-a}\Big(e^{b_n}\{1 + o(n^{-\frac{1}{2}})\} - 1 - b_n\Big)$ where $b_n = 0(n^{-\frac{1}{2}})$ hence

$$\widehat{g}_n(t) = e^{-a}\frac{b_n^2}{2} + o(n^{-\frac{1}{2}}) = o(n^{-\frac{1}{2}}),$$

the convergence rate of g_n is the same as $\widehat{g}_n$then $g_n = o(n^{-\frac{1}{2}})$.

Problem 4.2.6. Let $(N_t)_{t\geq 0}$ be a Poisson process with intensity λ, and with probabilities $P(N_t = k) = e^{-\lambda t}\frac{(\lambda t)^k}{k!}$. Prove that for every $t > 0$ the variable $t^{-1}N_t$ converges to λ in probability and that the characteristic function of $(\lambda t)^{-\frac{1}{2}}(N_t - \lambda t)$ converges to the characteristic function of a normal variable as t tends to infinity.

Answer. The characteristic function of $t^{-1}N_t$ is

$$E\left(e^{iut^{-1}N_t}\right) = e^{-\lambda t}\sum_{n\geq 0}\frac{\left(e^{iut^{-1}}\lambda t\right)^n}{n!} = e^{\lambda t\left(e^{iut^{-1}}-1\right)}$$

and by an expansion as t tends to infinity $\lambda t(e^{iut^{-1}} - 1) = iu\lambda + o(1)$ and $E(e^{iut^{-1}N_t})$ converges to $e^{iu\lambda}$. The characteristic function of $\frac{N_t-\lambda t}{\sqrt{\lambda t}}$ is

$$\widehat{f}_t(u) = e^{-iu(\lambda t)^{\frac{1}{2}}}Ee^{iuN_t(\lambda t)^{-\frac{1}{2}}} = e^{-iu(\lambda t)^{\frac{1}{2}}}e^{-\lambda t}\sum_{n\geq 0}\frac{\left\{e^{iu(\lambda t)^{-\frac{1}{2}}}\lambda t\right\}^n}{n!}$$

$$= e^{-iu(\lambda t)^{\frac{1}{2}}}e^{\lambda t\left(e^{iu(\lambda t)^{-\frac{1}{2}}}-1\right)},$$

by an expansion as t tends to infinity

$$\lambda t\left(e^{iu(\lambda t)^{-\frac{1}{2}}} - 1\right) = iu(\lambda t)^{\frac{1}{2}} - \frac{u^2}{2} + o(1)$$

and $\widehat{f}_t(u) = e^{-\frac{u^2}{2}+o(1)}$ converges to the characteristic function of the normal variable.

Problem 4.2.7. Let $X_1,\ldots,X_n$ be independent centered variables in $\mathbb{R}$, with respective distribution function $F_1,\ldots,F_n$ and variances $\sigma_1^2,\ldots,\sigma_n^2$, and let $s_n^2 = \sum_{k=1}^n \sigma_k^2$. Prove that under Lindeberg's condition

$$s_n^{-2}\sum_{k=1}^n E(X_k^2 1_{\{s_n^{-1}|X_k|\geq\varepsilon\}}) \to 0, \tag{4.4}$$

as n tends to infinity, for every $\varepsilon > 0$, the characteristic function of the variable $U_n = s_n^{-1}(X_1 + \cdots + X_n)$ converge to the characteristic function of a normal variable $\mathcal{N}(0,1)$.

Answer. The Lindeberg condition is equivalent to

$$s_n^{-2} \sum_{k=1}^{n} E(X_k^2 1_{\{s_n^{-1}|X_k|<\varepsilon\}}) \to 1,$$

it implies $\sigma_k < s_n\varepsilon$ for n sufficiently large, otherwise $|X_k| < s_n\varepsilon$ entails $|X_k| < \sigma_k$ and

$$s_n^{-2} \sum_{k=1}^{n} E(X_k^2 1_{\{|X_k|<s_n\varepsilon\}}) \leq s_n^{-2} \sum_{k=1}^{n} E(X_k^2 1_{\{|X_k|<\sigma_k\}}) < 1$$

which is contradictory. The characteristic function of U_n is

$$\varphi_n(t) = \prod_{k=1}^{n} \varphi_{Y_k}(s_n^{-1}t)$$

and by a second order expansion of $\varphi_{Y_k}(t_n)$ as t_n converges to zero, we have

$$\varphi_{Y_k}(t_n) = 1 - \frac{1}{2} t_n^2 \sigma_k^2 + o(\|t_n\|^2).$$

At $t_n = s_n^{-1}t$ and without the assumption that s_n tends to infinity, Lindeberg's condition implies that for every $\varepsilon > 0$ there exists n_0 such $n \geq n_0$ entails $t_n^2\sigma_k^2 = t^2 s_n^{-2}\sigma_k^2 < \varepsilon^2 t^2$ then, for every $t > 0$

$$\begin{aligned}\log \varphi_{U_n}(t) &= \sum_{k=1}^{n} \log \varphi_{Y_k}(s_n^{-1}t) \\ &= -\frac{1}{2s_n^2} \sum_{k=1}^{n} \sigma_k^2 t^2 + o(1) = -\frac{1}{2}t^2 + o(1)\end{aligned}$$

and $\varphi_{U_n}(t)$ converges to $e^{-\frac{t^2}{2}}$ which is the characteristic function of a centered Gaussian variable U with variance one.

Problem 4.2.8. Let $X_1, \ldots, X_n$ be independent centered variables in $\mathbb{R}$, with respective distribution function $F_1, \ldots, F_n$ and variances $\sigma_1^2, \ldots, \sigma_n^2$, and let $s_n^2 = \sum_{k=1}^{n} \sigma_k^2$. Prove that Lindeberg's condition (4.4) is satisfied under Lyapunov's conditions $E(|X_k|^{2+\delta})$ finite for $\delta > 0$ and

$$\sum_{k=1}^{n} E(|X_k|^{2+\delta}) = o(s_n^{2+\delta}).$$

Answer. The condition $E(|X_k|^{2+\delta})$ finite for $\delta > 0$ implies

$$\sigma_k = \|X_k\|_{L^2} \leq \|X_k\|_{L^{2+\delta}}$$

and σ_k is finite for every k. The variable $S_n = X_1 + \ldots + X_n$ has the variance s_n^2 and $s_n = \|S_n\|_{L^2} \leq \|S_n\|_{L^{2+\delta}}$.

The condition $\sum_{k=1}^n E(|X_k|^{2+\delta}) = o(s_n^{2+\delta})$ implies

$$1 = s_n^{-2} \sum_{k=1}^n E(X_k^2) \leq s_n^{-2} \sum_{k=1}^n E(|X_k|^{2+\delta}) = o(s_n^\delta),$$

therefore s_n tends to infinity as n tends to infinity. Schwarz's inequality entails

$$\begin{aligned}
\sum_{k=1}^n E(X_k^2 1_{\{|X_k|>s_n\varepsilon\}}) &\leq \sum_{k=1}^n \{E(|X_k|^{2+\delta})\}^{\frac{2}{2+\delta}} \{P(|X_k| > s_n\varepsilon)\}^{\frac{\delta}{2+\delta}} \\
&\leq \frac{1}{s_n^\delta \varepsilon^\delta} \sum_{k=1}^n \{E(|X_k|^{2+\delta})\}^{\frac{2}{2+\delta}} \{E(|X_k|^{2+\delta})\}^{\frac{\delta}{2+\delta}} \\
&= \frac{1}{s_n^\delta \varepsilon^\delta} \sum_{k=1}^n E(|X_k|^{2+\delta}),
\end{aligned}$$

under Lyapunov's condition $\sum_{k=1}^n E(X_k^2 1_{\{s_n^{-1}|X_k|\geq\varepsilon\}}) = o(s_n^{2+\delta})$, it follows that

$$s^{-2} \sum_{k=1}^n E(X_k^2 1_{\{|X_k|>s_n\varepsilon\}}) = o(1)$$

as n tends to infinity.

Problem 4.2.9. Let X and Y be independent positive variables with distribution function F such that $\bar{F}(x) = 1 - F(x) \sim x^{-a} L(x)$ with L varying slowly as x tends to infinity and $a > 1$, give a bound for the limit of $P(X > x \mid X + Y > x)$ as x tends to infinity.

Answer. For positive variables X and Y

$$P(X > x \mid X + Y > x) = \frac{P(X > x)}{P(X + Y > x)} = \frac{\bar{F}(x)}{1 - F^{\star 2}(x)}$$

where

$$\begin{aligned}
F^{\star 2}(x) &= \int \{1 - \bar{F}(x - y)\}\, dF(y) \\
&= 1 + \int \bar{F}(x - y)\, d\bar{F}(y),
\end{aligned}$$

hence $1 - F^{\star 2}(x) = -\bar{F}^{\star 2}(x)$. As x tends to infinity, the approximation $x^{-a}(x-y)^a = (1-x^{-1}y)^a = 1 - ax^{-1}y + o(1)$ implies

$$-\frac{\bar{F}^{\star 2}(x)}{F(x)} = L^{-1}(x)\int \Big\{1 - ax^{-1}y + o(1)\Big\} L(x-y)\, d\bar{F}(y)$$
$$= L^{-1}(x)\int L(x-y)\, d\bar{F}(y)$$
$$-ax^{-1}L^{-1}(x)\int yL(x-y)\, d\bar{F}(y) + o(1).$$

For an exponentially decreasing function $L(x) = e^{-bx}$ we have

$$L^{-1}(x)\int L(x-y)\, d\bar{F}(y) = L^{-1}(x)\int_0^x e^{-b(x-y)}\, d\bar{F}(y)$$
$$+L^{-1}(x)\int_x^\infty e^{-b(x-y)}\, d\bar{F}(y),$$

the last term is equivalent to

$$\int_x^\infty \{y^{-a}L^{-1}(y)L'(y) - ay^{-a-1}\}\, dy \le (b+1)x^{-a},$$

it converges to zero as x tends to infinity, and

$$\left|\int_0^x e^{-b(x-y)}\, d\bar{F}(y)\right| \le \int_0^x dF(y) \to 1.$$

By the same argument, for the first term $x^{-1}L^{-1}(x)\int_x^\infty yL(x-y)\, d\bar{F}(y)$ is equivalent to $x^{-1}\int_x^\infty \{y^{1-a}L^{-1}(y)L'(y) - ay^{-a}\}\, dy$, as x tends to infinity, and it is bounded by $x^{-a}\int_x^\infty L^{-1}(y)L'(y)\, dy - \int_x^\infty ay^{-a-1}\, dy$ which converges to zero, and

$$x^{-1}L^{-1}(x)\left|\int_0^x yL(x-y)\, d\bar{F}(y)\right| \le \int_0^x dF(y) \to 1.$$

It follows that $P(X > x \mid X + Y > x)$ converges to a limit $p \ge \frac{1}{2}$.

4.3 Asymptotic expansions

A real random variable X with a density and finite moments $E(X^k)$, for $k = 1, \ldots, K$, has a Fourier transform $\varphi(t) = E(e^{itX})$ with K derivatives

$$\varphi^{(k)}(t) = E\{(iX)^k e^{itX}\},\ k = 1, \ldots, K.$$

The kth cumulant κ_k is the coefficient of $\frac{(it)^k}{k!}$ in the Taylor series expansion of $\log\varphi$, if $E(X^k)$ is finite for every k, let

$$\varphi(t) = \sum_{k=0}^{\infty} \frac{(it)^k}{k!} E(X^k),$$
$$\log\varphi(t) = \sum_{k=1}^{\infty} \frac{(-1)^{k+1}}{k} \Big\{\sum_{k=1}^{\infty} \frac{(it)^k}{k!} E(X^k)\Big\}^k$$
$$= \sum_{k=1}^{\infty} \frac{(it)^k}{k!} \kappa_k,$$

κ_k is a homogeneous polynomial of the moments, of degree k, $\kappa_1 = E(X)$, $\kappa_2 = \mathrm{Var}(X)$, $\kappa_3 = E\{(X-\kappa_1)^3\}$, $\kappa_4 = E\{(X-\kappa_1)^4\} - 3\kappa_2^2$, etc.

For n independent and identically distributed variables X_i with moments of order k, the characteristic function $\varphi_n(t) = \varphi^n(n^{-\frac{1}{2}}t)$ of the normalized sum $U_n = n^{-\frac{1}{2}} \sum_{i=1}^n (X_i - EX_i)$ has the uniform series expansion

$$\varphi_n(t) = \exp\Big\{\sum_{j=2}^{k} \frac{(it)^j}{j!} n^{-\frac{j-2}{2}} \kappa_j + o\Big(n^{-\frac{k-2}{2}}\Big)\Big\}$$
$$= e^{-\frac{\sigma^2 t^2}{2}} \Big\{\sum_{j=1}^{k-2} n^{-\frac{j}{2}} r_{j-2}(it) + o\Big(n^{-\frac{k-2}{2}}\Big)\Big\},$$

where the polynomials r_j depend on the cumulants of X of order lower than $j+2$

$$r_1(s) = \frac{\kappa_3 s^3}{3!},\ r_2(s) = \frac{\kappa_4 s^4}{4!} + \frac{\kappa_2^2 s^6}{72}, \ldots$$

by a Taylor expansion in a neighborhood of zero. By inversion of the Fourier transform, the distribution function F_n of U_n is approximated by the distribution function Φ of a variable $\mathcal{N}(0,1)$. If $\lim_{|t|\to\infty} |\varphi_X(t)| < 1$ and the variables X_i have moments of order k, F_n has the uniform asymptotic expansion

$$P(U_n \le x) = \Phi(x) + \sum_{j=1}^{k-2} n^{-\frac{j}{2}} R_j(x) + o\Big(n^{-\frac{k-2}{2}}\Big), \tag{4.5}$$

where the polynomials R_j depend on the cumulants of X lower than j and their Fourier transforms are

$$\int_{\mathbb{R}} e^{itx}\, dR_j(x)\, dx = e^{-\frac{t^2}{2}} r_j(it).$$

Exercise 4.3.1. Let $(X_i)_{i=1,\dots,n}$ be independent variables such that $E|X_i|^3$ is finite, and let $V_n = \sum_{i=1}^n E|X_i - E(X_i)|^2$, prove the Berry–Essen inequality

$$\sup_{x\in\mathbb{R}} \Big| P\Big(V_n^{-\frac{1}{2}} \sum_{i=1}^n \{X_i - E(X_i)\} \le x\Big) - \Phi(x)\Big| \le K \frac{\sum_{i=1}^n E|X_i - EX_i|^3}{V_n^{\frac{3}{2}}}.$$

Answer. In the expansion (4.5), the first order term of the approximation of the distribution function $F_n(x)$ of the normalized sum $U_n = V_n^{-\frac{1}{2}}(\sum_{i=1}^n \{X_i - E(X_i)\}$ by $\Phi(x)$, for identically distributed variables, is $n^{-\frac{1}{2}} R_1(x)$ where

$$\int_{\mathbb{R}} e^{itx}\, dR_1(x)\, dx = -e^{-\frac{t^2}{2}} \frac{it^3\kappa_3}{6},$$

and by inversion of the Fourier transform

$$\sup_{x\in\mathbb{R}} |R_1(x)| = \frac{\kappa_3}{12\pi} \Big| \int_{\mathbb{R}} e^{isx} e^{-\frac{s^2}{2}}\, ds \Big| \le \frac{\kappa_3}{12\pi}$$

where $\kappa_3 = E|X_i - E(X_i)|^3$. The $k-3$ following terms have lower orders $n^{-\frac{j}{2}}$, the bound of $\sup_{x\in\mathbb{R}} |F_n(x) - \Phi(x)|$ is therefore a $O(n^{-\frac{1}{2}}\kappa_3\sigma^{-3})$ and it is asymptotically equivalent to the bound of the Berry–Essen inequality as $V_n = 0(n)$.

Under the conditions of independent variables, the Fourier transform of each variable $Y_i = V_n^{-\frac{1}{2}}\{X_i - E(X_i)\}$ has an expansion like $\varphi_n(t)$ and $R_1(x)$ is replaced by $n^{-1}R_{1n}(x)$ such that

$$\int_{\mathbb{R}} e^{itx}\, dR_{1n}(x)\, dx = -e^{-\frac{t^2}{2}} \frac{it^3\kappa_{3n}}{6},$$

with $\kappa_{3n} = \sum_{i=1}^n E|X_i - E(X_i)|^3$. By inversion of the Fourier transform

$$|R_{1n}(x)| \le \frac{\kappa_{3n}}{12\pi}$$

and $V_n^{-\frac{3}{2}} = 0(n^{-\frac{3}{2}})$ which yields a bound of the same order as the Berry–Essen inequality.

Exercise 4.3.2. Give an asymptotic expansion for the distribution of $U_n = n^{\frac{1}{2}}(\sigma_n^2 - \sigma^2)$ where $\sigma_n^2 = n^{-1}\sum_{i=1}^n X_i^2$ is the empirical variance of n independent and identically distributed centered variables X_i with moments of order k and variance σ^2.

Answer. The empirical variance σ_n^2 converges in probability to σ^2 and the Fourier transform of U_n is

$$Ee^{itU_n} = e^{-itn^{\frac{1}{2}}\sigma^2}\varphi_2^n(n^{-\frac{1}{2}}t)$$

where φ_2 is the characteristic function of the variables X_i^2, it has the expansion according to the cumulants $(\kappa_{2j})_{j\geq 1}$ of X_i^2, where $(\kappa_j)_{j\geq 1}$ are the cumulants of X_i

$$\begin{aligned}Ee^{itU_n} &= \exp\Big[\sum_{j=2}^{k}\frac{(it)^j n^{-\frac{j-2}{2}}}{j!}\kappa_{2j} + o\Big(n^{-\frac{k-2}{2}}\Big)\Big]\\ &= e^{-\frac{\mu_4 t^2}{2}}\Big\{\sum_{j=1}^{k-2} n^{-\frac{j}{2}} r_{j-2}^{(2)}(it) + o\Big(n^{-\frac{k-2}{2}}\Big)\Big\},\end{aligned}$$

where the polynomials $r_j^{(2)}(s)$ are defined by an expansion of the exponential with the even cumulants, $r_1^{(2)}(s) = \frac{\kappa_6 s^3}{3!}$, $r_2^{(2)}(s) = \frac{\kappa_8 s^4}{4!} + \frac{\kappa_4^2 s^6}{72}, \ldots$

The distribution function of U_n has the expansion

$$P(U_n \leq x) = \Phi(x) + \sum_{j=1}^{k-2} n^{-\frac{j}{2}} R_j^{(2)}(x) + o\Big(n^{-\frac{k-2}{2}}\Big)$$

where the polynomials $R_j^{(2)}$ have the Fourier transforms

$$\int_{\mathbb{R}} e^{itx}\, dR_j^{(2)}(x)\, dx = e^{-\frac{t^2}{2}} r_j^{(2)}(it).$$

Exercise 4.3.3. For n independent and identically distributed real variables X_i with moments of order k and for g a function of $C^2(\mathbb{R})$, give an asymptotic expansion for the distribution of $U_n = n^{\frac{1}{2}}\{T_n - g(\mu)\}$ where $T_n = n^{-1}\sum_{i=1}^n g(X_i)$ is the empirical mean of the variables X_i with expectation μ.

Answer. The cumulants of the variables $g(X_i)$ are denoted $\kappa_j^{(g)}$ and the polynomials $r_j^{(g)}$ and $R_j^{(g)}$ are deduced like above, then the distribution function of U_n has the expansion

$$P(U_n \leq x) = \Phi(x) + \sum_{j=1}^{k-2} n^{-\frac{j}{2}} R_j^{(g)}(x) + o\Big(n^{-\frac{k-2}{2}}\Big).$$

Exercise 4.3.4. Give an asymptotic expansion for the distribution of $U_n = n^{\frac{1}{2}}(\sigma_n^2 - \sigma^2)$ where $\sigma_n^2 = (n-1)^{-1}\sum_{i=1}^n (X_i - \bar{X}_n)^2$ is the empirical variance of n independent and identically distributed variables X_i with moments of order k and variance σ^2 and $\bar{X}_n$ is their empirical mean.

Answer. The variable σ_n^2 has the expectation σ^2, it is a function of the empirical mean of the variables $(Y, X) = (Y_i, X_i)_{i=1,\ldots,n}$ where $Y_i = X_i^2$ and their characteristic function is

$$\varphi(s,t) = E(e^{isY+itX}) = \int_{\mathbb{R}} e^{isx^2+itx}\, dF_X(x).$$

The derivatives of φ at zero are $\varphi'_s(0) = iE(X^2)$, $\varphi'_t(0) = iE(X)$ and

$$\varphi^{(k_1,k_2)}(0) = i^{k_1+k_2} E(X^{2k_1+k_2}) = i^{k_1+k_2} \mu_{2k_1+k_2}.$$

In a neighborhood of zero, $\varphi(s,t)$ has the expansion

$$\varphi(s,t) = \sum_{2j_1+j_2 \le k} \frac{(it)^{2j_1+j_2}}{(2j_1+j_2)!} \mu_{2j_1+j_2} + o(\|(s,t)\|^k),$$

and $\log \varphi(s,t)$ has the expansion

$$\begin{aligned}\log \varphi(s,t) &= \sum_{k=1}^{\infty} \frac{(-1)^{k+1}}{k} \{\varphi(s,t) - 1\}^k \\ &= \sum_{2j_1+j_2 \le k} \frac{(it)^{2j_1+j_2}}{(2j_1+j_2)!} \kappa_{2j_1+j_2} + o(\|(s,t)\|^k),\end{aligned}$$

with the cumulant of the variables X_i. Let $Z_i = (Y_i, X_i)$ and let g the function of $C^2(\mathbb{R}^{2n})$ defined as

$$g(y,x) = (n-1)^{-1} \sum_{i=1}^{n} y_i - \{n(n-1)\}^{-1} \Big(\sum_{i=1}^{n} x_i\Big)^2,$$

the function g has the first derivatives $g'_{1i}(y,x) = (n-1)^{-1}$ and $g'_{1i+n}(y,x) = n^{-1}(n-1)^{-1} \sum_{i=1}^{n} x_i$ and it has the second derivatives $g'_{1i+n}(y,x) = -\{n(n-1)\}^{-1}$.

The variable $U_n = n^{\frac{1}{2}}\{g(Y,X) - Eg(Y,X)\}$ has a Taylor expansion

$$\begin{aligned}U_n &= n^{\frac{1}{2}}(Z - EZ)^T Eg'(Z) \\ &\quad + \frac{n}{2}(Z - EZ)^T Eg''(Z)(Z - E(Z)\{1 + o_p(1)\},\end{aligned}$$

which is denoted $U_n = \widetilde{U}_n + o_p(1)$ and the cumulants κ_{nj} of U_n are approximated by those of $\widetilde{U}_n$

$$\kappa_{nj} = \widetilde{\kappa}_{nj} + o_p\Big(n^{-\frac{j-2}{2}}\Big).$$

Since $\widetilde{U}_n$ is a quadratic expression of Z, its cumulants $\widetilde{\kappa}_{nj}$ depends on the cumulants of Z of order $2j$ and of the first two derivatives of g

$$\widetilde{\kappa}_{n1} = \frac{n}{2(n-1)} \operatorname{Var}(Z), \ \widetilde{\kappa}_{n2} = \operatorname{Var}(\widetilde{U}_n), \ldots$$

$\widetilde{\kappa}_{n2}$ depends on $\kappa_j(Z)$, $j = 2, 3, 4$, etc.

The characteristic function $\varphi_n(t) = \varphi^n(n^{-\frac{1}{2}}t)$ of the variable U_n is approximated by the characteristic function $\widetilde{\varphi}_n$ of $\widetilde{U}_n$

$$\widetilde{\varphi}_n(t) = e^{-\frac{\sigma_Z^2 t^2}{2}} \Big\{ \sum_{j=0}^{k-2} n^{-\frac{j}{2}} \widetilde{r}_{n\,j-2}(it) + o\Big(n^{-\frac{k-2}{2}}\Big) \Big\}$$

where the polynomials $\widetilde{r}_{nj}$ depend on the cumulants of X of order lower than $4j+2$

$$r_{n1}(s) = \frac{\widetilde{\kappa}_{n3} s^3}{3!}, \ r_{n2}(s) = \frac{\widetilde{\kappa}_{n4} s^4}{4!} + \frac{\widetilde{\kappa}_{n2}^2 s^6}{72}, \ldots$$

The distribution function of U_n is approximated by the distribution function Φ of a variable $\mathcal{N}(0,1)$, it has the uniform asymptotic expansion

$$P(U_n \le x) = \Phi(x) + \sum_{j=1}^{k-2} n^{-\frac{j}{2}} \widetilde{R}_{nj}(x) + o\Big(n^{-\frac{k-2}{2}}\Big),$$

where the polynomials $\widetilde{R}_{nj}$ depend on the cumulants of Z lower than $2j$ and their Fourier transforms are

$$\int_{\mathbb{R}} e^{itx} \, d\widetilde{R}_{nj}(x) \, dx = e^{-\frac{t^2}{2}} \widetilde{r}_{nj}(it).$$

Chapter 5

Continuous distributions

5.1 Exponential variables

An exponential variable X with strictly positive parameter λ has the density $f(x) = \lambda e^{-\lambda x}$ for $x > 0$, its Laplace transform is

$$L(s) = E(e^{-sX}) = \lambda \int_0^\infty e^{-(\lambda+s)x}\,dx = \frac{\lambda}{\lambda+s}, \quad s > 0.$$

Its moments are obtained by derivation as $E(X) = -L'(0)\lambda^{-1}$, $E(X^2) = L''(0) = 2\lambda^{-2}$ and $E(X^k) = L^{(k)}(0) = k\lambda^{-k}$, for every integer k.

Exercise 5.1.1. Let X be a strictly positive random variable on $\mathbb{R}_+$ such that $P(X > s+t \mid X > s) = P(X > t)$ is strictly positive for all s and t of $\mathbb{R}_+$, find the distribution of X.

Answer. The property of the conditional distribution of X implies

$$P(X > s+t) = P(X > t)P(X > s)$$

for every s such that $h(s) = P(X > s)$ is strictly positive. For all integers k and n, we have $h(ks) = h^k(s)$ therefore $h(n^{-1}s) = h^{\frac{1}{n}}(s)$ and

$$h(n^{-1}ks) = h^{\frac{k}{n}}(s).$$

Every real number x is the limit of a sequence of rational numbers, it follows that $h(x) = h^x(1)$. Let $\lambda = P(X > 1)$, then $P(X > x) = \lambda^x = e^{x \log \lambda}$ where $\mu = \log \lambda$ is negative for λ in $]0,1[$, equivalently $P(X > x) = e^{-|\mu| x}$ and the variable X has an exponential distribution.

Exercise 5.1.2. For a random variable X on $\mathbb{R}_+$, calculate $E(t-X \mid X < t)$ and $E(X-t \mid X > t)$, prove that X has an exponential distribution if and only if $E(X \mid X > t)$ is constant.

Answer. Let $\bar{F}(x) = 1 - F(x)$, integrations by parts imply

$$\begin{aligned} E(t - X \mid X < t) &= t - \frac{\int_0^t x\, dF(x)}{F(t)} \\ &= \frac{\int_0^t F(x)\, dx}{F(t)}, \\ E(X - t \mid X > t) &= \frac{\int_t^\infty x\, d\bar{F}(x)}{\bar{F}(t)} - t \\ &= \frac{\int_t^\infty \bar{F}(x)\, dx}{\bar{F}(t)}. \end{aligned}$$

For an exponential variable with expectation λ^{-1} we have

$$E(X - t \mid X > t) = \lambda^{-1} = E(X).$$

Reversely let $E(X - t \mid X > t) = \alpha$ and $I_t = \int_t^\infty \bar{F}(x)\, dx$, then $I_t = \alpha \bar{F}(t)$ which is equivalent to $I_t = -\alpha I'(t)$, hence $I_t = \alpha e^{-\alpha^{-1}t}$ and $\bar{F}(t) = e^{-\alpha^{-1}t}$.

Exercise 5.1.3. Let $X_1, \ldots, X_n$ be independent real variables with exponential distribution function $F(x) = 1 - e^{-x}$ and let $X_n^* = \max_{k \leq n} X_k$. Let G_n be the distribution function of $a_n^{-1} X_n^*$, $a_n > 0$, prove that G_n converges to zero if $a_n = n$ and determine a_n for a non degenerated limit.

Answer. The distribution function of X_n^* is F^n and $G_n(x) = F^n(a_n x)$. For the exponential distribution, as $0 < 1 - e^{-nx} < 1$ for every $x > 0$, the limit

$$\lim_{n \to \infty} \log F^n(nx) = n \log(1 - e^{-nx}) = -\infty$$

implies that G_n converges to zero as n tends to infinity. With an arbitrary a_n tending to infinity as n tends to infinity, we have $\lim_n F(a_n x) = 1$ for every $x > 0$ and

$$e^{-a_n x} = 1 - \{G_n(x)\}^{\frac{1}{n}} = F(a_n x),$$

hence G_n converges to zero as n tends to infinity.

Exercise 5.1.4. Let X and Y be independent exponential variables with distinct parameters, find the distribution of XY^{-1}.

Answer. For every continuous and bounded function h on $\mathbb{R}_+$, integrable with respect to the distribution of XY^{-1}, the integral $I = \lambda\mu \int_{\mathbb{R}_+^2} h(xy^{-1}) e^{-\lambda x} e^{-\mu y}\, dx\, dy$ is written as

$$I = \lambda\mu \int_{\mathbb{R}_+^2} h(z) e^{-(\mu+\lambda z)y} y\, dz\, dy,$$

by a change of variable, the density of XY^{-1} is

$$\begin{aligned} f_Z(z) &= \lambda\mu \int_{\mathbb{R}_+} e^{-(\mu+\lambda z)y} y\, dy \\ &= \frac{\lambda\mu}{(\mu+\lambda z)^2} \int_{\mathbb{R}_+} e^{-s} s\, ds = \frac{\lambda\mu}{(\mu+\lambda z)^2} \end{aligned}$$

and its integral on $\mathbb{R}_+$ is 1.

Exercise 5.1.5. Let X and Y be independent exponential variables such that $X \sim \mathcal{E}_\lambda$ and $Y \sim \mathcal{E}_\mu$, $\lambda > \mu$, prove that $t^{-1}P(X \le t < X+Y)$ converges to λ as t converges to zero.

Answer. The probability of $\{X \le t < X+Y\}$ is

$$\begin{aligned} P(Y > t - X \ge 0) &= \lambda \int_0^t e^{-\mu(t-x)} e^{-\lambda x}\, dx \\ &= \lambda e^{-\mu t} \int_0^t e^{-(\lambda-\mu)x}\, dx \\ &= \frac{\lambda e^{-\mu t}}{\lambda-\mu} \{1 - e^{-(\lambda-\mu)t}\} \\ &= \frac{\lambda}{\lambda-\mu} \{e^{-\mu t} - e^{-\lambda t}\}. \end{aligned}$$

By an expansion of the exponential probabilities as t tends to zero, $e^{-\mu t} - e^{-\lambda t}$ is equivalent to $(\lambda-\mu)t + o(t)$ and the result follows.

Exercise 5.1.6. Let X be a variable in $\mathbb{R}_+^2$ with dependent exponential marginal densities and with joint density $f_X(x_1, x_2) = c \exp\{-\lambda x_1 - \eta x_2 - \theta(x_1 \wedge x_2)\}$, determine c, the expectation and the variance of X.

Answer. The constant is such that

$$\begin{aligned} c^{-1} &= \int_{\mathbb{R}_+} \int_0^{x_2} e^{-(\lambda+\theta)x_1} e^{-\eta x_2}\, dx_1\, dx_2 \\ &\quad + \int_{\mathbb{R}_+} \int_0^{x_1} e^{-(\eta+\theta)x_2} e^{-\lambda x_1}\, dx_1\, dx_2, \end{aligned}$$

$$\begin{aligned}
c^{-1} &= \frac{1}{\lambda+\theta}\int_{\mathbb{R}_+}\{1-e^{-(\lambda+\theta)x_2}\}e^{-\eta x_2}\,dx_2 \\
&\quad+\frac{1}{\eta+\theta}\int_{\mathbb{R}_+}\{1-e^{-(\eta+\theta)x_1}\}e^{-\lambda x_1}\,dx_1 \\
&= \frac{1}{\lambda+\theta}\Big\{\frac{1}{\eta}-\frac{1}{\lambda+\eta+\theta}\Big\}+\frac{1}{\eta+\theta}\Big\{\frac{1}{\lambda}-\frac{1}{\lambda+\eta+\theta}\Big\} \\
&= \frac{\eta+\lambda}{\eta\lambda(\lambda+\eta+\theta)}.
\end{aligned}$$

The Laplace transform of X is defined as $L(s) = E\{e^{-(s_1X_1+s_2X_2)}\}$ for $s=(s_1,s_2)$ in $\mathbb{R}_+^2$

$$\begin{aligned}
L(s) &= c\int_{\mathbb{R}_+}\int_0^{x_2} e^{-(\lambda+\theta+s_1)x_1}e^{-(\eta+s_2)x_2}\,dx_1\,dx_2 \\
&\quad+c\int_{\mathbb{R}_+}\int_0^{x_1} e^{-(\eta+\theta+s_2)x_2}e^{-(\lambda+s_1)x_1}\,dx_1\,dx_2 \\
&= \frac{c}{\lambda+\theta+s_1}\Big\{\frac{1}{\eta+s_2}-\frac{1}{\lambda+\eta+\theta+s_2}\Big\} \\
&\quad+\frac{c}{\eta+\theta+s_2}\Big\{\frac{1}{\lambda+s_1}-\frac{1}{\lambda+\eta+\theta+s_1}\Big\} \\
&= \frac{c(\lambda+\theta)}{(\lambda+\theta+s_1)(\eta+s_2)(\lambda+\eta+\theta+s_2)} \\
&\quad+\frac{c(\eta+\theta)}{(\eta+\theta+s_2)(\lambda+s_1)(\lambda+\eta+\theta+s_1)}.
\end{aligned}$$

The expectation of X_1 is equal to $-L_k'(0)$

$$\begin{aligned}
E(X_1) &= \frac{c}{\eta(\lambda+\theta)(\lambda+\eta+\theta)}+\frac{c}{\lambda^2(\lambda+\eta+\theta)}+\frac{c}{\lambda(\lambda+\eta+\theta)^2} \\
&= \frac{1}{\lambda+\eta}\Big\{\frac{\lambda}{\lambda+\theta}+\frac{\eta}{\lambda}+\frac{\eta}{\lambda(\lambda+\eta+\theta)}\Big\}.
\end{aligned}$$

The second derivative of the Laplace transform with respect to x_1 is $E(X_1^2)$ therefore

$$\begin{aligned}
E(X_1^2) &= \frac{2c}{\eta(\lambda+\theta)^2(\lambda+\eta+\theta)}+\frac{2c}{\lambda^3(\lambda+\eta+\theta)} \\
&\quad+\frac{2c}{\lambda(\lambda+\eta+\theta)^3}+\frac{2c}{\lambda^2(\lambda+\eta+\theta)^2},
\end{aligned}$$

therefore

$$E(X_1^2) = \frac{2}{\lambda+\eta}\Big\{\frac{\lambda}{(\lambda+\theta)^2}+\frac{\eta}{\lambda^2}+\frac{\eta}{(\lambda+\eta+\theta)^2}+\frac{\eta}{\lambda(\lambda+\eta+\theta)}\Big\},$$

and the variance is deduced. The expressions are similar for X_2 by exchanging λ and η.

Exercise 5.1.7. A two-dimensional random variable $T = (T_1, T_2)$ is defined by independent exponential variables Z_1, Z_2, Z_3 as $T_1 = \min(Z_1, Z_2)$ and $T_2 = \min(Z_1, Z_3)$. Prove that the survival function

$$\bar{F}_T(t_1, t_2) = P(T_1 \geq t_1, T_2 \geq t_2)$$

of $T = (T_1, T_2)$ satisfies $\bar{F}_T(s + t_1, s + t_2) = \bar{F}_T(s, s)\bar{F}_T(t_1, t_2)$ and that it is singular.

Answer. For exponential variables $\bar{F}(s + t) = \bar{F}(s)\bar{F}(t)$ and the survival function of (T_1, T_2) is

$$\bar{F}_T(t_1, t_2) = \bar{F}_1(t_1)\bar{F}_2(t_2)\bar{F}_3(t_1 \vee t_2),$$

its marginals are $\bar{F}_{T_1}(t_1) = e^{-(\lambda_1+\lambda_3)t_1}$ and $\bar{F}_{T_2}(t_2) = e^{-(\lambda_2+\lambda_3)t_2}$. The survival function of Z_3 satisfies

$$\begin{aligned}\bar{F}_3((s + t_1) \vee (s + t_2)) &= \bar{F}_3(s + t_1)1_{\{t_1 \geq t_2\}} + \bar{F}_3(s + t_2)1_{\{t_1 < t_2\}} \\ &= \bar{F}_3(s)\bar{F}_3(t_1 \vee t_2)\end{aligned}$$

and the expression of $\bar{F}_T(s + t_1, s + t_2)$ follows. The distribution function F_T is singular since the probability of equal components is not zero

$$\begin{aligned}P(T_1 = T_2) &= P(Z_3 < \min(Z_1, Z_2)) \\ &= \int_0^\infty \bar{F}_1(t)\bar{F}_2(t)\, dF_3(t).\end{aligned}$$

Exercise 5.1.8. The hazard function of a variable with density f and distribution function F is $h(t) = f(t)\{1 - F(t)\}^{-1}$, prove that a constant hazard function determine an exponential variable.

Answer. For an exponential variable with parameter λ, we have $r(t) \equiv \lambda$. Reversely let $h(t) = \lambda$, integrating the hazard function yields $\log\{1 - F(t)\} = -\int_0^t h(s)\, ds + c$ then

$$1 - F(t) = e^c \exp\{-\int_0^t h(s)\, ds\} = e^{-\lambda t}$$

it follows that $h(t) \equiv \lambda$ and $c = 0$.

Exercise 5.1.9. A Weibull variable $W(\lambda, \alpha)$ has the distribution function $F(t) = 1 - e^{-(\lambda t)^\alpha}$, for λ and $\alpha > 0$ and $t \geq 0$. Determine the hazard function of a Weibull variable and conditions for an increasing or decreasing hazard function. Let X be an exponential variable, find the distribution of $X^{\frac{1}{\alpha}}$.

Answer. The density of F is $f(t) = \alpha\lambda^\alpha t^{\alpha-1}e^{-(\lambda t)^\alpha}$ and its hazard function is $h(t) = \alpha\lambda^\alpha t^{\alpha-1}$. It is increasing if $\alpha > 1$ and decreasing if $\alpha < 1$. For an exponential variable $X^{\frac{1}{\alpha}}$ with parameter μ and for every integrable function g of X, we have

$$Eg(X^{\frac{1}{\alpha}}) = \mu \int_0^\infty g(x^{\frac{1}{\alpha}})e^{-\mu x}\,dx = \alpha\mu \int_0^\infty g(y)e^{-\mu y^\alpha} y^{\alpha-1}\,dy$$

and by the change of variable $\lambda^\alpha = \mu$, the variable $X^{\frac{1}{\alpha}}$ has a Weibull distribution $W(\lambda, \alpha)$.

Exercise 5.1.10. Let $X_1, \ldots, X_n$ be independent Weibull variables $W(\lambda, \alpha)$ find the distribution functions of their minimum and their maximum.

Answer. The distribution of the minimum variable $\underline{X_n} = \min_{k=1,\ldots,n} X_k$ is determined by

$$P(\underline{X_n} \geq t) = \prod_{k=1,\ldots,n} P(X_k \geq t) = \{1 - F(t)\}^n = e^{n(\lambda t)^\alpha},$$

this is a Weibull distribution $W(n^{\frac{1}{\alpha}}\lambda, \alpha)$.

The variable $\overline{X_n} = \max_{k=1,\ldots,n} X_k$ has the distribution function

$$P(\overline{X_n} \leq t) = \prod_{k=1,\ldots,n} P(X_k \leq t) = F^n(t).$$

Exercise 5.1.11. Find the limit of the Pareto distribution function $F(t) = 1 - (1 + \sigma^{-1}t)^{-\alpha}$, $\sigma > 0$, $t \geq 0$, and of its hazard function as σ tends to infinity and $\sigma^{-1}\alpha$ converges to a limit.

Answer. Let $\bar{F}(t) = 1 - F(t) = (1+\sigma^{-1}t)^{-\alpha}$, an expansion of the logarithm as σ tends to infinity and $\sigma^{-1}\alpha$ converges to λ yields

$$\bar{F}(t) = \exp\{-\alpha\log(1 + \sigma^{-1}t)) = e^{-\sigma^{-1}\alpha t} \sim e^{-\lambda t},$$

hence $F(t)$ converges to the exponential distribution function $1 - e^{-\lambda t}$ and the hazard function is

$$\lambda(t) = \frac{\alpha\sigma^{-1}}{1 + \sigma^{-1}t},$$

it converges to λ under the conditions.

Exercise 5.1.12. Let Y an exponential variable and let X an exponential variable with parameter λY, find its distribution.

Answer. Assuming that Y has an exponential distribution with parameter μ, the density of X is

$$f(x) = \lambda\mu \int_0^\infty y e^{-y(\mu+\lambda x)}\, dx = \frac{\lambda\mu}{(\mu+\lambda x)^2}$$

and its distribution function is

$$F(x) = 1 - \frac{1}{1+\lambda x}$$

under the condition $\mu = 1$.

Problem 5.1.1. Let $T_1, \ldots, T_n$ be independent exponential variables with the same distribution.

(a). Find the distribution of $S_k = \sum_{k=1}^k T_i$.

(b). Find the distribution of $N_t = \max\{n : S_n \le t\}$ and calculate $E\int_0^\infty h(N_t)\, dt$ as $\sum_{n\ge 0} h(n)$ converges.

Answer. (a). The distribution function of $S_2 = T_1 + T_2$ is

$$\begin{aligned} F^{\star 2}(t) &= \int_0^t P(T_1 \le t-s)\, dP(T_2 \le s) \\ &= \lambda \int_0^t \{1 - e^{-\lambda(t-s)}\} e^{-\lambda s}\, ds = 1 - e^{-\lambda t}\{1+\lambda t\}. \end{aligned}$$

Assuming that $S_k = \sum_{k=1}^k T_i$ has the distribution function

$$F^{\star k}(t) = 1 - e^{-\lambda t}\left\{1 + \lambda t + \cdots + \frac{\lambda^{k-1}t^{k-1}}{(k-1)!}\right\},$$

for S_{k+1} we have

$$\begin{aligned} F^{\star(k+1)}(t) &= 1 - e^{-\lambda t}\left\{1 + \lambda t + \frac{\lambda^2 t^2}{2} + \frac{\lambda^3}{2}\int_0^t (t-s)^2\, ds \right. \\ &\quad \left. + \cdots + \frac{\lambda^k}{(k-1)^{k-1}} \int_0^t (t-s)^{k-1}\, ds\right\} \\ &= 1 - e^{-\lambda t}\left\{1 + \lambda t + \cdots + \frac{\lambda^k t^k}{k!}\right\}. \end{aligned}$$

(b). The probabilities of N_t are given by

$$P(N_t \leq n) = P(S_n \leq t) = F^{\star n}(t)$$

and N_t has the distribution of a Poisson process

$$P(N_t = n) = F^{\star n}(t) - F^{\star(n+1)}(t) = e^{-\lambda t}\frac{\lambda^n t^n}{n!}.$$

The integral $E\int_0^\infty h(N_t)\,dt$ is sum

$$\sum_{n\geq 0} h(n)\int_0^\infty P(N_t = n)\,dt = \lambda^{-1}\sum_{n\geq 0} h(n).$$

5.2 Gamma and Beta variables

A variable X has a Gamma distribution $\Gamma(a,\lambda)$ with strictly parameters a and λ if its density is

$$f_{a,\lambda}(x) = \frac{\lambda^a}{\Gamma(a)}x^{a-1}e^{-\lambda x},\ x \geq 0, \tag{5.1}$$

its expectation is $E(X) = \lambda^{-1}a$ and its variance is $\mathrm{Var}(X) = \lambda^{-2}a$, its Laplace transform is

$$\varphi_X(t) = \frac{\lambda^a}{\Gamma(a)}\int_0^\infty x^{a-1}e^{-(\lambda+t)x}\,dx = \frac{\lambda^a}{(\lambda+t)^a}.$$

The sum of k Gamma variables X_i with parameters (a_i,λ) is Gamma variable with parameters $(\sum_{i=1}^k a_i, \lambda)$, as their Laplace transform is the product $\prod_{i=1}^k \varphi_{X_i}(t)$.

The Beta density on $[0,1]$ with strictly parameters a and b is

$$f_{a,b}(x) = B_{a,b}^{-1}x^{a-1}(1-x)^{b-1}, \quad B_{a,b} = \frac{\Gamma_a\Gamma_b}{\Gamma_{a+b}},$$

it has expectation $E(X) = (a+b)^{-1}a$ and its variance is

$$\mathrm{Var}(X) = \frac{ab}{(a+b)^2(a+b+1)}.$$

Under a Dirichlet distribution, a variable $X = (X_1,\ldots,X_k)$ on $S = \{x = (x_1,\ldots,x_k), \sum_{i=1}^k x_i = 1\}$ has the density with strictly positive parameters $a_1,\ldots,a_k$

$$f_a(x) = \frac{\Gamma_{\sum_{i=1}^k a_i}}{\prod_{i=1}^k \Gamma_{a_i}}\prod_{i=1}^{k-1} x_i^{a_i-1}\Big(1-\sum_{j=1}^{k-1} x_j\Big)^{a_k-1},$$

the expectation of X_i is $E(X_i) = (\sum_{i=1}^k a_i)^{-1} a_i$ and its variance is

$$\mathrm{Var}(X) = \frac{a_i \sum_{j \neq i} a_j}{(\sum_{i=1}^k a_i)^2 (\sum_{i=1}^k a_i + 1)},$$

the covariance of X_i and X_j is

$$\mathrm{Cov}(X_i, X_j) = \frac{a_i a_j}{(\sum_{i=1}^k a_i)^2 (\sum_{i=1}^k a_i + 1)}.$$

Exercise 5.2.1. Prove that $\Gamma_{\frac{1}{2}} = \sqrt{\pi}$ and $\Gamma_{\frac{3}{2}} = \frac{\sqrt{\pi}}{2}$.

Answer. By the change of variable $x = \frac{y^2}{2}$, the density $f_{\frac{1}{2}}(x)$ has the integral

$$\begin{aligned}
\int_0^\infty f_{\frac{1}{2}}(x)\, dx &= \frac{1}{\Gamma(\frac{1}{2})} \int_0^\infty x^{-\frac{1}{2}} e^{-x}\, dx \\
&= \frac{\sqrt{2}}{\Gamma(\frac{1}{2})} \int_0^\infty e^{-\frac{y^2}{2}}\, dy \\
&= \frac{1}{\sqrt{2}\Gamma(\frac{1}{2})} \int_{\mathbb{R}} e^{-\frac{y^2}{2}}\, dy = 1
\end{aligned}$$

where $\int_{\mathbb{R}} e^{-\frac{y^2}{2}}\, dy = \sqrt{2\pi}$, it follows $\Gamma_{\frac{1}{2}} = \sqrt{\pi}$. Integrating by parts yields $\Gamma_x = (x-1)\Gamma_{x-1}$ hence $\Gamma_{\frac{3}{2}} = \frac{\sqrt{\pi}}{2}$.

Exercise 5.2.2. Calculate the convolution of two Gamma densities $f_{\lambda_1,a}$ and $f_{\lambda_2,b}$, for a integer.

Answer. Gamma densities with distinct parameters λ_1 and λ_2 have convolutions

$$\begin{aligned}
f^{\star 2}(x) &= \frac{\lambda_1^a \lambda_2^b}{\Gamma(a)\Gamma(b)} \int_0^x e^{-\lambda_1(x-y)-\lambda_2 y} (x-y)^{a-1} y^{b-1}\, dy \\
&= \frac{\lambda_1^a \lambda_2^b}{\Gamma(a)\Gamma(b)} e^{-\lambda_1 x} \sum_{k=0}^a \binom{a}{k} x^{a-1-k} \int_0^x e^{-(\lambda_2-\lambda_1)y} y^{k+b-1}\, dy \\
&= \frac{e^{-\lambda_1 x} \lambda_1^a \lambda_2^b}{\Gamma(a)\Gamma(b)(\lambda_2-\lambda_1)^b} \sum_{k=0}^a \binom{a}{k} x^{a-1-k} \frac{1}{(\lambda_2-\lambda_1)^k} \int_0^x e^{-u} u^{k+b-1}\, du
\end{aligned}$$

where $\int_0^x e^{-u} u^{k+b-1}\, du = k! - k! e^{-x} \Big\{ \sum_{j=0}^k \frac{x^{k-j}}{(k-j)!} \Big\}$.

For Gamma densities with $\lambda_1 = \lambda_2 = \lambda$, we have

$$f^{\star 2}(x) = \frac{\lambda^{a+b}}{\Gamma(a)\Gamma(b)} e^{-\lambda x} \sum_{k=0}^{a} \binom{a}{k} x^{a-1-k} \int_0^x y^{k+b-1}\, dy$$
$$= \frac{\lambda^{a+b}}{\Gamma(a)\Gamma(b)} e^{-\lambda x} \sum_{k=0}^{a} \binom{a}{k} \frac{x^{a+b-1}}{k+b}.$$

Exercise 5.2.3. Let X be a $\Gamma(2, \lambda)$ variable with a random parameter λ following a $\Gamma(2, \alpha)$ distribution, find the density of X.

Answer. The variable X has the density

$$f_\alpha(x) = \alpha^2 x \int_0^\infty y^3 e^{-(x+\alpha)y}\, dy$$
$$= \frac{6\alpha^2 x}{(x+\alpha)^4}$$

where the integrand is proportional to the $\Gamma(4, x+\alpha)$ density. Writing

$$\frac{x}{(x+\alpha)^4} = \frac{1}{(x+\alpha)^3} - \frac{\alpha}{(x+\alpha)^4}$$

the integral of f_α on $\mathbb{R}_+$ is

$$\int_0^\infty f_\alpha(x)\, dx = 6\alpha^2 \int_0^\infty \Big\{ \frac{1}{(x+\alpha)^3} - \frac{\alpha}{(x+\alpha)^4} \Big\}\, dx$$
$$= 6\alpha^2 \Big\{ \frac{1}{2\alpha^2} - \frac{\alpha}{3\alpha^3} \Big\} = 1$$

and the function f_α is a density.

Exercise 5.2.4. Let X be a binomial variable with a random probability p following a Beta distribution, find the density of X and the density of p conditionally on X.

Answer. If $p \sim B_{a,b}$, the joint density of $X \sim \mathcal{B}(n,p)$ and p and the probabilities of X are

$$P(X=k, p=y) = \frac{\binom{n}{k}}{B_{a,b}} y^{a+k-1}(1-y)^{n-k+b-1},$$
$$\pi_k = P(X=k) = \int_0^1 P(X=k, p=y)\, dy = \frac{B_{a+k,n-k+b}\binom{n}{k}}{B_{a,b}}.$$

The conditional density of p is the Beta density with parameters $a+k$ and $n-k+b$.

Exercise 5.2.5. Let X be a geometric variable with a random probability p following a Beta distribution, find the unconditional distribution of X and the distribution of p conditionally on X.

Answer. Let p have a Beta(a,b) distribution, the probabilities of X are

$$\pi_k = \frac{1}{B_{a,b}} \int_0^1 y^a (1-y)^{b+k-2}\, dy = \frac{B_{a+1,b+k-1}}{B_{a,b}}$$

and p has a Beta distribution $B_{a+1,b+k-1}$ conditionally on X, the conditional distribution of p is

$$f(p \mid X = k) = \frac{1}{B_{a+1,b+k-1}} p^a (1-p)^{b+k-2}.$$

Exercise 5.2.6. Let X be an exponential variable with a random parameter following a $\Gamma_{a,\lambda}$ distribution, find the distribution of X.

Answer. The unconditional density of X is

$$\begin{aligned} f(x) &= \frac{\lambda^a}{\Gamma(a)} \int_0^\infty y^a e^{-y(\lambda+x)}\, dy \\ &= \frac{a\lambda^a}{(\lambda+x)^{a+1}}, \end{aligned}$$

on $\mathbb{R}_+$ and its distribution function is

$$F(x) = 1 - \frac{\lambda^a}{(\lambda+x)^a}.$$

Exercise 5.2.7. Let X be a Poisson variable with a random intensity λ, find the distribution of X if λ has a $\Gamma_{p,\alpha}$ distribution.

Answer. The probabilities of X are

$$\begin{aligned} \pi_k &= \frac{\alpha^p}{k!(p-1)!} \int_0^\infty e^{-x(\alpha+1)} x^{k+p-1}\, dx \\ &= \frac{\alpha^p (k+p-1)!}{(\alpha+1)^{k+p} k!(p-1)!} \\ &= \binom{k+p-1}{k} \frac{\alpha^p}{(\alpha+1)^{k+p}}, \end{aligned}$$

the unconditional distribution of X is therefore binomial with parameters $N_k = k+p-1$ and $p_\alpha = (\alpha+1)^{-1}$.

Exercise 5.2.8. Let Y be a $\Gamma_{\alpha+X,\beta}$ variable conditionally to a Poisson variable X, find the distribution of Y, its expectation and its variance.

Answer. The density of Y is

$$\begin{aligned} f(y) &= e^{-\lambda-\beta y} \sum_{k \geq 0} \frac{\lambda^k \beta^{\alpha+k}}{k!(\alpha+k-1)!} y^{\alpha+k-1} \\ &= e^{-\lambda-\beta y} \beta^{\frac{\alpha+1}{2}} y^{\frac{\alpha-1}{2}} I_{2k\alpha-1+}(2\sqrt{\beta y}) \end{aligned}$$

where I_n is the nth Bessel function. The expectation of Y is

$$E(Y) = \beta^{-1} E(\alpha + X) = \beta^{-1}(\alpha + \lambda),$$

the variance of Y conditionally on X is

$$\text{Var}(Y \mid X) = \beta^{-2}(\alpha + X)$$

and its variance is

$$\begin{aligned} \text{Var}(Y) &= E\{\text{Var}(Y \mid X)\} + \text{Var}\{E(Y \mid X)\} \\ &= \frac{E(\alpha+X)}{\beta^2} + \frac{\text{Var}(X)}{\beta^2} \\ &= \frac{\alpha+2\lambda}{\beta^2}. \end{aligned}$$

Exercise 5.2.9. Let $f = pf_1 + (1-p)f_2$ be a mixture density with a random probability p, find f and the variance of a variable with density f as p has a uniform or a Beta distribution on $[0, 1]$.

Answer. If p has a uniform distribution, $\int_0^1 p\,dp = \frac{1}{2}$ and f is a mixture density with the probabilities $\frac{1}{2}$. Let $\mu_k = \int_0^1 x f_k(x)\,dx$ and let $\sigma_k^2 = \int_0^1 (x-\mu_k)^2 f_k(x)\,dx$, for $k = 1, 2$, the expectation and the variance of a variable X with density f are

$$\begin{aligned} E(X) &= \frac{1}{2}(\mu_1 + \mu_2), \\ \text{Var}(X) &= \frac{1}{12}(\mu_1 + \mu_2)^2 + \frac{1}{2}(\sigma_1^2 + \sigma_2^2). \end{aligned}$$

If p has a Beta distribution $B_{a,b}$, its expectation is $E(p) = (a+b)^{-1}a$ and f is a mixture density with the probabilities $(a+b)^{-1}a$ and $(a+b)^{-1}b$. The variables of p and $1-p$ have the same distribution, their variance is denoted $v_{a,b}$. The expectation and the variance of a variable X are

$$\begin{aligned} E(X) &= \frac{a\mu_1 + b\mu_2}{a+b}, \\ \text{Var}(X) &= v_{a,b} \frac{(a\mu_1 + b\mu_2)^2}{(a+b)^2} + \frac{a\sigma_1^2 + b\sigma_2^2}{a+b}. \end{aligned}$$

Exercise 5.2.10. Let Y be a multinomial variable in $\mathbb{R}^d$ with random probability under a Dirichlet distribution, find the distribution of Y.

Answer. Let $Y = (Y_1, \ldots, Y_k)$ the multinomial variable with random probability $p = (p_1, \ldots, p_{k-1}, 1 - \sum_{i=1}^{k-1} p_i)$, under a Dirichlet distribution, with density f_a, then Y has the probability distribution

$$P(Y = y) = \frac{\Gamma_{\sum_{i=1}^k a_i}}{\prod_{i=1}^k \Gamma_{a_i}} \prod_{i=1}^{k-1} \int_{\sum_{i=1}^k p_i = 1} p_i^{y_i + a_i - 1} \Big(1 - \sum_{j=1}^{k-1} p_j\Big)^{y_k + a_k - 1} \prod_{i=1}^k dp_i,$$

therefore

$$P(Y = y) = \frac{\Gamma_{\sum_{i=1}^k a_i} \prod_{i=1}^k \Gamma_{y_i + a_i}}{\Gamma_{\sum_{i=1}^k (y_i + a_i)} \prod_{i=1}^k \Gamma_{a_i}}.$$

Problem 5.2.1. Prove that the ratio $U = Y^{-1}X$ of independent variables X and Y with Gamma distributions with parameters a and b has a Beta density $f_{a,b}$ and find the distribution of the variable $V = (X + Y)^{-1}X$. Prove the independence of V and $X + Y$.

Answer. The density of the variable $U = Y^{-1}X$ is defined for every function $h(U)$ in $L^1(\mathbb{R}_+)$ by the change of variable $x = uy$ with Jacobian $|J| = y$

$$\begin{aligned}
Eh(U) &= \int_{\mathbb{R}_+} \int_{\mathbb{R}_+^*} h(y^{-1}x) f_a(x) f_b(y)\, dx\, dy \\
&= \int_{\mathbb{R}_+} \int_{\mathbb{R}_+^*} h(u) f_a(uy) f_b(y) y\, du\, dy \\
&= \frac{1}{\Gamma(a)\Gamma(b)} \int_{\mathbb{R}_+} \int_{\mathbb{R}_+^*} h(u) e^{-y(u+1)} u^{a-1} y^{a+b-1}\, du\, dy,
\end{aligned}$$

the density of U is therefore

$$\begin{aligned}
f_U(u) &= \frac{u^{a-1}}{(1+u)^{a+b-1}\Gamma(a)\Gamma(b)} \int_{\mathbb{R}_+} e^{-v} v^{a+b-1}\, dv \\
&= \frac{\Gamma(a+b)}{\Gamma(a)\Gamma(b)} \frac{u^{a-1}}{(1+u)^{a+b}}.
\end{aligned}$$

Denoting

$$B_{a,b} = \int_{\mathbb{R}_+} \frac{u^{a-1}}{(1+u)^{a+b}}\, du = \frac{\Gamma(a)\Gamma(b)}{\Gamma(a+b)},$$

and $1-x=(1+u)^{-1}$ in $]0,1]$ yields the Beta density

$$f_U(u)\,du = B_{a,b}^{-1}x^{a-1}(1-x)^{b-1}\,dx.$$

By the same argument, the variable $V=(X+Y)^{-1}X$ has a Beta distribution with parameter $B(a,a+b)$.

With distributions $\Gamma(a,\lambda)$ and $\Gamma(b,\lambda)$, we have

$$Eh(U) = \frac{\lambda^{a+b}}{\Gamma(a)\Gamma(b)}\int_{\mathbb{R}_+}\int_{\mathbb{R}_+^*} h(u)e^{-\lambda y(u+1)}u^{a-1}y^{a+b-1}\,du\,dy$$

and U has the $B_{a,b}$ distribution with the density

$$\begin{aligned} f_U(u) &= \frac{\lambda^{a+b}u^{a-1}}{(1+u)^{a+b-1}\Gamma(a)\Gamma(b)}\int_{\mathbb{R}_+} e^{-v}v^{a+b-1}\,dv \\ &= \frac{\Gamma(a+b)}{\Gamma(a)\Gamma(b)}\,\frac{u^{a-1}}{(1+u)^{a+b}}. \end{aligned}$$

The variable $S=X+Y$ has a $\Gamma(a+b,\lambda)$ distribution and $V=S^{-1}X$ has still a $B_{a,a+b}$ distribution.

The independence of S and V is proved like the independence of Y and U, the joint density of (U,Y) if

$$f_{U,Y}(u,y) = \frac{\lambda^{a+b}}{\Gamma(a)\Gamma(b)}e^{-\lambda y(u+1)}u^{a-1}y^{a+b-1}$$

and by the same change of variable as previously, the density of Y conditionally on U is the density of a $\Gamma(a+b)$ distribution which does not depend on U .

Problem 5.2.2. Let $(X_i)_{i=1,\dots,k}$ be a vector of independent Γ_{a_i} variables, prove that the vector Z with components $Z_j=(\sum_{i=1}^k X_i)^{-1}X_j$ is independent of $Y=\sum_{i=1}^k X_i$ and has a Dirichlet distribution.

Answer. For $i=1,\dots,k$, let f_{a_i} be the density of X_i, the density of Y is f_A, with $A=\sum_{i=1}^k a_i$. The density of the variable Z is defined for every function $h(Z)$ in $C_b(S)$ for $S=\{z=(z_1,\dots,z_k),\sum_{i=1}^k z_i=1\}$, by the change of variable $x_i=z_iy$ for $i=1,\dots,k-1$, and $x_k=y-\sum_{i=1}^{k-1}x_i$ where $y=\sum_{i=1}^k x_i$, with the Jacobian $|J|=y^{k-1}$, it follows that

$$\begin{aligned} Eh(Z) &= \int_{\mathbb{R}_+^k}\int_{\mathbb{R}_+^*} h(y^{-1}x)\prod_{i=1}^k f_{a_i}(x_i)\,dx_i\} \\ &= \int_{\mathbb{R}_+^k} h(z)\prod_{i=1}^{k-1} f_{a_i}(z_iy)f_{a_k}(y(1-\sum_{i=1}^{k-1} z_i))y^{k-1}\prod_{i=1}^{k-1}dz_i\,dy, \end{aligned}$$

therefore

$$Eh(Z) = \frac{\Gamma(A)}{\prod_{i=1}^k \Gamma(a_i)} \int_S h(z) \prod_{i=1}^{k-1} z_i^{a_i-1} z_k^{a_k-1} \prod_{i=1}^k dz_i,$$

the variable Z is thus independent of Y and it has a Dirichlet distribution with parameter $a = (a_1, \dots, a_k)$.

Problem 5.2.3. Let Y be a Gamma distributed variable with mean 1 and variance a, prove that the Poisson variable N with conditional intensity λY has a negative binomial distribution $NB(\lambda, a)$. Calculate the generating function, the expectation and the variance of N.

Answer. A Gamma distributed variable with density $f_{\alpha,\beta}$ defined by (5.1) has the expectation $\beta^{-1}\alpha$ and the variance $\beta^{-2}\alpha$, the variable Y has therefore a $\Gamma(a^{-1}, a^{-1})$ distribution with $\alpha = \beta$. The expectation and the variance of N are $E(N \mid Y) = \lambda Y$, $E(N) = \lambda$ and $\text{Var}(N \mid Y) = \lambda Y$, $\text{Var}(N) = \text{Var}\{E(N \mid Y)\} + E\{\text{Var}(N \mid Y)\}$ therefore

$$\text{Var}(N) = \lambda^2(a-1) + \lambda.$$

The generating function of N is the expectation of a conditional Poisson variable with intensity λY

$$G(s) = E\{G_N(s \mid Y)\} = E\{e^{\lambda Y(s-1)}\},$$

it is the Laplace transform of Y at $-\lambda(s-1))$ given in Section 5.2

$$L_Y(-\lambda(s-1)) = \frac{1}{\{1 - \lambda(s-1)a\}^{\frac{1}{a}}}.$$

The conditional probabilities of N are

$$\pi_k(Y) = e^{-\lambda Y} \frac{(\lambda Y)^k}{k!},$$

denoting $\theta = a^{-1}$, the probabilities $\pi_k = P(N = k) = E\pi_k(Y)$ are deduced

$$\begin{aligned}
\pi_k &= \frac{\theta^\theta \lambda^k}{k!\Gamma(\theta)} \int_0^\infty e^{-(\theta+\lambda)y} y^{k+\theta-1}\, dy \\
&= \frac{\Gamma(k+\theta)}{k!\Gamma(\theta)} \frac{\theta^\theta \lambda^k}{(\theta+\lambda)^{k+\theta}},
\end{aligned}$$

they are the probabilities of the negative binomial distribution $NB(\lambda, a)$.

5.3 Gaussian variables

Let X a centered Gaussian variable with variance σ^2, and with density ϕ, its Fourier transform is

$$E(e^{itX}) = \frac{e^{-\frac{t^2}{2\sigma^2}}}{\sqrt{2\pi}\sigma}\int_{\mathbb{R}} e^{-\frac{(x-it\sigma^2)^2}{2\sigma^2}}\,dx = e^{-\frac{\sigma^2t^2}{2}}.$$

Exercise 5.3.1. Find the affinity and the Kullback information of the probabilities of Gaussian variables.

Answer. Let μ_1 and μ_2 and respectively σ_1^2 and σ_2^2 be the distinct expectations and variances of Gaussian variables with densities f_1 and respectively f_2 and let

$$\mu = \frac{\sigma_2^2\mu_1 + \sigma_1^2\mu_2}{(\sigma_1^2+\sigma_2^2)^{\frac{1}{2}}}.$$

The affinity of Gaussian variables with densities f_1 and f_2 is

$$\begin{aligned}
\rho(f_1,f_2) &= \int_{\mathbb{R}}(f_1f_2)^{\frac{1}{2}} = \frac{1}{(2\pi\sigma_1\sigma_2)^{\frac{1}{2}}}\int_{\mathbb{R}} e^{-\frac{(x-\mu_1)^2}{4\sigma_1^2}-\frac{(x-\mu_2)^2}{4\sigma_2^2}}\,dx\\
&= \frac{\exp\Big\{-\frac{(\mu_1-\mu_2)^2}{4(\sigma_1^2+\sigma_2^2)}\Big\}}{(2\pi\sigma_1\sigma_2)^{\frac{1}{2}}}\int_{\mathbb{R}}\exp\Big\{-\frac{\{(\sigma_1^2+\sigma_2^2)^{\frac{1}{2}}x-\mu\}^2}{4\sigma_1^2\sigma_2^2}\Big\}\,dx\\
&= \sqrt{2\sigma_1\sigma_2}\exp\Big\{-\frac{(\mu_1-\mu_2)^2}{4(\sigma_1^2+\sigma_2^2)}\Big\}.
\end{aligned}$$

If $\sigma_1=\sigma_2$, denoting $\mu = \frac{\mu_1+\mu_2}{\sqrt{2}}$, the affinity of the variables is

$$\begin{aligned}
\rho(f_1,f_2) &= \frac{1}{\sigma\sqrt{2\pi}}e^{-\frac{(\mu_1-\mu_2)^2}{8\sigma^2}}\int_{\mathbb{R}} e^{-\frac{(\sqrt{2}x-\mu)^2}{4\sigma^2}}\,dx\\
&= \sqrt{2}e^{-\frac{(\mu_1-\mu_2)^2}{8\sigma^2}},
\end{aligned}$$

and if $\mu_1=\mu_2$ and $\sigma_1\neq\sigma_2$

$$\rho(f_1,f_2) = \frac{\sqrt{2\sigma_1\sigma_2}}{\sqrt{\sigma_1^2+\sigma_2^2}}.$$

Their information is

$$\begin{aligned}
K(f_1,f_2) &= -\int_{\mathbb{R}}\Big\{\frac{(x-\mu_1)^2}{2\sigma_1^2}-\frac{(x-\mu_2)^2}{2\sigma_2^2}\Big\}f_1(x)\,dx + \log\frac{\sigma_2}{\sigma_1}\\
&= -\frac{\sigma_2^2-\sigma_1^2}{2\sigma_1^2\sigma_2^2}(\sigma_1^2+\mu_1^2) + \log\frac{\sigma_2}{\sigma_1}\\
&\quad + \Big\{\frac{\mu_1}{\sigma_1^2}-\frac{\mu_2}{\sigma_2^2}\Big\}\mu_1 - \Big\{\frac{\mu_1^2}{2\sigma_1^2}-\frac{\mu_2^2}{2\sigma_2^2}\Big\},
\end{aligned}$$

therefore

$$K(f_1, f_2) = \frac{1}{2}\Big(\frac{\sigma_1^2}{\sigma_2^2} - 1\Big) + \log\frac{\sigma_2}{\sigma_1} + \frac{(\mu_1 - \mu_2)^2}{2\sigma_2^2}.$$

Exercise 5.3.2. Let Y be a Gaussian variable $\mathcal{N}(\theta, \sigma^2)$ conditionally on θ, a Gaussian variable $\mathcal{N}(\mu, \tau^2)$, find the distribution of Y.

Answer. The density of Y is

$$\begin{aligned} f(y) &= \frac{1}{2\pi\sigma\tau}\int_{\mathbb{R}} e^{-\frac{(y-\theta)^2}{2\sigma^2}} e^{-\frac{(\theta-\mu)^2}{2\tau^2}}\, d\theta \\ &= \frac{1}{2\pi\sigma\tau} e^{-\frac{(y-\mu)^2}{2(\sigma^2+\tau^2)}} \int_{\mathbb{R}} \exp\Big\{-\frac{(\sqrt{\sigma^2+\tau^2}\theta - \frac{\tau^2 y + \sigma^2\mu}{\sqrt{\sigma^2+\tau^2}})^2}{2\sigma^2\tau^2}\Big\}\, d\theta \\ &= \frac{1}{\sqrt{2\pi(\sigma^2+\tau^2)}} e^{-\frac{(y-\mu)^2}{2(\sigma^2+\tau^2)}}, \end{aligned}$$

it is the density of a Gaussian variable $\mathcal{N}(\mu, \sigma^2 + \tau^2)$.

Exercise 5.3.3. Let Y be a Gaussian variable $\mathcal{N}(0, \tau^{-1})$ conditionally on τ, where τ has an exponential distribution, find the distribution of Y.

Answer. Let $\tau \sim \mathcal{E}_{\frac{\lambda}{2}}$ and $h(y) = \frac{(y-\theta)^2+\lambda}{2}$, the density of Y is

$$\begin{aligned} f(y) &= \frac{\lambda}{2\sqrt{2\pi}}\int_0^\infty e^{-\{\frac{(y-\theta)^2+\lambda}{2}\}\tau}\sqrt{\tau}\, d\tau \\ &= \frac{\lambda\Gamma_{\frac{3}{2}}}{2\sqrt{2\pi}h^{\frac{3}{2}}(y)} = \frac{\lambda}{4h^{\frac{3}{2}}(y)}. \end{aligned}$$

Exercise 5.3.4. Let Y be a Gaussian variable $\mathcal{N}(\theta, \tau^{-1})$ conditionally on τ, where τ is a Gamma variable $\Gamma_{a,\lambda}$, find the distribution of Y and the distribution of τ conditionally on Y.

Answer. The density of Y is

$$\begin{aligned} f(y) &= \frac{\lambda^a}{\sqrt{2\pi}\Gamma_a}\int_0^\infty e^{-\{\frac{(y-\theta)^2}{2}+\lambda\}\tau}\tau^{a-\frac{1}{2}}\, d\tau \\ &= \frac{\lambda^a\Gamma_{a+\frac{1}{2}}}{\{h(y)\}^{a+\frac{1}{2}}\sqrt{2\pi}\Gamma_a} \end{aligned}$$

where

$$h(y) = \frac{(y-\theta)^2}{2} + \lambda.$$

The conditional density of τ is

$$f(\tau \mid y) = \frac{\tau^{a-\frac{1}{2}}}{\Gamma_{a+\frac{1}{2}}} e^{-h(y)\tau} \{h(y)\}^{a+\frac{1}{2}},$$

it is proportional to a Gamma distribution with parameters $a + \frac{1}{2}$ and $h(y)$.

Exercise 5.3.5. Let X be a Gaussian variable $\mathcal{N}(0, \tau)$ where τ has the inverse Gaussian density

$$f_a(t) = \frac{a}{\sqrt{2\pi t^3}} e^{-\frac{a^2}{2t}}, \ t \geq 0,$$

find the distribution of X and the distribution function of τ^{-1} conditionally on X.

Answer. The density of X is

$$\begin{aligned} f(x) &= \int_0^\infty \frac{a}{2\pi t^2} e^{-\frac{a^2+x^2}{2t}} \, dt \\ &= \frac{a}{2\pi} \int_0^\infty e^{-\frac{(a^2+x^2)y}{2}} \, dy \\ &= \frac{a}{\pi(a^2+x^2)}, \end{aligned}$$

it is the Cauchy density with parameter a on $\mathbb{R}$. The density of τ conditionally on X is

$$f(t; X) = \frac{(a^2 + X^2)}{2t^2} e^{-\frac{a^2+X^2}{2t}}$$

and the conditional distribution function of τ^{-1} is

$$\begin{aligned} F(t; X) &= \int_{\frac{1}{t}}^\infty \frac{(a^2 + X^2)}{2s^2} e^{-\frac{a^2+X^2}{2s}} \, ds \\ &= \frac{(a^2 + X^2)}{2} \int_0^t e^{-\frac{(a^2+X^2)u}{2}} \, du \\ &= 1 - e^{-\frac{(a^2+X^2)t}{2}}, \end{aligned}$$

it is the exponential distribution function with parameter $a^2 + X^2$.

Exercise 5.3.6. A variable X has a log-normal distribution if $\log(X)$ has a Gaussian variable $\mathcal{N}(\mu, \sigma^2)$, find the expectation and the variance of X.

Answer. The expectation of X is

$$E(X) = E(e^{\log(X)}) = \frac{1}{\sigma\sqrt{2\pi}} \int_{\mathbb{R}} e^{x - \frac{(x-\mu)^2}{2\sigma^2}} \, dx$$
$$= \frac{e^{\mu+\frac{\sigma^2}{2}}}{\sigma\sqrt{2\pi}} \int_{\mathbb{R}} e^{x - \frac{(x-\mu-\sigma^2)^2}{2\sigma^2}} \, dx = e^{\mu+\frac{\sigma^2}{2}},$$

in the same way

$$E(X^2) = E(e^{2\log(X)}) = \frac{1}{\sigma\sqrt{2\pi}} \int_{\mathbb{R}} e^{2x - \frac{(x-\mu)^2}{2\sigma^2}} \, dx$$
$$= \frac{e^{2(\mu+\sigma^2)}}{\sigma\sqrt{2\pi}} \int_{\mathbb{R}} e^{x - \frac{(x-\mu-2\sigma^2)^2}{2\sigma^2}} \, dx = e^{2(\mu+\sigma^2)}$$

and the variance of X is $e^{2\mu+\sigma^2}(e^{\sigma^2} - 1)$.

Problem 5.3.1. Write the density and the Fourier transform of a Gaussian vector $X_{(p)} = (X_1, \ldots, X_p)^T$ with distribution $\mathcal{N}(0, \Sigma)$ such that the inverse matrix Σ^{-1} exists.

Answer. The matrix Σ is symmetric then there exists a variable Y such that $X_{(p)} = YP$, with an orthogonal matrix P, and the variance of Y is

$$D = P^T \Sigma P = \text{diag}(\sigma_i^2)$$

with determinant $|D| = |\Sigma|$. The density of Y is Gaussian

$$\phi_p(y) = \prod_{i=1}^{p} \frac{1}{\sigma_i\sqrt{2\pi}} e^{-\frac{y_i^2}{2\sigma_i^2}} = \frac{1}{|D|^{\frac{1}{2}}(2\pi)^{\frac{n}{2}}} e^{-\frac{y^T D^{-1} y}{2}}$$

and its Fourier transform is

$$\widehat{\phi}_Y(t) = E(e^{it^T Y}) = e^{-\frac{t^T D t}{2}}.$$

The density of $X_{(p)}$ is

$$\frac{1}{(2\pi|\Sigma|)^{\frac{n}{2}}} e^{-\frac{x^T \Sigma^{-1} x}{2}}$$

and its Fourier transform is

$$\widehat{\phi}_{(p)}(t) = E(e^{it^T X(p)}) = e^{-\frac{t^T \Sigma t}{2}}.$$

Problem 5.3.2. Let S_n be the sum of n independent and identically distributed Gaussian variables $\mathcal{N}(0, \sigma^2)$, find the limit of the Fourier transform of $n^{-\frac{1}{2}} S_n$.

Answer. The Fourier transform of S_n is $E(e^{itS_n}) = \prod_{i=1}^n E(e^{itX_i}) = \widehat{\phi}_X^n(t)$ such that $\widehat{\phi}_X(0) = 1$, $\widehat{\phi}'_X(t) = 0$ as $EX = 0$ and $\widehat{\phi}'_X(t) = -\sigma^2$. For $n^{-\frac{1}{2}} S_n$ we have $\widehat{\phi}_X^n(n^{-\frac{1}{2}}t)$. As n tends to infinity, $\widehat{\phi}_X^n(n^{-\frac{1}{2}}t)$ converges to zero and $\widehat{\phi}_X(0)$ By an expansion as n tends to infinity

$$\begin{aligned}
\widehat{\phi}_X(n^{-\frac{1}{2}}t) &= 1 - n^{-1}t^2\sigma^2 + o(n^{-1}), \\
\log \widehat{\phi}_X^n(n^{-\frac{1}{2}}t) &= n \log\{\widehat{\phi}_X(n^{-\frac{1}{2}}t)\} \\
&= n\{\widehat{\phi}_X(n^{-\frac{1}{2}}t) - 1\} - \frac{n}{2}\{\widehat{\phi}_X(n^{-\frac{1}{2}}t)\}^2\{1 + o(1)\} \\
&= -\frac{\sigma^2}{2} + o(1)
\end{aligned}$$

and $\widehat{\phi}_X^n(n^{-\frac{1}{2}}t)$ converges to the transform $\widehat{\phi}$ of a Gaussian variable with distribution $\mathcal{N}(0, \sigma^2)$.

Problem 5.3.3. (a). Write the density of a centered Gaussian vector (X, Y) in $\mathbb{R}^2$ with correlation ρ and variances $E(X^2) = \sigma_X^2$ and $E(Y^2) = \sigma_Y^2$, and the density of their sum.

(b). Write the densities of X conditionally on $Y = y$ and of Y conditionally on $X = x$, their conditional expectations and variances.

(c). Determine the distribution function of their ratio and their product under the condition $E(X^2) = E(Y^2) = 1$.

Answer. (a). The density of (X, Y) is

$$\begin{aligned}
f(x,y) &= \frac{1}{2\pi\sigma_X\sigma_Y\sqrt{(1-\rho^2)}} \exp\Big\{-\frac{\sigma_Y^2x^2 - 2\rho\sigma_X\sigma_Y xy + \sigma_X^2y^2}{2\sigma_X^2\sigma_Y^2(1-\rho^2)}\Big\} \\
&= \frac{1}{2\pi\sigma_X\sigma_Y\sqrt{(1-\rho^2)}} \exp\Big\{-\frac{(\sigma_Y x - \rho\sigma_X y)^2}{2\sigma_X^2\sigma_Y^2(1-\rho^2)} - \frac{y^2}{2\sigma_Y^2}\Big\} \\
&= \frac{1}{2\pi\sigma_X\sigma_Y\sqrt{(1-\rho^2)}} \exp\Big\{-\frac{(\sigma_X y - \rho\sigma_Y x)^2}{2\sigma_X^2\sigma_Y^2(1-\rho^2)} - \frac{x^2}{2\sigma_X^2}\Big\}.
\end{aligned}$$

The density of their sum is $\int_{\mathbb{R}} f(x-y, y)\, dy$

$$\begin{aligned}
&\frac{1}{2\pi\sigma_X\sigma_Y\sqrt{(1-\rho^2)}} \exp\Big\{-\frac{\sigma_Y^2(x-y)^2 - 2\rho\sigma_X\sigma_Y(x-y)y + \sigma_X^2y^2}{2\sigma_X^2\sigma_Y^2(1-\rho^2)}\Big\}\, dy \\
&= \frac{1}{\sqrt{2\pi}} \exp\Big\{-\frac{x^2}{2\sigma^2}\Big\}
\end{aligned}$$

with the variance

$$\sigma^2 = \sigma_X^2 + \sigma_Y^2 + 2\rho\sigma_X\sigma_Y.$$

(b). The density of X conditionally on $Y = y$ is

$$\begin{aligned} f(x \mid Y = y) &= f_Y^{-1}(y) f(x, y) \\ &= \frac{1}{\sigma_X\sqrt{2\pi(1-\rho^2)}} \exp\Big\{-\frac{(\sigma_Y x - \rho\sigma_X y)^2}{2\sigma_X^2\sigma_Y^2(1-\rho^2)}\Big\} \end{aligned}$$

and the density of Y conditionally on $X = x$ is

$$\begin{aligned} f(y \mid X = x) &= f_X^{-1}(x) f(x, y) \\ &= \frac{1}{\sigma_Y\sqrt{2\pi(1-\rho^2)}} \exp\Big\{-\frac{(\sigma_X y - \rho\sigma_Y x)^2}{2\sigma_X^2\sigma_Y^2(1-\rho^2)}\Big\}. \end{aligned}$$

The conditional expectations are

$$\begin{aligned} E(X \mid Y = y) &= \sigma_X\sigma_Y^{-1}\rho y, \\ E(Y \mid X = x) &= \sigma_Y\sigma_X^{-1}\rho x \end{aligned}$$

and the conditional variances are

$$V_{X|Y} = \sigma_X^2\sigma_Y^2(1-\rho^2) = V_{Y|X}.$$

(c). If $E(X^2) = E(Y^2) = 1$, the variable $X^{-1}Y$ has the distribution function

$$\begin{aligned} F(t) &= \frac{1}{2\pi(1-\rho^2)} \int_{\mathbb{R}^2} 1_{\{y \le xt\}} e^{-\frac{x^2-2\rho xy+y^2}{2(1-\rho^2)}}\, dx\, dy \\ &= \frac{1}{2\pi(1-\rho^2)} \int_{\mathbb{R}^2} 1_{\{y \le xt\}} e^{-\frac{(y-\rho x)^2}{2(1-\rho^2)}} e^{-\frac{x^2}{2}}\, dx\, dy \\ &= \frac{1}{\sqrt{1-\rho^2}} \int_{\mathbb{R}} \Phi_\rho(x(t-\rho))\, d\Phi(t) \end{aligned}$$

where Φ_ρ is the density of a Gaussian real centered variable with variance $1-\rho^2$ and Φ is the density of a normal Gaussian real variable. By the same arguments, the variable XY has the distribution function

$$\begin{aligned} F(t) &= \frac{1}{2\pi(1-\rho^2)} \int_{\mathbb{R}^2} 1_{\{y \le x^{-1}t\}} e^{-\frac{x^2-2\rho xy+y^2}{2(1-\rho^2)}}\, dx\, dy \\ &= \frac{1}{\sqrt{1-\rho^2}} \int_{\mathbb{R}} \Phi_\rho(x^{-1}t - x\rho)\, d\Phi(t). \end{aligned}$$

Problem 5.3.4. Let X_1 and X_2 be independent Gaussian variables and let $X = (X_1, X_2)^T$. For a symmetric matrix A, give conditions for $Y = AX$ to have independent components and for Y to be a Gaussian variable in $\mathbb{R}^2$.

Answer. The variance matrix of $Y = AX$ is

$$\Sigma_Y = A^T \mathrm{Var}(X) A = \mathrm{Var}(X) A^T A$$

and its components are independent if $A^T A$ is a diagonal matrix, then Y is Gaussian in $\mathbb{R}^2$. The variable Y is Gaussian in $\mathbb{R}^2$ if and only if every linear combination of its components is Gaussian in $\mathbb{R}$, and this is true by the independence of X_1 and X_2.

Problem 5.3.5. Let $(X_i)_{i\leq n}$ be independent centered Gaussian variables with respective variances σ_i^2.

(a). Let $S_n = \sum_{i=1}^n X_i$, write the density of (S_n, S_m), its marginal densities and the density of S_n conditionally on S_m, for $m < n$.

(b). Let $U_n = \sum_{i=1}^n X_i^2$, write the density of (U_n, U_m), its marginal densities and the density of U_m conditionally on U_n, for $m < n$.

Answer. (a). The variable S_n is Gaussian, centered with variance $s_n = \sum_{i=1}^n \sigma_i^2$ for every n. For $n > m$, $S_n - S_m$ is independent of S_m and $E\{S_m(S_n - S_m)\} = 0$, the covariance of S_n and S_m is

$$E(S_m S_n) = E S_{m\wedge n}^2 = s_{m\wedge n}$$

and their correlation is

$$\rho = s_{m\wedge n}(s_n s_m)^{-\frac{1}{2}}.$$

The variable (S_n, S_m) has then the density

$$f(x,y) = \frac{1}{2\pi\sqrt{(1-\rho^2)s_n s_m}} \exp\Big\{-\frac{s_m x^2 - 2\rho\sqrt{s_n s_m}xy + s_n y^2}{2 s_n s_m (1-\rho^2)}\Big\}$$

and the marginal densities

$$\begin{aligned} f(y) &= \frac{1}{2\pi\sqrt{(1-\rho^2)s_n s_m}} \exp\Big\{-\frac{y^2}{2s_m}\Big\} \int_{\mathbb{R}} \exp\Big\{-\frac{(\sqrt{s_m}x - \rho\sqrt{s_n}y)^2}{2s_n s_m(1-\rho^2)}\Big\}\, dx \\ &= \frac{1}{\sqrt{2\pi s_m}} \exp\Big\{-\frac{y^2}{2s_m}\Big\} \\ f(x) &= \int_{\mathbb{R}} f(x,y)\, dy = \frac{1}{\sqrt{2\pi s_n}} \exp\Big\{-\frac{x^2}{2s_n}\Big\}, \end{aligned}$$

they are Gaussian densities. For all scalars a and b and for $n > m$, $aS_n + bS_m = (a+b)S_m + b(S_n - S_m)$ where S_m and $S_n - S_m$ are independent

Gaussian variables, aS_n+bS_m is therefore a Gaussian variable and (S_n, S_m) is a Gaussian vector.

The conditional density of S_n given S_m is

$$\begin{aligned} f(x \mid y) &= \frac{1}{\sqrt{2\pi(1-\rho^2)s_n}} \exp\Big\{-\frac{s_m x^2 - 2\rho\sqrt{s_n s_m}xy + s_n y^2}{2s_n s_m(1-\rho^2)}\Big\} \exp\Big\{\frac{y^2}{2s_m}\Big\} \\ &= \frac{1}{\sqrt{2\pi(1-\rho^2)s_n}} \exp\Big\{-\frac{(\sqrt{s_m}x - \rho\sqrt{s_n}y)^2}{2s_n s_m(1-\rho^2)}\Big\}, \end{aligned}$$

its expectation is $\rho\sqrt{s_n}y$ and its variance is $(1-\rho^2)s_n s_m$.

(b). The density of (U_m, U_n) at (x, y) is the product of the densities of the independent variables U_m at x and $U_n - U_m$ at $y - x$, they are densities of $\Gamma(\frac{m}{2}, \frac{1}{2})$, and respectively $\Gamma(\frac{n-m}{2}, \frac{1}{2})$ distributions. For $m < n$, the density of U_m conditionally on U_n is the ratio of the density of (U_m, U_n) with the density of the $\Gamma(\frac{n-m}{2}, \frac{1}{2})$ distribution

$$f(x \mid U_n) = \frac{\Gamma(\frac{n}{2})}{U_n\Gamma(\frac{m}{2})\Gamma(\frac{n-m}{2})}\Big(\frac{x}{U_n}\Big)^{\frac{n}{2}-1}\Big(1-\frac{x}{U_n}\Big)^{\frac{n-m}{2}-1}.$$

Problem 5.3.6. Let X be a normal variable $N(0,1)$, prove that the variable $\frac{X^2}{2}$ has a Gamma distribution $\Gamma(\frac{1}{2})$ with density $f(y) = \pi^{-\frac{1}{2}}y^{-\frac{1}{2}}e^{-y}$.

Answer. By a change of variable, for every function h of $\frac{X^2}{2}$ we have

$$\begin{aligned} \int_{\mathbb{R}_+} h\Big(\frac{x^2}{2}\Big)\, dF_X(x) &= \frac{\sqrt{2}}{\sqrt{\pi}} \int_{\mathbb{R}_+} h\Big(\frac{x^2}{2}\Big)\, e^{-\frac{x^2}{2}}\, dx \\ &= \frac{1}{\sqrt{\pi}} \int_{\mathbb{R}_+} h(y) y^{-\frac{1}{2}} e^{-y}\, dy \end{aligned}$$

and $\Gamma(\frac{1}{2}) = \int_{\mathbb{R}_+} y^{-\frac{1}{2}} e^{-y}\, dy = 2\int_{\mathbb{R}_+} e^{-x^2}\, dx = \int_{\mathbb{R}} e^{-x^2}\, dx = \sqrt{\pi}$, therefore $\frac{X^2}{2}$ has a distribution $\Gamma(\frac{1}{2})$.

5.4 χ^2 variables

A χ^2 variable Z is the square of a normal variable $X \sim \mathcal{N}(0,1)$, its characteristic function is

$$E(e^{itZ}) = \frac{1}{(1-2it)^{\frac{1}{2}}}.$$

Exercise 5.4.1. Write the density and the characteristic function of a non centered χ^2 variable $Z = (X+a)^2$.

Answer. The distribution function of Z is

$$F(x) = P((X+a)^2 \leq x) = P(X \leq \sqrt{x} - a) - P(X \geq -\sqrt{x} - a)$$

and it density is deduced from the normal density ϕ

$$\begin{aligned} fF(x) &= \frac{1}{2\sqrt{x}}\{\phi(\sqrt{x}-a) + \phi(-\sqrt{x}-a)\} \\ &= \frac{1}{\sqrt{2\pi x}} e^{-\frac{x+a^2}{2}} \cosh(a\sqrt{x}). \end{aligned}$$

The characteristic function of Z is

$$\begin{aligned} E(e^{itZ}) &= \frac{e^{ita^2}}{\sqrt{2\pi}} \int_{\mathbb{R}} e^{-\frac{(1-2it)x^2}{2} + 2itax}\, dx \\ &= \frac{1}{(1-2it)^{\frac{1}{2}}} e^{\frac{2ta^2}{1-2it}}. \end{aligned}$$

Problem 5.4.1. (a). Let $X_1, \ldots, X_n$ be independent normal variables $\mathcal{N}(0,1)$, prove that the variable $Z = X_1^2 + \cdots + X_n^2$ has a $\chi_n^2 = \Gamma(\frac{n}{2}, \frac{1}{2})$ distribution, write its expectation and its variance. Prove the inequality

$$P(Z - n \geq 2x) \leq e^{-x}, \ \forall x > 0.$$

(b). For independent variables X_i with Gaussian distributions $\mathcal{N}(\mu_i, 1)$, for $i = 1, \ldots, n$, prove that the variable Z has an uncentered $\chi_{n,\mu}^2$ distribution where $\lambda = \sum_{i=1}^n \mu_i^2$.

(c). Let $X = (X_1, \ldots, X_n)$ be a normal vector $\mathcal{N}(0, \Sigma)$ with a definite positive variance matrix Σ, prove that the variable $X^T\Sigma^{-1}X$ has a χ_n^2 distribution. Find the distribution of $X^T\Sigma^{-1}X$ when $E(X) = \mu$.

Answer. (a). The Laplace transform of X_1^2 is

$$\begin{aligned} \varphi(t) = E(e^{-tX^2}) &= \frac{1}{\sqrt{2\pi}} \int_0^\infty e^{-\frac{x^2}{2}(2t+1)}\, dx \\ &= (2t+1)^{-\frac{1}{2}} = \frac{2^{-\frac{1}{2}}}{(t+\frac{1}{2})^{\frac{1}{2}}}, \end{aligned}$$

it has therefore a $\Gamma(\frac{1}{2}, \frac{1}{2})$ distribution and the sum $X_1^2 + \cdots + X_n^2$ has the distribution $\Gamma(\frac{n}{2}, \frac{1}{2})$. The expectation of X_1^2 is 1 and its variance is 2, then $E(\chi_n^2) = n$ and $\mathrm{Var}(\chi_n^2) = 2n$.

For all integers $0 < s < \frac{1}{2}$ and $y > 0$

$$
\begin{aligned}
P(Z-n \geq ny) &= P(e^{(Z-n)s} \geq e^{nsy}) \leq e^{-ns(y+1)}\varphi^n(-s), \\
\log P(Z-n \geq ny) &\geq -ns(y+1) - \frac{n}{2}\log(1-2s) \\
&= -nsy - ns + \frac{n}{2}\sum_{k\geq 1}(2s)^k \\
&\leq -nsy + \frac{ns^2}{1-2s},
\end{aligned}
$$

the bound is minimum at s such that

$$y = \frac{2s(1-s)}{(1-2s)^2}$$

and this equation has a unique solution in $]0, \frac{1}{2}[$

$$s = \frac{1}{2}(1 - \frac{1}{\sqrt{2y+1}})$$

where the inequality becomes

$$P(Z-n \geq ny) \leq e^{-\frac{ny}{2}}.$$

(b). If the variables X_i have expectations μ_i, there exists an orthogonal matrix of rotation R which maps $\mu = E(X)$ to a vector $(a, 0, \ldots, 0)^T$ of the first axis and the $n-1$ last components of the vector $S = RX$ are centered, the variance matrix of S is $RE(XX^T)R^T = I_n$ so the components of S are independent. The expectation of the first components is $E(S_1) = R_1\mu$ where the first line R_1 of R is chosen as the vector $\lambda^{-\frac{1}{2}}\mu$ therefore $E(S_1) = \lambda^{\frac{1}{2}}$ and S_1 is a Gaussian variable $\mathcal{N}(\lambda^{\frac{1}{2}}, 1)$.

The quadratic form Z is

$$Z = \sum_{i=1}^n X_i^2 = X^T X = RS^T SR^T$$

where $S^T S$ is the sum of the uncentered $\chi^2_{1,\lambda}$ variable S_1^2 with expectation λ and of the independent χ^2_{n-1} variable $\sum_{i=1}^{n-1} S_i^2$. The distribution of $S^T S$ and Z is the convolution $\chi^2_{n,\mu}$ of the χ^2_{n-1} distribution with the density of S_1^2.

(c). The matrix Σ is symmetric so there exists an orthogonal matrix P of change of basis and a diagonal matrix D such that $\Sigma = PDP^T$ and $PP^T = P^T P = I_n$ the identity matrix of $\mathbb{R}^n$, the vector $Y = P^T X$ with components $(Y_1, \ldots, Y_n)$ is a centered Gaussian vector with the variance matrix

$$E(YY^T) = E(P^T XX^T P) = P^T E(XX^T)P = P^T \Sigma P = D$$

where the elements of the diagonal matrix D are the variances σ_i^2 of the variables Y_i which are independent. By the change of basis $X = PY$ we have $\Sigma = PDP^T$ and its inverse is $\Sigma^{-1} = PD^{-1}P^T$. The quadratic form $X^T\Sigma^{-1}X$ develops as

$$X^T\Sigma^{-1}X = Y^TP^T\Sigma^{-1}PY = Y^TD^{-1}Y = \sum_{i=1}^n \sigma_i^{-2}Y_i^2,$$

its has a χ_n^2 distribution.

When the variable X is a vector of uncentered and dependent Gaussian variables, the distribution of Z is a $\chi_{n,\mu}^2$ obtained by diagonalization of its variance matrix and a rotation like in (b) for a projection of the expectation of Y on the first axis.

Problem 5.4.2. Let U and V be independent normal variables $N(0,1)$.

(a). Write the joint density of $X = U + V$ and $Y = (U+V)^{-1}V$ and the density of X conditionally on Y.

(b). Find the densities of $Z = (U^2 + V^2)^{\frac{1}{2}}$ and of $T = UVZ^{-1}$.

Answer. (a). By the change of variable $x = u + v$ and $y = (u+v)^{-1}v$, we have $u = x(1-y)$ and $v = xy$, $du = dx$ and $dv = (1-y)^{-1}ydy$. For normal variables U and V, let $k^{-2}(y) = (1-y)^2 + y^2$ and the density of (X,Y) is

$$\begin{aligned} h(x,y) &= (2\pi)^{-1}(1-y)^{-1}y\phi(xy)\phi(x(1-y)) \\ &= (2\pi)^{-1}(1-y)^{-1}y\exp\Big\{-\frac{x^2}{2k^2(y)}\Big\}. \end{aligned}$$

The marginal density of Y is

$$h_Y(y) = (2\pi)^{-\frac{1}{2}}(1-y)^{-1}yk(y)$$

and the density of X conditionally on $Y = y$ is a centered Gaussian variable with variance $k^2(y)$ and with density

$$h_{X|Y}(x) = (2\pi)^{-\frac{1}{2}}k^{-1}(y)\exp\Big\{-\frac{x^2}{2k^2(y)}\Big\}.$$

(b). The variable Z^2 has a χ_2^2 distribution with a $\Gamma_{\frac{1}{2}}$ density, then Z has the distribution function $P(Z \le t) = P(Z^2 \le t^2)$ and the $\Gamma_{\frac{3}{2}}$ density

$$f_Z(t) = \frac{2}{\sqrt{\pi}}e^{-t}x^{\frac{1}{2}}.$$

For all $a > 0$ and $|t| < 1$, the distribution function

$$P(X(X^2+a)^{-\frac{1}{2}} \le t) = P\Big\{|X| \le t\Big(\frac{a}{1-t^2}\Big)^{\frac{1}{2}}\Big\}$$

has the density

$$\frac{\sqrt{2a}}{\sqrt{\pi(1-t^2)^3}} \exp\Big\{-\frac{at^2}{2(1-t^2)}\Big\}.$$

It follows $P(T \leq t) = \int_{\mathbb{R}} P(Xy(X^2+y^2)^{-\frac{1}{2}} \leq t)\varphi(y)\,dy$ where φ is the normal Gaussian density and the density of T is

$$\begin{aligned} f_T(t) &= \frac{\sqrt{2}}{\sqrt{\pi}}(1-t^2)^{-\frac{3}{2}} \int_{\mathbb{R}} y \exp\Big\{-\frac{y^2t^2}{2(1-t^2)}\Big\}\varphi(y), dy \\ &= \frac{1}{\pi}(1-t^2)^{-\frac{3}{2}} \int_{\mathbb{R}} y \exp\Big\{-\frac{y^2}{2(1-t^2)}\Big\}\,dy \\ &= \frac{\sqrt{2}}{\sqrt{\pi}(1-t^2)}, \end{aligned}$$

T has the arcsine distribution function $P(T \leq t) = \frac{\sqrt{2}}{\sqrt{\pi}} \arcsin(t)$.

Problem 5.4.3. Let $X = (X_1, \ldots, X_n)$ be a vector of independent variables with Gaussian distribution $\mathcal{N}(\mu, \sigma^2)$, prove that the empirical mean $\bar{X}_n$ and the empirical variance $s_n = (n-1)^{-1}\sum_{i=1}^n (X_i - \bar{X}_n)^2$ are independent and find their distributions.

Answer. The empirical mean $\bar{X}_n$ is a Gaussian variable with expectation μ and variance $n^{-1}\sigma^2$, the empirical variance s_n has the expectation σ^2. The vector Y with components $X_i - \bar{X}_n$ is a linear combination of the components of X

$$(X_i - \bar{X}_n)_{i=1,\ldots,n} = A_n X \tag{5.2}$$

with the symmetric matrix

$$A_n = \begin{pmatrix} 1-n^{-1} & -n^{-1} & \cdots & -n^{-1} \\ -n^{-1} & 1-n^{-1} & \cdots & -n^{-1} \\ \vdots & \vdots & \vdots & \vdots \\ -n^{-1} & \cdots - n^{-1} & 1-n^{-1} & \end{pmatrix},$$

Y is a Gaussian vector with expectation $E(Y) = A_n E(X) = \mu A_n I_n = 0$ and with variance $\mathrm{Var}(Y) = A_n \mathrm{Var}(X) A_n = \sigma^2 A_n^{\otimes 2}$. For $i = 1, \ldots, n$, the covariance of the Gaussian variables $\bar{X}_n$ and $X_i - \bar{X}_n$ is

$$E\{(X_i - \bar{X}_n)(\bar{X}_n - \mu)\} = E(n^{-1}X_i^2) + (1-n^{-1})\mu^2 - E(\bar{X}_n^2) = 0$$

and the covariance of the Gaussian variables $\bar{X}_n$ and Y is zero then Y is independent of $\bar{X}_n$. As Y and $\bar{X}_n$ belong to the Gaussian space H_n generated by the components of X, their independence implies that Y belongs to a subspace of H_n with dimension $n-1$ and the weighted sum s_n of the Y_i^2 has a χ_{n-1}^2 distribution.

Problem 5.4.4. Let $X = (X_1, \dots, X_n)^T$ be a vector of Gaussian variables with distribution $\mathcal{N}(0, \Sigma)$, prove that $\sum_{i=1}^n X_i^2$ has a χ_r^2 distribution, for $r \leq n$ is the rank of Σ.

Answer. The symmetric matrix Σ has the rank $k \leq n$, it has k non zero eigen values λ_i, $i \leq k$, and $n-k$ zero eigen values. There exists an orthogonal matrix P such that $\Sigma = E(XX^T) = PE(Y^T)P^T = D$, the diagonal matrix of the eigen values where $X = PY$ and $Y = P^T X$, then the $n-r$ components of the vector Y are zero and

$$\sum_{i=1}^n X_i^2 = X^T X = Y^T Y = \sum_{i=1}^r Y_i^2,$$

it is a χ_r^2 variable.

5.5 Ordered variables

Let $X_1, \dots, X_n$ be a sample of variables with distribution function F, the ordered sample is $X_{(1)} \leq X_{(2)} \leq \dots \leq X_{(n)}$ and the inequalities are strict if F is not a discrete distribution. If F has a density, the ordered sample has the density $n! \prod_{i=1}^n f(X_{(i)})$ and $X_{(i)}$ has the density

$$\begin{aligned} f_{(i)}(x) &= n! \int_{\infty}^{x_{(2)}} \cdots \int_{\infty}^{x} \int_{x}^{\infty} \cdots \int_{x_{(n-1)}}^{\infty} \prod_{i=1}^{n} f(X_{(i)}) \\ &\quad . \, dx_{(1)} \dots dx_{(i-1)} \, dx_{(i+1)} \dots dx_{(n)} \\ &= n \binom{n-1}{i-1} \{F(x)\}^{i-1} \{1 - F(x)\}^{n-i} f(x). \end{aligned} \tag{5.3}$$

Exercise 5.5.1. Let $X_1, \dots, X_{2n}$ be independent uniform variables on $[0,1]$, calculate the characteristic function of the variable $X_{(n)} - E(X_{(n)})$.

Answer. The median of the uniform distribution on $[0,1]$ is $M = \frac{1}{2}$ such that $P(M > \frac{1}{2}) = \frac{1}{2} = P(M < \frac{1}{2})$, the order statistic $X_{(n)}$ is the empirical median with density

$$f_{n,2n}(x) = \frac{(2n)!}{n!(n-1)!} (1-x)^n x^{n-1},$$

its expectation is

$$\frac{(2n)!}{n!(n-1)!}\int_0^1 (1-x)^n x^n \, dx = \frac{(2n)!}{n!(n-1)!} B_{n+1,n+1} = \frac{n}{2n+1}$$

and it converges to $\frac{1}{2}$ as n tends to infinity. The variance of $X_{(n)}$ is equivalent to $E(X_{(n)}^2) - \frac{1}{4}$ where

$$E(X_{(n)}^2) = \frac{(2n)!}{n!(n-1)!} B_{n+1,n+2} = \frac{n(n+1)}{(2n+1)(2n+2)} = \frac{n}{2(2n+1)},$$

then

$$\mathrm{Var}(X_{(n)}) = \frac{n}{2n+1}\Big(\frac{1}{2} - \frac{n}{2n+1}\Big) = \frac{n}{2(2n+1)^2} \sim \frac{1}{8n}.$$

The higher order moments of the variable $X_{(n)}$ and $X_{(n)} - \frac{n}{2n+1}$ are

$$\begin{aligned}
E(X_{(n)}^3) &= \frac{(2n)!}{n!(n-1)!} B_{n+1,n+3} = \frac{n(n+2)}{2(2n+1)(2n+3)}, \\
E\Big\{(X_{(n)} - \frac{n}{2n+1})^3\Big\} &= \frac{n(n+2)}{2(2n+1)(2n+3)} - \frac{3n^2}{2(2n+1)^2} + \frac{2n^3}{(2n+1)^3} \\
&= \frac{2}{2(2n+1)^3(2n+3)} \sim \frac{1}{16n^4},
\end{aligned}$$

and

$$\begin{aligned}
E(X_{(n)}^4) &= \frac{(2n)!}{n!(n-1)!} B_{n+1,n+4} = \frac{n(n+3)}{4(2n+1)(2n+3)}, \\
E\Big\{(X_{(n)} - \frac{n}{2n+1})^4\Big\} &= \frac{n(n+3)}{4(2n+1)(2n+3)} - 2\frac{n^2(n+2)}{(2n+1)^2(2n+3)} \\
&\quad +3\frac{n^3}{(2n+1)^3} - 2\frac{n^4}{(2n+1)^4} \sim \frac{1}{16},
\end{aligned}$$

then the characteristic function of the variable $X_{(n)} - \frac{n}{2n+1}$ has the expansion

$$\varphi_n(t) = 1 - \frac{t^2}{16n} + \sum_{k\geq 4} \frac{(it)^k}{k!} E\{(X_{(n)} - \frac{n}{2n+1})^k\} + o(n^{-1}).$$

Exercise 5.5.2. Let $X_{(1)}, \ldots, X_{(n)}$ be an ordered sample of uniform variables on $[0,1]$, find the expectation of $Y_k = \log(1 - X_{(k)})$ and $Z_k = -\log(X_{(k)})$.

Answer. The distribution function of $Y_k = \log(1 - X_{(k)})$ is defined on $\mathbb{R}_+$ as

$$\begin{aligned} P(-\log(1 - X_{(k)}) \geq x) &= P(X_{(k)} \leq 1 - e^{-x}) \\ &= \sum_{j=k}^{n} P(X_{(j)} \leq 1 - e^{-x} < X_{(j+1)}) \\ &= \sum_{j=k}^{n} \binom{n}{j} e^{-(n-j)x}(1 - e^{-x})^j, \end{aligned}$$

by (5.3) it has the density

$$f_k(x) = n\binom{n-1}{k-1}(1 - e^{-x})^{k-1}e^{-(n-k)x}$$

and its expectation is

$$E(Y_k) = n\binom{n-1}{k-1}\int_{\mathbb{R}_+} xe^{-(n-k)x}(1 - e^{-x})^{k-1}\, dx$$

where the integral develops as

$$\begin{aligned} I &= \sum_{j=0}^{k-1} \binom{k-1}{j}(-1)^j \int_{\mathbb{R}_+} xe^{-(n-k+j)x}\, dx \\ &= \sum_{j=0}^{k-1} \frac{\binom{k-1}{j}(-1)^j}{(n-k+j)^2}, \end{aligned}$$

it follows

$$E\{-\log(1 - X_{(k)})\} = \frac{n!}{(n-k)!}\sum_{j=0}^{k-1} \frac{(-1)^j}{(k-j-1)!j!(n-k+j)^2}.$$

The distribution function of Z_k on $\mathbb{R}_+$ is

$$\begin{aligned} P(-\log(X_{(k)}) \geq x) &= P(X_{(k)} \leq e^{-x}) \\ &= \sum_{j=k}^{n} P(X_{(j)} \leq e^{-x} < X_{(j+1)}) \\ &= \sum_{j=k}^{n} \binom{n}{j} e^{-jx}(1 - e^{-x})^{n-j}, \end{aligned}$$

it has the density

$$f_k(x) = n\binom{n-1}{k-1}e^{-(k-1)x}(1 - e^{-x})^{n-k}$$

and its expectation is

$$E(Y_k) = n\binom{n-1}{k-1}\sum_{j=0}^{n-k}\binom{n-k}{j}(-1)^j\int_{\mathbb{R}_+} xe^{-(k+j-1)x}\,dx$$
$$= \frac{n!}{(k-1)!}\sum_{j=0}^{n-k}\frac{(-1)^j}{(n-k-j)!j!(k+j-1)^2}.$$

Exercise 5.5.3. Let $X_1,\ldots,X_n$ be independent uniform variables on $[0,1]$, prove that for the ordered variables $X_{(k)}$, $P(X_{(k)} \le n^{-1}x)$ converges to $\Gamma_{k-1}(x)$, the Gamma distribution function, as n tends to infinity. Find the expectation and the variance of $X_{(k)}$. Deduce the limit of the distribution function $P\{X_{(k)} \le F^{-1}(n^{-1}x)\}$ for independent real variables $X_1,\ldots,X_n$ with continuous and strictly increasing distribution function F.

Answer. The order statistic $X_{(k)}$ has the density

$$f_k(x) = n\binom{n-1}{k-1}(1-x)^{n-k}x^{k-1},$$
$$f_k(n^{-1}x) = n\binom{n-1}{k-1}(1-n^{-1}x)^{n-k}n^{1-k}x^{k-1}$$

where $(n-k)!n^k \sim n!$ and $(1-n^{-1}x)^{n-k} \sim e^{-x}$ as n tends to infinity, hence $f_k(n^{-1}x)$ is asymptotically equivalent to the Γ_{k-1} density. The expectation and the variance of $X_{(k)}$ are

$$E(X_{(k)}) = n\binom{n-1}{k-1}\int_0^1 (1-x)^{n-k}x^k\,dx$$
$$= \frac{n!}{(n-k)!(k-1)!}B_{k,n-k} = \frac{k}{n+1},$$
$$E(X_{(k)}^2) = n\binom{n-1}{k-1}\int_0^1 (1-x)^{n-k}x^{k+1}\,dx$$
$$= \frac{n!}{(n-k)!(k-1)!}B_{k+1,n-k}$$
$$= \frac{k(k+1)}{(n+1)(n+2)},$$
$$\mathrm{Var}(X_{(k)}) = \frac{k(n-k+1)}{(n+1)^2(n+2)}.$$

Under a continuous strictly increasing distribution function F, the variables $U_i = F(X_i)$ have the distribution function

$$\begin{aligned} P(F(X) \leq t) &= P(X \leq F^{-1}(t)) \\ &= F \circ F^{-1}(t) = t, \end{aligned}$$

they are uniform on $[0, 1]$ and the limit of

$$P\{X_{(k)} \leq F^{-1}(n^{-1}x)\} = P\{F(X_{(k)}) \leq n^{-1}x\}$$

is the $\Gamma_{k-1}(x)$ distribution function deduced from the limit for the ordered uniform variable.

Chapter 6

Empirical processes and weak convergence

On a metric space S, endowed with the Borel sigma-algebra $\mathcal{A}$, a sequence of distribution functions $(F_n)_{n\geq 1}$ converges to a distribution function F if

$$F_n(I) \to F(I)$$

for every interval of continuity of F.

Distribution functions F_n with discontinuities on $\mathbb{R}$ converge to a distribution function F on $\mathbb{R}$ if and only if for all $\varepsilon > 0$ and $h > 0$ there exist n_0 such that for all integer $n > n_0$ and real t

$$F(t-h) - \varepsilon < F_n(t) < F(t-h) - \varepsilon.$$

Let $(X_k)_{k=1,\ldots,n}$ be n independent and identically distributed real random variables with distribution function F on $(S, \mathcal{A})$, their empirical distribution

$$\widehat{F}_n(x) = n^{-1} \sum_{k=1,\ldots,n} 1_{\{X_k \leq x\}}$$

is an increasing right-continuous step function with left-limits. On $\mathbb{R}^d$, $\widehat{F}_n$ converges a.s. to F at every point of continuity of F, if F is continuous then $\widehat{F}_n$ converges a.s. uniformly to F.

The convergence in distribution of a sequence of variables $(X_n)_n$ to the distribution of a variable X is equivalent to the convergence of their characteristic functions $\varphi_n(t)$ to the characteristic function $\varphi_X(t)$ of X. By the strong (respectively weak) law of large number, the empirical mean of n independent and identically distributed random variables converges a.s. (respectively in probability) to their expectation.

The empirical process of a n sample of independent and identically distributed random variables $(X_k)_{k=1,\ldots,n}$ with distribution function F is

$$\nu_n(t) = n^{-\frac{1}{2}}\{\widehat{F}_n(t) - F(t)\}.$$

6.1 Weak convergence

A probability measure P on a metric space $(\mathcal{S}, \mathcal{B})$ is tight if for every $\varepsilon > 0$ there exists a compact set K_ε of $\mathcal{S}$ such that $P(K_\varepsilon) > 1-\varepsilon$. If $\mathcal{S}$ is separable and complete, every probability measure P on $(\mathcal{S}, \mathcal{B})$ is tight.
A sequence of variables $(X_n)_n$ on $(\mathcal{S}, \mathcal{B})$ is tight if for every $\varepsilon > 0$ there exists a compact set K_ε of $\mathcal{S}$ such that $P(X_n \in K_\varepsilon) \geq 1-\varepsilon$ for every n.

On a metric space $(\mathcal{S}, \mathcal{B})$, probability measures P_n converge weakly to a probability measure P if

$$\lim_n \int_{\mathcal{S}} h\, dP_n = \int_{\mathcal{S}} h\, dP$$

for every function h of $\mathcal{C}_b(\mathcal{S}) = \{h : (\mathcal{S}, \mathcal{A}) \mapsto (\mathbb{R}, \mathcal{B})$, continuous and bounded$\}$. The weak convergence of P_n to P is equivalent to

$$\limsup_{n\to\infty} P_n(A) \leq P(A)$$

for every closet set A of S and it is denoted $P_n \Rightarrow P_X$.

The weak convergence of a sequence of variables $(X_n)_n$ to the distribution of a variable X is the weak convergence of the probability distributions P_n of the variables X_n to the probability distributions P of the variable X, it is denoted $X_n \stackrel{\mathcal{D}}{\to} X$. Then for every function h of $\mathcal{C}_b(\mathcal{S})$

$$\lim_n E\{h(X_n)\} = E\{h(X)\}.$$

Let $(X_k)_{k=1,\ldots,n}$ be n independent and identically distributed random variables in $\mathbb{R}^p$, with expectation μ and variance σ^2, the central limit theorem states that $\sigma^{-1}n^{-\frac{1}{2}}(\bar{X}_n-\mu)$ converges weakly to a normal Gaussian variable $\mathcal{N}(0,1)$.

Let (X_n, X_n') be random variables on a product space $\mathcal{S} \times \mathcal{S}'$, if they converge weakly to a variable (X, X') then $X_n \stackrel{\mathcal{D}}{\to} X$ and $X_n' \stackrel{\mathcal{D}}{\to} X'$. Reversely if X_n and X_n' are independent, $X_n \stackrel{\mathcal{D}}{\to} X$ and $X_n' \stackrel{\mathcal{D}}{\to} X'$ then $(X_n, X_n') \stackrel{\mathcal{D}}{\to} (X, X')$.

Continuous mapping: Let $X_n \stackrel{\mathcal{D}}{\to} X$ and let h be a continuous function from $(S, \mathcal{S})$ to $(S', \mathcal{S}')$ then $h(X_n) \stackrel{\mathcal{D}}{\to} h(X)$.

Exercise 6.1.1. Let $(X_n)_{n\geq 1}$ be a sequence of variables of L^p, $p > 0$, such that $\sup_n E(|X_n|^p)$ is finite, prove the tightness of $(X_n)_{n\geq 1}$.

Answer. For every $\varepsilon > 0$, there exists a such that for every n

$$P(|X_n| > a) \leq a^{-p} \sup_n E(|X_n|^p) \leq \varepsilon$$

then $P(|X_n| \leq a) > 1-\varepsilon$ and $(X_n)_n$ is tight.

Exercise 6.1.2. Let $(X_n)_{n\geq 1}$ and $(Y_n)_{n\geq 1}$ be tight sequences of variables, prove that $(X_n, Y_n)_{n\geq 1}$ is tight.

Answer. For every $\varepsilon > 0$, there exist a and b such that $P(|X_n| \leq a) > 1-\varepsilon$ and $P(|Y_n| \leq b) > 1-\varepsilon$ for every n, it follows that $P(|X_n| > a) \leq \varepsilon$ and $P(|X_n| \leq a, |Y_n| > b) \leq P(|Y_n| > b) \leq \varepsilon$, hence

$$\begin{aligned} P(|X_n| \leq a, |Y_n| \leq b) &= 1 - P(|X_n| > a) - P(|X_n| \leq a, |Y_n| > b) \\ &\geq 1 - 2\varepsilon \end{aligned}$$

and (X_n, Y_n) belongs to a compact for every n with a probability larger than $1 - 2\varepsilon$.

Exercise 6.1.3. Let Φ be the distribution function of the normal distribution $\mathcal{N}(0,1)$, and let $x_n = n^{\frac{1}{2}}x$ for $x > 0$, prove that $1 - \Phi(x_n) = O(x_n^{-2})$ as n is large.

Answer. Let S_n be the sum of n independent and identically distributed variables with expectation μ and variance σ^2, and let $c > \mu$ and $x_n = n^{\frac{1}{2}}(c-\mu)$. There is equivalence between $n^{-1}S_n > c$ and $n^{-\frac{1}{2}}(S_n - \mu) > n^{\frac{1}{2}}(c-\mu)$ hence by Bienaymé–Chebychev inequality we obtain

$$P(n^{-1}S_n > c) = P(n^{-\frac{1}{2}}(S_n - \mu) > x_n) \leq x_n^{-2}\sigma^2.$$

The weak convergence of $n^{-\frac{1}{2}}(S_n - \mu)$ to a normal variable as n tends to infinity implies that for n large enough $1 - \Phi(x_n) \leq x_n^{-2}\sigma^2$.

Exercise 6.1.4. Let h be a function of $C^2(\mathbb{R})$ and let $(X_n)_{n\geq 1}$ be a sequence of real variables such that $n^{\frac{1}{2}}(X_n - \mu)$ converges weakly to a Gaussian variable $\mathrm{N}(0, \sigma^2)$, find the limiting distribution of $n^{\frac{1}{2}}\{h(X_n) - h(\mu)\}$.

Answer. By a second order expansion of h we have

$$h(X_n) - h(\mu) = (X_n - \mu)h'(\mu) + O_p((X_n - \mu)^2),$$

the weak convergence of $n^{\frac{1}{2}}(X_n - \mu)$ implies that $n^{\frac{1}{2}}(X_n - \mu)^2$ converge in probability to zero and

$$n^{\frac{1}{2}}\{h(X_n) - h(\mu)\} = n^{\frac{1}{2}}(X_n - \mu)h'(\mu) + o_p(1)$$

it converges to a centered Gaussian variable with variance $h'^2(\mu)\sigma^2$.

Problem 6.1.1. On a metric space $(\mathcal{S}, \mathcal{B})$ let X_n converge weakly to X and Y_n converge in probability to a constant a.

(a) Prove that $X_n + Y_n$ converges weakly to $X + a$.

(b) Prove that $X_n Y_n$ converges weakly to aX.

(c) Let (X_n, Y_n) converges weakly to (X, Y), prove that $X_n Y_n$ converges weakly to XY.

Answer. (a) For all δ_1 and $\delta_2 > 0$ there exists $\varepsilon > 0$ such that for every function h of $\mathcal{C}_b(\mathcal{S})$

$$|h(X_n + Y_n) - h(X_n + a)| \leq \delta_2 \ \text{ a.s.}$$

on $\{\|Y_n - a\| \leq \varepsilon\}$ and

$$P(\|Y_n - a\| > \ \varepsilon) \leq \delta_1.$$

We have

$$\begin{aligned} |E\{h(X_n + Y_n)\} - E\{h(X + a)\}| &\leq |E\{h(X_n + a)\} - E\{h(X + a)\}| \\ &\quad + |E\{h(X_n + Y_n)\} - E\{h(X_n + a)\}| \end{aligned}$$

where $\limsup_n E\{h(X_n + a)\} = E\{h(X + a)\}$ by the weak convergence of X_n, and the second term has the bound

$$\begin{aligned} E|h(X_n + Y_n) - h(X_n + a)| &\leq E\{|h(X_n + Y_n) - h(X_n + a)|1_{\{\|Y_n - a\| \leq \varepsilon\}}\} \\ &\quad + E\{|h(X_n + Y_n) - h(X_n + a)|1_{\{\|Y_n - a\| > \varepsilon\}}\} \\ &\leq \delta_2 + 2\|h\|_\infty P(\|Y_n - a\| > \varepsilon) \\ &\leq \delta_2 + 2\|h\|_\infty \delta_1, \end{aligned}$$

as δ_1 and δ_2 are arbitrary, the second term converges to zero.

(b) For every function h of $\mathcal{C}_b(\mathcal{S})$, $|h(X_n Y_n) - h(aX_n)| \leq \delta_2$ a.s. on $\{\|Y_n - a\| \leq \varepsilon\}$ and $Eh(aX_n)$ converges to $Eh(aX)$, then $|E\{h(X_n Y_n)\} - \{Eh(aX_n)\}|$ is bounded by

$$\begin{aligned} &E\{|h(X_n Y_n) - h(aX_n)|1_{\{\|Y_n - a\| \leq \varepsilon\}}\} \\ &+ E\{|h(X_n Y_n) - h(aX_n)|1_{\{\|Y_n - a\| > \varepsilon\}}\} \\ &\leq \delta_2 + 2\|h\|_\infty P(\|Y_n - a\| > \varepsilon), \end{aligned}$$

therefore $|E\{h(X_n Y_n)\} - \{Eh(aX)\}|$ converges to zero.

(c) Let F_n be the distribution function of (X_n, Y_n), the Fourier transform of $X_n + Y_n$ is

$$\varphi_{X_n + Y_n}(t) = \int_{\mathcal{S}^2} e^{it(x+y)} \, dF_n(x, y)$$

where $\int_{\mathcal{S}^2} h((x,y)\,dF_n(x,y)$ converges to $\int_{\mathcal{S}^2} h((x,y)\,dF(x,y)$ for every function h of $\mathcal{C}_b(\mathcal{S}^2)$, therefore $\limsup_n \varphi_{X_n+Y_n}(t) = \varphi_{X+Y}(t)$.

Problem 6.1.2. Let $(X_i)_{i=1,\dots,n}$ and $(Y_i)_{i=1,\dots,n}$ be independent sequences of independent and identically distributed real variables with distribution functions F_X and respectively F_Y, with empirical mean and variance $\bar{X}_n$ and s^2_{nX}, and respectively $\bar{Y}_n$ and s^2_{nY}. Prove the weak convergence of $T_n = (\bar{X}_n - \bar{Y}_n)(s^2_{nX} + s^2_{nY})^{-\frac{1}{2}}$.

Answer. The empirical means have the variances $n^{-1}\sigma^2_X$ and respectively $n^{-1}\sigma^2_Y$, and the empirical variances converges a.s. to σ^2_X and respectively σ^2_Y. By the central limit theorem, $n^{-\frac{1}{2}}\{\bar{X}_n - E(X_1)\}$ and $n^{-\frac{1}{2}}\{\bar{Y}_n - E(Y_1)\}$ converge weakly to independent centered Gaussian variables, then $n^{-\frac{1}{2}}(\bar{X}_n - \bar{Y}_n)$ converges weakly to a Gaussian variable with expectation $E(X_1) - E(Y_1)$ and variance $\sigma^2_X + \sigma^2_Y$, the weak convergence of T_n to a Gaussian variable with expectation $E(X_1) - E(Y_1)$ and variance 1 follows.

Problem 6.1.3. Let $(X_i)_{i=1,\dots,n}$ be a sequence of independent and identically distributed real variables and let $h : \mathbb{R}^2 \mapsto \mathbb{R}$ be a symmetric function such that $\sigma^2 = \operatorname{Var} h(X_1, X_2)$ is finite, prove the convergence in probability of the variable $S_n = \{n(n-1)\}^{-1} \sum_{i\neq j=1}^n h(X_i, X_j)$ and the weak convergence of

$$U_n = \frac{1}{\sqrt{n(n-1)}} \sum_{i\neq j=1}^{n} h(X_i, X_j) - E\{h(X_i, X_j)\}.$$

Answer. Let F be the distribution function of the variables X_i, the variable S_n has the expectation $m = E\{h(X_i, X_j) = \int_{\mathbb{R}^2} h(x,y)\,dF(x)\,dF(y)$. The characteristic function of $h(X_i, X_j) - m$ is

$$\phi(t) = E[e^{it\{h(X_i,X_j)-m\}}] = \int_{\mathbb{R}} e^{it\{h(x,y)-m\}}\,dF(x)\,dF(y).$$

For every i, let $m(X_i)$ be the expectation of $h(X_i, X_j)$ conditionally on X_i and let $\phi_{n,X_i}(t) = E[e^{it\{h(X_i,X_j)-m(X_i)\}} \mid X_i]$, the characteristic function of

$$H_n(X_i) = \frac{1}{n-1} \sum_{j\neq i, j=1}^{n} \{h(X_i, X_j) - m(X_i)\}$$

conditionally on X_i is

$$\phi_{n,X_i}(t) = e^{-it\{m(X_i)-m\}} \phi^{n-1}((n-1)^{-1}t, X_i)$$

and the logarithm of $\phi^{n-1}((n-1)^{-1}t, X_i)$ has a second order expansion

$$(n-1)\log\phi((n-1)^{-1}t, X_i) = -\frac{t^2}{2(n-1)}\sigma_i^2\{1+o(1)\},$$
$$\sigma_i^2 = Var\{h(X_i, X_j) \mid X_i\}.$$

Conditionally on X_i, the convergence to zero as n tends to infinity of $\log\phi_{n,X_i}(t)$ for every t implies the convergence of $\phi_{n,X_i}(t)$ to 1 and $H_n(X_i)$ converges in probability to zero, it follows that $n^{-1}\sum_{i=1}^n H_n(X_i)$ converges in probability to zero. The variable S_n is written as

$$S_n = \frac{1}{n}\sum_{i=1}^n H_n(X_i) + \frac{1}{n}\sum_{i=1}^n m(X_i),$$

it converges therefore to $m = E\{m(X_i)\}$. The variable

$$U_{ni} = (n-1)^{-\frac{1}{2}} \sum_{j\neq i, j=1}^n \{h(X_i, X_j) - m\}$$

has the conditional characteristic function

$$\phi_{U_{ni}}(t) = E(e^{itU_{ni}} \mid X_i) = \phi^{n-1}((n-1)^{-\frac{1}{2}}t, X_i)$$

by a second order expansion of its logarithm we have

$$\log\phi^{n-1}((n-1)^{-\frac{1}{2}}t, X_i) = -\frac{1}{2}t^2\sigma_i^2 + o(1)$$

and $\phi^{n-1}((n-1)^{-\frac{1}{2}}t, X_i)$ converges to the characteristic function of a Gaussian variable with random variance σ_i^2, conditionally on X_i. The variable U_n has the variance $\sigma^2 = E\sigma_i^2 + \mathrm{Var}\{m(X_i)\}$ and

$$U_n = \frac{1}{\sqrt{n}}\sum_{i=1}^n U_{ni},$$

it is the normalized sum of asymptotically independent Gaussian variables with expectation zero and variances σ_i^2 therefore it converges weakly to a centered Gaussian variables with variance σ^2.

Problem 6.1.4. Let $(X_i)_{i\geq 1}$ be a sequence of independent and identically distributed centered variables with finite variance $\sigma^2 > 0$ and let $S_n = \sum_{i=1}^n X_i$. Find the joint limiting distribution of the minimum and the maximum of the variables $Y_{n,k} = S_k - n^{-1}kS_n$, $k = 0, \ldots, n$.

Answer. As n tends to infinity we assume that n is even then $Y_{n,0} = 0$ and $Y_{n,n} = 0$

$$\begin{aligned} Y_{n,1} &= S_1 - n^{-1}S_n = (1-n^{-1})X_1 - n^{-1}(S_n - S_1), \\ Y_{n,k} &= (1-kn^{-1})S_k - kn^{-1}(S_n - S_k), \\ Y_{n,n-1} &= S_{n-1} - (1-n^{-1})X_n \end{aligned}$$

where $S_n - S_k$ and S_{n-k} have the same distribution so $Y_{n,k}$ and $Y_{n,n-k}$ have the same distribution and it is sufficient to consider the variables $Y_{n,k}$, for $k = 0, \ldots, \frac{n}{2}$.

The variance of $Y_{n,k}$ is

$$v_{n,k} = \{(1-kn^{-1})^2k + n^{-2}k^2(n-k)\}\sigma^2 = k(1-kn^{-1})\sigma^2$$

and it is maximum at the integer part k_0 of $\frac{n}{2}$, where

$$Y_{n,k_0} = \frac{1}{2}(S_{[\frac{n}{2}]} - (S_n - S_{[\frac{n}{2}]}).$$

The maximum $\overline{Y}_n$ and the minimum $\underline{Y}_n$ of the variables $Y_{n,k}$, $k = 0, \ldots, \frac{n}{2}$, are therefore reached at k_0, by symmetry of the Gaussian variables and according to the sign of Y_{n,k_0}.

The variable $(n^{-\frac{1}{2}}\sigma^{-1}\overline{Y}_n, n^{-\frac{1}{2}}\sigma^{-1}\underline{Y}_n)$ converges weakly to

$$\begin{aligned} &\Big(\frac{1}{2\sqrt{2}}(Z_2 - Z_1), \frac{1}{2\sqrt{2}}(Z_1 - Z_2)\Big) \quad \text{if } Z_1 > Z_2, \\ &\Big(\frac{1}{2\sqrt{2}}(Z_1 - Z_2), \frac{1}{2\sqrt{2}}(Z_2 - Z_1), 0\Big) \quad \text{if } Z_1 < Z_2, \end{aligned}$$

where Z_1 and Z_2 are independent normal Gaussian variables hence $P(Z_1 = Z_2) = 0$ and the variance of $Z_1 - Z_2$ is 2. Then the variable $(n^{-\frac{1}{2}}\sigma^{-1}\overline{Y}_n, n^{-\frac{1}{2}}\sigma^{-1}\underline{Y}_n)$ converges weakly to

$$\Big(-\frac{|Z|}{2}, \frac{|Z|}{2}\Big)$$

where Z is a normal Gaussian variable.

6.2 Weak convergence of processes

Let I a subset of $\mathbb{R}$ a measurable process $X = (X(t))_{t\in I}$ is defined on a probability space $(\Omega, \mathcal{A}, P)$, with values in a functional metric space $(\mathcal{S}, d)$. The space C_I of the continuous real functions on $I \subseteq \mathbb{R}$ is endowed with the uniform norm $\|x\| = \sup_{t\in I}|x(t)|$, for every function x of C_I. With the uniform distance

$$d(x, y) = \|x - y\|, \, x, y \in C_I,$$

(C_I, d) is a separable and complete space.

The space D_I of the right-continuous real functions with left limits on I is not complete for the uniform distance d, but it is for the Skorohod distance which allows a small change according to the jumps of the functions. The empirical distribution of a sequence $(X_i)_{i=1,\dots,n}$ of real random variables is a process in $D_{\mathbb{R}}$. Let x et $y \in D_{[0,1]}$, the Skorohod distance π is defined as the smallest $\varepsilon > 0$ such that there exists a strictly increasing and continuous transformation g from $[0,1]$ onto $[0,1]$ satisfying $g(0) = 0$, $g(1) = 1$ and

$$\sup_{t\in[0,1]} |g(t) - t| \le \varepsilon, \qquad \sup_{t\in[0,1]} |x(t) - y(g(t))| \le \varepsilon.$$

If $(x_n)_n$ is a sequence of $D_{[0,1]}$ that converges for the Skorohod distance to $x \in C_{[0,1]}$, then x_n converges uniformly to x.

Proposition 6.1. *The distribution of a process X is determined by its finite dimensional distributions $(X(t_1), \dots, X(t_k))$, for all integer k and $(t_1, \dots, t_k)$ in S^k.*

A sequence of distributions $(P_{X_n})_n$ is relatively compact on a metric space if one extract from every sub-sequence of $(P_{X_n})_n$ a sub-sequence $(P_{X_{n'}})_{n'}$ that converges to a limit Q. The weak convergence of the finite dimensional distributions of a sequence of processes $X_n = (X_n(t))_{t\in I}$ to the finite dimensional distributions of a process $X = (X(t))_{t\in I}$ is not sufficient the weak convergence of X_n to X. If the sequence of distributions P_{X_n} is relatively compact, every sub-sequence $P_{X_{n'}}$ converging to a probability Q has its finite dimensional distributions that converge to the finite dimensional distributions of Q which are the same as P_X and the limit is unique. Reversely, every converging sequence $(P_{X_n})_n$ is relatively compact, if its limit is a probability P_X, then $P_{X_n} \Rightarrow P_X$.

Theorem 6.1 (Prokhorov's theorem). *Let $(P_n)_n$ a sequence of probabilities on a measurable $(S, \mathcal{S})$. If $(P_n)_n$ is tight then it is relatively compact. If (S, d) is a separable and complete metric space and if $(P_{X_n})_n$ is relatively compact then it is tight.*

Theorem 6.2. *Let (S, d) be a separable and complete metric space, if the finite dimensional distributions of a sequence of processes $X_n = (X_n(t))_{t\in I}$ converge to the finite dimensional distributions of a process X and if P_{X_n} is tight then $X_n \xrightarrow{\mathcal{D}} X$.*

Theorem 6.3. *A sequence of processes $(X_n)_n$ of C_I on a separable and complete metric space (S,d) is tight if*

(1) for every t in I, the sequence $(X_n(t))_n$ is tight,
(2) for all $\eta > 0$ and $\varepsilon > 0$, there exists $\delta > 0$ such that

$$\limsup_{n\to\infty} P\{\sup_{|t-s|<\delta} \|X_n(t) - X_n(s)\| > \varepsilon\} < \eta.$$

The Wiener process or Brownian motion B is a Gaussian process of $C_{[0,1]}$ such that for all s and t in $[0,1]$

$$EB_t = 0, \quad E\{B_s B_t\} = s \wedge t.$$

The Brownian bridge B^0 on $[0,1]$ is a Gaussian process of $C_{[0,1]}$ such that for all s and t in $[0,1]$

$$EB_t^0 = 0, \quad E\{B_s^0 B_t^0\} = s \wedge t - st,$$

it has the same distribution as $B_t - tB_1$.

Theorem 6.4 (Donsker's theorem). *Let $(X_i)_{i\geq 1}$ be a sequence of independent and identically distributed centered variables with finite variance $\sigma^2 > 0$ and let $S_n = \sum_{i=1}^n X_i$, then the process*

$$X_n(t) = \frac{1}{\sigma n^{\frac{1}{2}}} S_{[nt]} + \frac{nt - [nt]}{\sigma n^{\frac{1}{2}}} X_{[nt]+1} \tag{6.1}$$

converges weakly to the Brownian motion on $[0,1]$.

Theorem 6.5. *Let F be the uniform distribution function on $[0,1]$ and let $\widehat{F}_n$ be the empirical distribution function of an uniform sample $(X_i)_{i=1,\dots,n}$, the empirical process $\nu_n = \sqrt{n}(\widehat{F}_n - F)$ converges weakly to the Brownian bridge B^0 on $[0,1]$.*

Exercise 6.2.1. Prove that the process $V_t = (1+t)B^0_{\frac{t}{1+t}}$ is a Brownian motion on $\mathbb{R}_+$.

Answer. The process V is a centered Gaussian process and for $s < t$ in $[0,1]$, $\frac{s}{1+s} < \frac{t}{1+t}$ and the variables V_s and V_t have the covariance $E(V_s V_t) = s(1+t) - st = s$, it is therefore a Brownian motion on $\mathbb{R}_+$.

Exercise 6.2.2. Deduce from Donsker's theorem that the Brownian motion has independent increments and provide a direct proof.

Answer. Under the conditions of Donsker's theorem and as

$$\sup_{t\in[0,1]} \left| X_n(t) - \frac{1}{\sigma n^{\frac{1}{2}}} S_{[nt]} \right| \leq \frac{1}{\sigma n^{\frac{1}{2}}} |X_1|$$

converges to zero in L^2, it is sufficient that the variables $S_{[ns]} = \sum_{i=1}^{[ns]} X_i$ and $S_{[nt]} - S_{[ns]} = \sum_{i=[ns]+1}^{[nt]} X_i$ are independent.

A direct proof is deduced from the properties of the Brownian motion. Let $s < t$ in $[0,1]$, the Gaussian variables B_s and $B_t - B_s$ have the covariance $E\{B_s(B_t - B_s) = 0$, they are therefore independent.

Exercise 6.2.3. Let $(X_t)_{t\in[0,1]}$ be a process on a metric space $(\mathcal{X}, d)$ such that $E \sup_{|t-s|<\delta} d^a(X_t, X_s) \leq |t-s|^b$ for s, t in $[0,1]$, $a > 0$ and $b > 1$, prove that $P(X \in C) = 1$.

Answer. For $\varepsilon > 0$ there exists a partition $t_0 = 0 < t_1 < \cdots < t_m = 1$ such that $t_i - t_{i-1} \leq \delta$ and for every $\delta > 0$

$$\begin{aligned} &P(\max_{i=1,\ldots,m} \sup_{|t-t_i|<\delta} d(X_t, X_{t_i}) > \varepsilon) \\ &\quad \leq \sum_{i=1,\ldots,m} P(\max_{i=1,\ldots,m} \sup_{|t-t_i|<\delta} d(X_t, X_{t_i}) > \varepsilon) \\ &\quad \leq \frac{1}{\varepsilon^{\frac{1}{a}}} \sum_{i=1,\ldots,m} E\{\sup_{|t-t_i|<\delta} d^a(X_t, X_{t_i})\} \\ &\quad \leq \frac{1}{\varepsilon^{\frac{1}{a}}} \sum_{i=1,\ldots,m} |t_i - t_{i-1}|^b \\ &\quad \leq \frac{1}{\varepsilon^{\frac{1}{a}}} m\delta^b \leq \frac{1}{\varepsilon^{\frac{1}{a}}} \delta^{b-1} \end{aligned}$$

where $m\delta \leq 1$. For every $\eta > 0$, the bound is smaller than η for δ small enough, this proves the a.s. convergence of $\sup_{|t-s|<\delta} d(X_t, X_s)$ to zero as δ tends to zero.

Exercise 6.2.4. Let $(P_n)_{n\geq 1]}$ and P be probability measures on a complete metric space $(\mathcal{X}, d)$ such that $P_n(f)$ converges to $P(f)$ for every real function f of $C_b(X)$. For every $\varepsilon > 0$, let $w_f(\varepsilon, x) = \sup\{|f(x) - f(y)| : d(x,y) < \varepsilon\}$, and let $L(k_0) = \{f : |f(x) - f(y)| \leq k_0 d(x,y)\}$. Prove that $\sup_{f\in L(k_0)} P_n(x : w_f(\varepsilon, x) > \eta)$ converges to zero as η converges to zero.

Answer. The convergence of $P_n(f)$ to $P(f)$ for every real function f of $C_b(X)$ implies that P_n converges weakly to P. Every real function f of

$C_b(X)$ is uniformly continuous then for every $\delta > 0$ there exists $\eta > 0$ such that $\sup_{d(x,y)<\eta} |f(x) - f(y)| < \varepsilon$, therefore

$$\sup_{d(x,y)\leq\delta} |w_f(\eta,x) - w_f(\eta,y)| = |\sup_{d(x,z)\leq\delta} |f(x) - f(z)| - \sup_{d(y,z)\leq\eta} |f(y) - f(z)|| \leq w_f(\eta + 2\delta, x)$$

and the function $x \mapsto w_f(\eta, x)$ is uniformly continuous on $(\mathcal{X}, d)$. For every f in $L(k_0)$ we have $\sup_{d(x,y)\leq\eta} |f(x) - f(y)| \leq k_0\eta$ and $\varepsilon = k_0\eta$.

The set $A = \{x : w_f(\varepsilon, x) < \eta\}$ is open in $(\mathcal{X}, d)$ and its complementary $A^c = \{x : w_f(\varepsilon, x) \geq \eta\}$ is closed hence $\limsup_{n\to\infty} P_n(A_\varepsilon) \leq P(A_\varepsilon)$ where

$$\lim_{\varepsilon\to 0} \sup_{f\in L(k_0)} P(x : w_f(\varepsilon, x) > \eta) = 0,$$

hence the limit of $\sup_{f\in L(k_0)} P(x : w_f(\varepsilon, x) > \eta)$ as ε tends to zero and n tends to infinity is zero.

Exercise 6.2.5. Using the process $X_n(t)$ defined by (6.1), find the joint limiting distribution of the minimum and the maximum of the variables $Y_{n,k} = S_k - n^{-1}kS_n$, $k = 0, \ldots, n$.

Answer. We have

$$\max_{k=0,\ldots,n} Y_{n,k} = \sup_{t\in[0,1]} X_n(t), \qquad \min_{k=0,\ldots,n} Y_{n,k} = \inf_{t\in[0,1]} X_n(t)$$

where the process X_n converges weakly to the Brownian motion W on $[0, 1]$. By continuous mapping, the variable $(\min_{k=0,\ldots,n} Y_{n,k}, \max_{k=0,\ldots,n} Y_{n,k})$ converges weakly to $(\inf_{t\in[0,1]} W_t, \sup_{t\in[0,1]} W_t)$, as n tends to infinity. The symmetry of the Brownian motion along the line $x = 0$ implies the symmetry of the limit

$$(\inf_{t\in[0,1]} W_t, \sup_{t\in[0,1]} W_t) = (-\sup_{t\in[0,1]} |W_t|, \sup_{t\in[0,1]} |W_t|).$$

Exercise 6.2.6. Let $(X_i)_{i=1,\ldots,n}$ a sample of independent and identically distributed variables with a continuous distribution function F on $I \subset \mathbb{R}$, prove the weak convergence of its empirical process $\nu_n = \sqrt{n}(\widehat{F}_n - F)$.

Answer. The variables $U_i = F(X_i)$ have an uniform distribution on $[0, 1]$. The inverse of the distribution function F is

$$F^{-1}(s) = \inf\{t : s \leq F(t)\},$$

then $\{s \le F(t)\} = \{F^{-1}(s) \le t\}$ and the variables X_i have the empirical distribution function $\widehat{F}_n(x) = n^{-1}\sum_{i=1}^n 1_{\{U_i \le F(x)\}}$ and their empirical process is written as

$$\nu_n = \sqrt{n}(\widehat{F}_n - F) = \nu_n^0 \circ F$$

where ν_n^0 is the empirical process of the variables U_i. The function

$$h : D_{[0,1]} \to D_{[0,1]}, \quad h(\varphi)(t) = \varphi \circ F(t)$$

is continuous on $D_{[0,1]}$ at every ψ of $C_{[0,1]}$: for every $\varepsilon > 0$, let φ in $D_{[0,1]}$ and let ψ in $C_{[0,1]}$ such that $\pi(\varphi, \psi) \le \varepsilon$ for the Skorohod distance π of $D_{[0,1]}$, then $\pi(\varphi \circ F, \psi \circ F) \le \varepsilon$ in $D_{\mathbb{R}}$. It follows that $B_n = h(B_n^0)$ converges weakly to $B = h(B^0) = B^0 \circ F$ with covariance function $F(s \wedge t) - F(s)F(t)$ for all s and t in I.

Exercise 6.2.7. Let $\widehat{F}_n$ be the empirical distribution function of independent variables $(X_1, \ldots, X_n)$ with distribution function F and let h be a real function of $C_b^1([0,1])$, prove the weak convergence of $n^{\frac{1}{2}}\{h(\widehat{F}_n) - h(F)\}$ and give the variance of its limit.

Answer. The weak convergence of $\nu_n = n^{\frac{1}{2}}(\widehat{F}_n - F)$ entails

$$n^{\frac{1}{2}}\{h(\widehat{F}_n) - h(F)\} = h'(F)\nu_n + o_p(1)$$

and it converges weakly to $h'(F)B \circ F$, its limit variance is $h^{'2}(F)\,F(1-F)$.

Exercise 6.2.8. Let $(X_{1n}, \ldots, X_{nn})$ be independent real variables with distribution function F_n such that F_n converges F_n uniformly to a distribution function F and $G_n = n^{\frac{1}{2}}(F_n - F)$ converges to G. Let h be a real function of $C_b^1([0,1])$, define the limit of $n^{\frac{1}{2}}\{h(\widehat{F}_n) - h(F)\}$, for the empirical distribution function of $(X_{1n}, \ldots, X_{nn})$.

Answer. The empirical distribution function $\widehat{F}_n$ of $(X_{1n}, \ldots, X_{nn})$ converges uniformly to F and the empirical process $\nu_n = n^{\frac{1}{2}}(\widehat{F}_n - F_n)$ converges weakly to the Gaussian process $B \circ F$, therefore $n^{\frac{1}{2}}(\widehat{F}_n - F)$ converges weakly to $G \circ F + G$. By the convergence of F_n and the first order expansion $n^{\frac{1}{2}}\{h(\widehat{F}_n) - h(F_n)\} = h'(F_n)\nu_n + o_p(1)$, it follows that

$$n^{\frac{1}{2}}\{h(\widehat{F}_n) - h(F)\} = h'(F_n)\nu_n + h'(F)G_n + o_p(1),$$

it converges weakly to the uncentered Gaussian variable $h'(F)\{B \circ F + G\}$ with variance $h^{'2}(F)\,F(1-F)$.

Exercise 6.2.9. Let $(X_{ni})_{i\geq 1}$ be a sequence of independent and centered variables with finite variances $\sigma_{ni}^2 > 0$ and let $S_{nk} = \sum_{i=1}^k X_{ni}$ and $s_{nk}^2 = \sum_{i=1}^k \sigma_{ni}^2$, prove that the process

$$X_n(t) = \frac{1}{s_{nk}} S_{[nt]}$$

converges weakly to the Brownian motion on $[0,1]$ as $n^{-1}s_{nk}^2$ converges.

Answer. For all $a_1, \ldots, a_k$ in $\mathbb{R}^k$ and $t_1, \ldots, t_k$ in $[0,1^k$, the variable $\sum_{i=1}^k a_i X_n(t_i)$ converges weakly to a centered Gaussian variable with finite variance (Problem 4.2.3). By definition of the process X_n, it has independent increments and the variance of $X_n(t)$ is $n^{-1}[nt](1-n^{-1}[nt])$, it converges to $t(1-t)$. The tightness of the process X_n is proved like in Donsker's Theorem for identically distributed variables $(X_i)_{i\geq 1}$.

Exercise 6.2.10. Let $(X_{1n}, \ldots, X_{nn})$ be independent real variables with density f_{θ_n} such that θ_n belongs to a subset I of $\mathbb{R}^p$, $\theta \mapsto f_\theta$ belongs to $C_b^2(\Theta)$ uniformly on $\mathbb{R}$, and $n^{\frac{1}{2}}(\theta_n - \theta)$ converges weakly to a limit ϑ. Using a second order expansion of $\log f_\theta$ with respect to θ at θ_n, prove the weak convergence of $T_n = \sum_{i=1}^n \{\log f_{\theta_n}(X_i) - \log f_\theta(X_i)\}$.

Answer. Let $\dot{f}_\theta$ and $\ddot{f}_\theta$ denote the first and second order derivatives of f_θ with respect to θ, a second order expansion of T_n is written

$$\begin{aligned} T_n &= (\theta_n - \theta)^T \sum_{i=1}^n \frac{\dot{f}_\theta(X_{in})}{f_\theta(X_{in})} \\ &\quad + (\theta_n - \theta)^T \sum_{i=1}^n \frac{\ddot{f}_\theta(X_{in})}{f_\theta(X_{in})} (\theta_n - \theta) + o_p(1). \end{aligned}$$

The expectation of $\frac{\dot{f}_\theta(X_{in})}{f_\theta(X_{in})}$ under f_{θ_n} is $\int_{\mathbb{R}} \dot{f}_\theta(x) f_\theta^{-1}(x) f_{\theta_n}(x)\,dx$ and it converges to $\int_{\mathbb{R}} \dot{f}_\theta(x)\,dx = 0$ as θ_n converges to θ, the limit of its variance matrix is

$$\Sigma_\theta^2 = \int_{\mathbb{R}} \frac{\dot{f}_\theta^{\otimes 2}}{f_\theta},$$

then the variable $n^{-\frac{1}{2}} \sum_{i=1}^n \frac{\dot{f}_\theta(X_{in})}{f_\theta(X_{in})}$ converges weakly to a centered Gaussian variable with variance matrix Σ_θ^2. The second term of the expansion of T_n is written as

$$\frac{1}{2} n^{\frac{1}{2}} (\theta_n - \theta) n^{-1} \sum_{i=1}^n \frac{\ddot{f}_\theta(X_{in})}{f_\theta(X_{in})} n^{\frac{1}{2}} (\theta_n - \theta),$$

the second order derivative of $\log f_\theta$ is $h_\theta = f_\theta^{-2}\{f_\theta \ddot{f}_\theta - \dot{f}_\theta^{\otimes 2}\}$ such that $Eh_\theta(X)$ converges to zero, it follows that

$$\lim_{n\to\infty} \int_{\mathbb{R}} \frac{\ddot{f}_\theta}{f_\theta} f_{\theta_n} = \lim_{n\to\infty} \int_{\mathbb{R}} \frac{\dot{f}_\theta^{\otimes 2}}{f_\theta^2} f_{\theta_n} = 0$$

and the variable $n^{-1}\sum_{i=1}^n \frac{\ddot{f}_\theta(X_{in})}{f_\theta(X_{in})}$ converges in probability to Σ_θ^2.

It follows that the second term of the expansion converges weakly to $\frac{1}{2}\vartheta^T\Sigma_\theta^2\vartheta$ where $n^{\frac{1}{2}}(\theta_n - \theta)$ converges weakly to ϑ, so the variable T_n converges weakly to a Gaussian variable with expectation $\frac{1}{2}\vartheta^T\Sigma_\theta^2\vartheta$ and with variance matrix $\vartheta^T\Sigma_\theta^2\vartheta$.

Exercise 6.2.11. Extend Exercise 4 to independent variables $(X_{1n}, \ldots, X_{nn})$ densities f_{θ_n,η_n} such that $f_{\theta,\eta} = g(x;\theta,\eta)h(\eta)$ with g in $C_b^2(\Theta \times A)$ and h in $C_b^2(A)$ uniformly on $\mathbb{R}$, $n^{\frac{1}{2}}(\eta_n - \eta)$ converges weakly to ζ and $n^{\frac{1}{2}}(\theta_n - \theta)$ converges weakly to ϑ.

Answer. The derivative of $\log f(x;\theta,\eta)$ with respect to η is

$$f^{-1}(x;\theta,\eta)\{\dot{g}_\eta(x;\theta,\eta)h(x;\eta) + g(x;\theta,\eta)h'(x;\eta)\} = \frac{\dot{g}_\eta(x;\theta,\eta)}{g_\eta(x;\theta,\eta)} + \frac{h'(x;\eta)}{h(x;\eta)}$$

the expectation of $\frac{\dot{f}_\eta(X_{in};\theta,\eta)}{f(X_{in};\theta,\eta)}$ converges to zero and the variance matrix of the derivative of $\log f(x;\theta,\eta)$ with respect to (θ,η) is

$$\Sigma = \begin{pmatrix} \Sigma_\theta & \Sigma_{\theta,\eta} \\ \Sigma_{\theta,\eta} & \Sigma_\eta + S_\eta \end{pmatrix}$$

where

$$\Sigma_{\theta,\eta} = \int_{\mathbb{R}} \frac{\dot{g}_\theta(x;\theta,\eta)\dot{g}_\eta(x;\theta,\eta)}{g(x;\theta,\eta)} h(x;\eta)\,dx,$$
$$S_\eta = \int_{\mathbb{R}} \frac{h'^2(x;\eta)}{h(x;\eta)}\,dx$$

$$\Sigma_\eta = \int_{\mathbb{R}} \frac{\dot{g}_\eta^2(x;\theta,\eta)}{g(x;\theta,\eta)} h(x;\eta)\,dx,$$
$$\Sigma_\theta = \int_{\mathbb{R}} \frac{\dot{g}_\theta^2(x;\theta,\eta)}{g(x;\theta,\eta)} h(x;\eta)\,dx$$

and

$$\lim_{n\to\infty} E\Big\{\frac{\dot{g}(X_{in};\theta,\eta)}{g(X_{in};\theta,\eta)}\frac{h'(X_{in};\eta)}{h(X_{in};\eta)}\Big\} = 0.$$

The expansion of T_n has the same form as previously and it converges to a Gaussian variable with expectation $\mu(\vartheta,\zeta) = \frac{1}{2}(\vartheta^T,\zeta^T)\Sigma^2(\vartheta^T,\zeta^T)^T$ and with variance matrix $2\mu(\vartheta,\zeta)$.

Problem 6.2.1. Let P_n be the empirical probability of n independent and identically distributed variables with distribution function F under P_0, and let $K(P_n,P_0)$ be the Kullback–Leibler information of P_n and P_0. Prove the weak convergence of $nK(P_n,P_0)$.

Answer. By a second order expansion of the logarithm, we have

$$\begin{aligned}\log\frac{dP_n}{dP_0} &= \log\Big\{1+\frac{d(P_n-P_0)}{dP_0}\Big\}\\ &= \frac{d(P_n-P_0)}{dP_0}+\frac{1}{2}\Big\{\frac{d(P_n-P_0)}{dP_0}\Big\}^2\{1+o_p(1)\},\end{aligned}$$

the distance $K(P_n,P_0)$ satisfies

$$\begin{aligned}nK(P_n,P_0) &= n\int\log\frac{dP_n}{dP_0}\,dP_n\\ &= \frac{1}{2}\int\Big\{\frac{d\nu_n}{dP_0}\Big\}^2 dP_0+o(1),\end{aligned}$$

and it converges to $\frac{1}{4}\int\Big\{\frac{dP_F}{dP_0}\Big\}^2 dP_0$ where P_F is the probability distribution of $B\circ F$ where B is the standard Brownian bridge on $[0,1]$.

Problem 6.2.2. Let $S_n = \sum_{i=1}^n X_i$ for a sequence of independent and identically distributed centered variables $(X_i)_{i\geq 1}$ with finite variance $\sigma^2>0$, and let $\tau_n = \min\{k : S_k = \max_{i\leq n} S_i\}$, prove the convergence of $P(n^{-1}\tau_n\leq a)$ to $\frac{2}{\pi}\arcsin\sqrt{a}$.

Answer. With the notation (6.1), we have

$$\begin{aligned}n^{-1}\tau_n &= \inf\{s : S_{[ns]} = \sup_{t\in[0,1]} S_{[nt]}\}\\ &= \inf\{s : X_n(s) = \sup_{t\in[0,1]} X_n(t)\}\end{aligned}$$

and it converges weakly to $\tau = \inf\{s : B(s) = \sup_{t\in[0,1]} B(t)\}$ where B is the Brownian motion, by Donsker's theorem. The distribution of τ is given by

$$\begin{aligned}P(T\leq t) &= P(\sup_{s\in[0,t]} B_s \geq \sup_{x\in[t,1]} B_x)\\ &= P(\sup_{u\in[0,t]} B_{t-u}-B_t \geq \sup_{v\in[0,1-t]} B_{t+v}-B_t).\end{aligned}$$

Let $X_t = \sup_{u \in [0,t]} B_{t-u} - B_t$ and $Y_t = \sup_{v \in [0,1-t]} B_{t+v} - B_t > 0$ a.s., they are independent and the variables $X = t^{-\frac{1}{2}} X_t$ and $Y = (1-t)^{-\frac{1}{2}} Y_t$ have the same distribution as $\sup_{t \in [0,1]} B_t - B_0$, then

$$P(T \le t) = P(X_t \ge Y_t) = P\Big(\frac{Y}{(X^2+Y^2)^{\frac{1}{2}}} \le t^{\frac{1}{2}}\Big)$$

where the variable $Y(X^2+Y^2)^{-\frac{1}{2}}$ is the sine of a uniform variable on $[0, \frac{\pi}{2}]$, hence

$$P(T \le t) = \frac{2}{\pi} \arcsin(t^{\frac{1}{2}}).$$

Problem 6.2.3. Let X a centered Gaussian process on I with covariance function Γ on I^2. Prove the equivalence between the continuity of the function Γ and the continuity of X in probability.

Answer. By the continuity in probability of X if for every t in I and for every $\varepsilon > 0$

$$\lim_{h \to 0} \sup_{|t-s|<h} P(|X_t - X_s| > \varepsilon) = 0$$

where $P(|X_t - X_s| > \varepsilon) \le \varepsilon^{-2} E(|X_t - X_s|^2)$ and

$$E(|X_t - X_s|^2) = \Gamma_{t,t} - 2\Gamma_{s,t} + \Gamma_{s,s},$$

therefore $\lim_{h \to 0} \sup_{|t-s|<h} E(|X_t - X_s|^2) = 0$ so the continuity of the function Γ implies the continuity of X in probability. Reversely, if X is continuous in probability, for every finite integer set $\mathcal{K}$ and for all $(t_j)_{j \in \mathcal{K}}$, the vector $\{X(t_j)\}_{j \in \mathcal{K}}$ is continuous in probability on $\mathbb{R}^{\mathcal{K}}$ and it has a Gaussian distribution with a continuous covariance function $\{\Gamma(t_i, t_j)\}_{i,j \in \mathcal{K}^2}$ on $\mathbb{R}^{\mathcal{K}} \times \mathbb{R}^{\mathcal{K}}$ then $P(|X_{t_i} - X_{t_j}| > \varepsilon, (t_i, t_j)\}_{i,j \in \mathcal{K}^2})$ is continuous and X is continuous in probability.

Problem 6.2.4. Let $(X_i)_{i=1,\dots,n}$ a sample of independent and identically distributed variables with a distribution function F of $C^1(I)$, for an interval $I \subset \mathbb{R}$, prove the weak convergence of the empirical processus $\sqrt{n}(\widehat{F}_n^{-1} - F^{-1})$, for the inverse distribution functions $\widehat{F}_n^{-1}$ and F^{-1}.

Answer. For every differentiable function ψ on $[0,1]$, the empirical process $\zeta_n = n^{1/2}\{\psi(\widehat{F}_n) - \psi(F)\}$ has the approximation

$$\zeta_n(x) = \{\psi \circ F(x)\}' \,.\, \nu_n(x)$$

and it converges weakly to $(\psi \circ F)' \,.\, B_0 \circ F$ where B_0 is a Brownian bridge. Using the inverse function $\psi(F(x)) = F^{-1}(x)$, it follows that the process $n^{1/2}(\widehat{F}_n^{-1} - F^{-1})$ converges weakly to the process

$$\zeta_0 = \frac{B_0 \circ F}{f \circ F^{-1}}.$$

6.3 Inequalities

For the sum S_n of n independent and identically distributed variables $(X_i)_{i\leq n}$, with expectation zero and characteristic function φ_X, Chernoff's theorem states the inequality

$$\log P(n^{-1}S_n > a) = \inf_{t>0}\{n \log \varphi_X(n^{-1}t) - at\}. \tag{6.2}$$

Exercise 6.3.1. For the empirical process of a sample $(X_1, \ldots, X_n)$ with distribution function F, prove the inequality

$$\lim_{n\to\infty} P(\sup_x \nu_n(x) \geq a) = e^{-2a^2}, \ a > 0.$$

Answer. Let $\tau_n = \inf\{y : \nu_n(y) = \sup_x \nu_n(x)\}$, the events $\{\tau_n > y\}$ and $\{\nu_n(\tau_n) < \sup_x \nu_n(x)\}$ are equivalent and τ_n is a stopping time for the filtration generated by the empirical process ν_n. As n tends to infinity, the process ν_n converges weakly to the Brownian bridge ν_0 and τ_n converges weakly to the stopping time $\tau_0 = \inf\{y : \nu_0(y) = \sup_x \nu_0(x)\}$ of the Brownian bridge.

By Chernoff's theorem the probability $P(\sup_x \nu_n(x) \geq a)$ has the limit

$$\begin{aligned}\lim_n P(\sup_x \nu_n(x) \geq a) &= \lim_n P(\nu_n(\tau_n) \geq a)\\ &= \inf_{t>0} \exp\{\log \varphi_{\nu_0(\tau_0)}(t) - at\},\end{aligned}$$

where the characteristic function of the Brownian bridge at x is

$$\varphi_{\nu_{0x}}(t) = e^{\frac{v_x t^2}{2}}$$

with the variance $v_x = x(1-x)$ of the Brownian bridge. The minimum of the bound is

$$\lim_n P(\sup_x \nu_n(x) \geq a) = e^{-\frac{a^2}{2v_0}} = e^{-2a^2}$$

as v_x is maximum at $x = \frac{1}{2}$ where $v_0 = \frac{1}{4}$.

Exercise 6.3.2. Let $(X_i)_{i\geq 1}$ be a sequence of independent random variables with the same distribution function F on $\mathbb{R}$ and with empirical process ν_n and let $Y_n(t) = \int_0^t A_s \, d\nu_n(s)$ be the integral of a predictable process A with respect to the empirical process ν_n. Extend Chernoff's theorem to $\lim_{n\to\infty} P(|Y_n(t)| > x)$.

Answer. The empirical process is $\nu_n(t) = n^{-\frac{1}{2}}\{S_n(t) - \mu_n(t)\}$ defined by $S_n(t) = \sum_{i=1}^n 1_{\{X_i\leq t\}}$ and $\mu_n(t) = ES_n(t) = nF(t)$, it converges weakly to the process $W_F = W \circ F$, where W is the standard Brownian bridge, its covariance function is $v_F(s,t) = F(s \wedge t) - F(s)F(t)$ and its variance is $v_F(x) = F(x)\{1 - F(x)\}$. The Laplace transform transformed Brownian bridge W_F at x is

$$L_{W_F(x)}(u) = e^{-\frac{u^2 v_F(x)}{2}}$$

and for every real number $a > 0$, Chernoff's theorem asserts

$$\lim_{n\to\infty} P(|\nu_n(x)| > a) = 2\exp\Bigl(-\frac{a^2}{2v_F(x)}\Bigr).$$

The maximum value v_0 of the variance v_F is achieved as $F(x) = \frac{1}{2}$ hence $v_0 = \frac{1}{4}$ and

$$\log P(\sup_x \nu_n(x) > a) = e^{-2a^2}.$$

The process Y_n has the variance function

$$V_A(x) = E\Bigl\{\int_0^x A_s^2\, dF(s) - \Bigl(\int_0^x A_s\, dF(s)\Bigr)^2\Bigr\},$$

and Chernoff's theorem implies

$$\lim_{n\to\infty} P(|Y_n(x)| > a) = 2\exp\Bigl\{-\frac{a^2}{2V_A(x)}\Bigr\},$$

$$\lim_n P(\sup_x Y_n(x) \geq a) = 2e^{-\frac{a^2 v_0}{2}}$$

where v_0 is the maximum value of the function $V_A(x)$.

Exercise 6.3.3. Let ν_n be the empirical process of independent variables $(X_1, \ldots, X_n)$ of $L^2(\mathbb{R})$, such that $E|\nu_n| \geq b_n > 0$. Prove that for every $a_n < b_n$

$$P(|\nu_n| > a_n) \geq \frac{(b_n - a_n)^2}{E(|\nu_n|^2)}.$$

Answer. Writting $|\nu_n| = |\nu_n| 1_{\{|\nu_n| > a\}} + |\nu_n| 1_{\{|\nu_n| \leq a\}}$, we obtain

$$\begin{aligned} E|\nu_n| &= E\{|\nu_n| 1_{\{|\nu_n| > a\}} + E\{|\nu_n| 1_{\{|\nu_n| \leq a\}} \\ &\leq \{E(\nu_n^2)\, P(|\nu_n| > a)\}^{\frac{1}{2}} + E\{|\nu_n| 1_{\{|\nu_n| \leq a\}} \end{aligned}$$

where $V_n = E(\nu_n^2) = n^{-1} \sum_{i=1}^n \mathrm{Var}(X_i)$ and

$$E|\nu_n| - E(|\nu_n| 1_{\{|\nu_n| \leq a\}}) \geq b_n - a_n,$$

the result follows.

6.4 Contiguity

On measurable spaces $(\Omega_n, \mathcal{A}_n)$, probabilities Q_n are contiguous to P_n if for every A_n in $\mathcal{A}_n$, the convergence of $P_n(A_n)$ to zero implies the convergence of $Q_n(A_n)$ to zero. Equivalently, for every sequence of variables $(Y_n)_{n\geq 1}$, the convergence of Y_n to zero in probability under P_n implies the convergence of Y_n to zero in probability under Q_n.

Let μ be probabilities on $\Omega_n, \mathcal{A}_n)$ such that P_n and Q_n are absolutely continuous with respect to μ, with densities f_n and respectively g_n, and let

$$\begin{aligned} L_n &= f_n^{-1} g_n \text{ if } f_n \neq 0, \\ &= 1 \text{ if } f_n = g_n = 0, \\ &= \infty \text{ if } f_n = 0,\ g_n \neq 0, \end{aligned}$$

by LeCam's first Lemma, if L_n converges weakly under P_n to a normal variable $\mathcal{N}(-\frac{\sigma^2}{2}, \sigma^2)$, then Q_n is contiguous to P_n.

Exercise 6.4.1. Let f_n and g_n be the densities of Gaussian variables with expectations α_n, and respectively β_n, and with variances σ_n^2, and respectively τ_n^2, give conditions on the parameters for the contiguity of Q_n to P_n.

Answer. The Gaussian densities are strictly positive, under the condition $\delta_n^2 = \sigma_n^2 - \tau_n^2 > 0$, the ratio $L_n = f_n^{-1} g_n$ is written as

$$\begin{aligned} L_n &= \frac{\tau_n}{\sigma_n} \exp\Big\{-\frac{(\delta_n x - \mu_n)^2}{2\sigma_n^2\tau_n^2}\Big\} \exp\Big\{-\frac{\beta_n^2}{2\tau_n^2} + \frac{\alpha_n^2}{2\sigma_n^2} + \frac{\mu_n^2}{2\sigma_n^2\tau_n^2}\Big\} \\ &= \frac{\tau_n}{\sigma_n} \exp\Big\{-\frac{\delta_n^2(x - \delta_n^{-1}\mu_n)^2}{2\sigma_n^2\tau_n^2}\Big\} \exp\Big\{-\frac{(\beta_n - \alpha_n)^2}{2\delta_n^2}\Big\}, \\ \mu_n &= \frac{\beta_n\sigma_n^2 - \alpha_n\tau_n^2}{\delta_n^2}. \end{aligned}$$

If α_n and β_n converge to the same limit α and δ_n converge to $\delta > 0$, μ_n converges to α, $\delta_n^{-1}\mu_n$ converges to $\delta^{-1}\alpha$ and $\delta_n^{-2}\sigma_n^2\tau_n^2$ converges to $\delta^{-2}\tau^2\sigma^2$. By LeCam's first Lemma, a condition for the contiguity is $\delta^{-1}\alpha = -\frac{1}{2}\delta^{-2}\tau^2\sigma^2$, i.e.

$$\alpha = -\frac{\tau^2\sigma^2}{2\delta}.$$

Furthermore, the scaling ratio $\tau^{-1}\sigma$ must be equal to $\sqrt{2\pi}\tau\sigma$ which entails $\tau^{-2} = \sqrt{2\pi}$.

Exercise 6.4.2. Let $P_n << \mu$ with density f_n and let $N_n = \{f_n = 0\}$, prove that $P_n(N_n) = 0$ and, if Q_n is contiguous to P_n, prove that $\lim_{n\to\infty} Q_n(N_n) = 0$.

Answer. Every subset A of N_n has the probability $P_n(A) = \int_A f_n \, d\mu_n = 0$, therefore $P_n(N_n) = 0$ and $Q_n(N_n)$ converges to zero by contiguity.

Exercise 6.4.3. Let $F_n(x) = P_n(L_n < x)$, prove that $\int_0^\infty x \, dF_n(x) \leq 1$ and converges to one if Q_n is contiguous to P_n.

Answer. By a change of variable, we have

$$\int_0^\infty x \, dF_n(x) = E_{P_n}(L_n) = \int_{N_n^c} dQ_n + \int_{N_n} \frac{dQ_n}{dP_n} \, dP_n$$

where N_n^c is the complementary set of N_n, the second integral is $Q_n(N_n)$ and it converges to zero, the second integral $Q_n(N_n^c)$ is lower than 1 and it converges to one.

Exercise 6.4.4. Let $F_n(x) = P_n(L_n < x)$ converge to a distribution function F and let $P_n(A_n)$ converge to zero, prove that if Q_n is contiguous to P_n then $\overline{\lim}_{n\to\infty} Q_n(A_n) = 1 - \int_0^\infty x \, dF(x)$.

Answer. As $L_n \leq 1$ under P_n, the probability $Q_n(A_n) \geq 0$ has the bound $Q_n(A_n) = \int_{A_n} L_n \, dP_n \leq P_n(A_n)$ and it converges to zero, moreover

$$Q_n(A_n) = 1 - \int_{A_n^c} L_n \, dP_n,$$

$$\limsup_{n\to\infty} Q_n(A_n) = 1 - \liminf_{n\to\infty} \int_{A_n^c} L_n \, dP_n,$$

where $\int_{A_n^c} L_n \, dP_n = \int_\Omega L_n \, dP_n - Q_n(A_n) = \int_0^\infty x \, dF_n(x) - Q_n(A_n)$ and

$$\liminf_{n\to\infty} \int_{A_n^c} L_n \, dP_n = \int_0^\infty x \, dF(x).$$

Chapter 7

Discrete martingales and stopping times

7.1 Martingales and supermartingales

A filtration $\mathbb{F} = (\mathcal{F}_n)_{n\geq 0}$ on a probability space $(\Omega, \mathcal{A}, P)$ is an increasing, complete, right-continuous with right-limits family of sub-sigma algebras of $\mathcal{A}$. A stopping time τ with respect to $\mathbb{F}$ is an integer random variable such that the event $\{\tau \leq n\}$ belongs to $\mathcal{F}_n$, for every integer n. Then

$$\mathcal{F}_\tau = \{B \in \mathcal{A} : B \cap \{\tau \leq n\} \in \mathcal{F}_n\}$$

is a sub-sigma algebra of $\mathcal{A}$. If τ' is a stopping time such that $\tau \leq \tau'$, then $\mathcal{F}_\tau \subset \mathcal{F}_{\tau'}$. On a filtered probability space $(\Omega, \mathcal{F}, P, \mathbb{F})$, a martingale $M = (M_n)_{n\geq 0}$ with respect to $\mathbb{F}$ is a sequence of variables such that

$$E(M_m|\mathcal{F}_n) = M_n, m > n \in \mathbb{N},$$

then $E(M_n) = E(M_0)$ for every integer n, and for all stopping times such that $\tau \leq \tau'$

$$E(M_{\tau'}|\mathcal{F}_\tau) = M_\tau.$$

A sequence of variables $M = (M_n)_{n\geq 0}$ is a local martingale with respect to $\mathbb{F}$ if there exists an increasing sequence of stopping times $(\tau_n)_n$ tending to infinity such that $(M_{n\wedge\tau_n})_{n\leq 0}$ is a martingale. A sequence $(X_n)_{n\geq 0}$ is a submartingale with respect to $\mathbb{F}$ if $E(X_m|\mathcal{F}_n) \geq X_n$ for every $m > n \geq 0$ then for all stopping times $\tau \leq \tau'$

$$E(X_{\tau'}|\mathcal{F}_\tau) \geq X_\tau.$$

The sequence $(X_n)_{n\geq 0}$ is a supermartingale if $E(X_m|\mathcal{F}_n) \leq X_n$.

Theorem 7.1. *Maximal inequality for a positive supermartingale $(X_n)_{n\geq 0}$: for every $x \geq 0$*

$$P(\max_{k=0,\dots,n} X_k > x \mid \mathcal{F}_0) \leq \frac{X_0}{x} \wedge 1.$$

Proof. The variable $T_n = \min\{k = 0, \ldots, n : X_k > x\}$ is a stopping time and the event $\{T = k\}$ is $\mathcal{F}_k$-measurable, then

$$E(X_n \mid \mathcal{F}_0) = \sum_{k=0,\ldots,n} E\{E(X_n \mid T = k) \mid \mathcal{F}_0\} P(T = k \mid \mathcal{F}_0)$$

where $E(X_n \mid T = k) \leq X_k > x$, it follows

$$\begin{aligned} E(X_n \mid \mathcal{F}_0) &\leq \sum_{k=0,\ldots,n} x P(T = k \mid \mathcal{F}_0) \\ &\geq x P(\max_{k=0,\ldots,n} X_k > x \mid \mathcal{F}_0). \end{aligned}$$

□

The maximal inequality extends to L^p norms

$$P(\max_{k=0,\ldots,n} X_k > x \mid \mathcal{F}_0) \leq x^{-p} E(|X_n|^p \mid \mathcal{F}_0).$$

Theorem 7.2. *Every positive supermartingale $(X_n)_{n\geq 0}$ converges a.s. to a limit X_∞ such that $E(X_\infty|\mathcal{F}_n) \leq X_n$ for every integer n. Every positive submartingale $(X_n)_{n\geq 0}$ such that $\sup_{n\in\mathbb{N}} E(X_n^+)$ is finite converges a.s. to a limit X_∞ in L^1 such that $E(X_\infty|\mathcal{F}_n) \geq X_n$ for every integer n.*

Examples of discrete martingales and supermartingales

(1) For every positive variable X of L^p, $p \geq 1$, the sequence of the variables $X_n = E(X \mid \mathcal{F}_n)$ is a positive martingale of L^p and it converges a.s.
(2) A random walk $X_n = \sum_{i=1}^n \zeta_i$ is defined by a sequence of independent and identically distributed random variables $(\zeta_i)_i = (X_i - X_{i-1})_{i=1,\ldots,n}$. Let $\mu = E\zeta_i$ be the mean increment of X_n, if $\mu = 0$ then X_n is a martingale, if $\mu > 0$ then X_n is a submartingale and if $\mu < 0$ then X_n is a supermartingale.
(3) Let $(X_n)_{n\geq 0}$ be a real martingale and let φ be a real convex, respectively concave, function defined from $\mathbb{R}$, then $\varphi(X_n)$ is a submartingale, respectively supermartingale.
(4) For every increasing sequence of random variables $(A_n)_{n\geq 1}$ such that A_n is $\mathcal{F}_n$-measurable and $E(A_\infty|\mathcal{F}_0) < \infty$ a.s. then $X_n = E(A_\infty|\mathcal{F}_n) - A_n$ is a supermartingale.

For a martingale $(X_n)_n$, the following properties are equivalent

(1) $(X_n)_n$ converges in L^1 to X,
(2) $\sup E|X_n|$ finite and $X_n = E(X \mid \mathcal{F}_n)$,
(3) $(X_n)_n$ is equi-integrable and X_n converges in probability to X.

Exercise 7.1.1. On a probability space $(\Omega, \mathcal{F}, P)$ let $(A_n)_{n\geq 1}$ a partition of Ω with probabilities $p_k = P(A_k)$, let $B_n = \cup_{k>n} A_k$ and let $\mathcal{F}_n$ the sigma-algebra generated by $(A_k)_{k\leq n}$ and B_n. For a positive sequence $(a_n)_{n\geq 1}$ such that $\sum_{n\geq 1} a_k$ is finite, find the sequence $(b_n)_{n\geq 0}$ such that $M_n = \sum_{k\leq n} a_k 1_{A_k} + b_n 1_{B_n}$ is a martingale.

Answer. The probability of B_n is $r_n = \sum_{k>n} p_k$ and for integers $m < n$

$$E(M_n \mid \mathcal{F}_m) = \sum_{k\leq m} a_k 1_{A_k} + \sum_{k=m+1}^{n} a_k p_k + b_n E(1_{B_n} \mid \mathcal{F}_m),$$

M_n is martingale if $E(M_n \mid \mathcal{F}_m) = M_m$ which requires

$$\begin{aligned}
b_m 1_{B_m} &= \sum_{k=m+1}^{n} a_k p_k + b_n E(1_{B_n} \mid \mathcal{F}_m), \\
\sum_{k=m+1}^{n} a_k p_k &= E(b_m 1_{B_m} - b_n 1_{B_n}) \\
&= E(b_m 1_{B_m \setminus B_n}) + (b_m - b_n) P(B_n) \\
&= b_m(r_m - r_n) + (b_m - b_n) r_n = b_m r_m - b_n r_n,
\end{aligned}$$

this provides a recursive formula for every $n \geq 1$

$$b_n = b_{n-1} + r_n^{-1}(b_{n-1} - a_n) p_n.$$

Exercise 7.1.2. Let $(X_n)_{n\geq 0}$ and $(Y_n)_{n\geq 0}$ be a real submartingales on a space $(\Omega, \mathcal{F}, P, \mathbb{F})$, prove that $X_n \wedge Y_n$ is a $\mathbb{F}$-martingale.

Answer. The events $\{X_n \leq Y_n\}$ and $\{X_n \geq Y_n\}$ are $\mathcal{F}_n$-measurable and we have

$$\begin{aligned}
E\{X_{n+1} \wedge Y_{n+1} \mid \mathcal{F}_n\} &= E\{X_{n+1} 1_{\{X_{n+1} \leq Y_{n+1}\}} \mid \mathcal{F}_n\} \\
&\quad + E\{Y_{n+1} 1_{\{X_{n+1} \geq Y_{n+1}\}} \mid \mathcal{F}_n\} \\
&= [E\{X_{n+1} \mid \mathcal{F}_n\}] \wedge [E\{Y_{n+1} \mid \mathcal{F}_n\}] \\
&= X_n \wedge Y_n.
\end{aligned}$$

Exercise 7.1.3. Let τ be a stopping time on a filtered probability space $(\Omega, \mathcal{F}, P, \mathbb{F})$ and or every integer n let $\phi(n) = \min\{p \geq 1 : \{\tau = n\} \in \mathcal{F}_p\}$. Prove that $\phi(\tau)$ is a stopping time and $\phi(\tau) \leq \tau$.

Answer. For all integers n and k, $\phi(n) = k$ entails $\{\tau = n\}$ belongs to $\mathcal{F}_k \cap \mathcal{F}_n \subset \mathcal{F}_{k\wedge n}$, therefore $\phi(\tau)$ is a stopping time and $k \leq n$, on $\{\tau = n\} \cap \{\phi(n) = k\}$ this is equivalent to $\phi(\tau) \leq \tau$.

Exercise 7.1.4. Let $X_n^* = \max_{k=0,\dots,n} X_k$, for a positive supermartingale $(X_n, \mathcal{F}_n)_n$ of L^p, $p \geq 1$, prove the inequalities

$$P(X_n^* > x) \leq x^{-1}E(X_n),\ x > 0$$

and $E\|X_n^*\|_p \leq \frac{p}{p-1}E\|X_n\|_p$, for every $x > 0$.

Answer. Let $\tau = \inf\{n : X_n > x\}$, if $\tau \leq n$ then $E(X_\tau) \leq E(X_n \mid \mathcal{F}_\tau)$ by the supermartingale property and

$$\begin{aligned} E(X_\tau 1_{\{\tau \leq n\}}) &\leq E\{(X_n \mid \mathcal{F}_\tau) 1_{\{\tau \leq n\}})\} \\ &= E(X_n 1_{\{\tau \leq n\}}) \leq E(X_n). \end{aligned}$$

Let $X_\tau > x$, $\tau \leq n$ is equivalent to $X_n^* > x$, this entails

$$\begin{aligned} P(X_n^* > x) = P(\tau \leq n) &\leq x^{-1}E(X_\tau 1_{\{\tau \leq n\}}) \\ &\leq E(X_n). \end{aligned}$$

For $p \geq 1$, $x > 0$ and $(X_n)_n$ in L^p, we have

$$\lim_{x\to\infty} x^p P(X_n^* \geq x) \leq \lim_{x\to\infty} E(|X_n^*|^p 1_{\{X_n^* \geq x\}}) = 0,$$

with an integration by parts we obtain

$$\begin{aligned} \|X_n^*\|_p &= p\int_0^\infty x^{p-1} P(X_n^* \geq x)\, dx \\ &\leq pE\int_0^\infty x^{p-2} X_n 1_{\{X_n^* \geq x\}})\, dx \\ &= \frac{p}{p-1} E\{X_n\, (X_n^*)^{p-1}\} \\ &\leq \frac{p}{p-1} \|X_n^*\|_p^{\frac{p-1}{p}}\, \|X_n\|_p \end{aligned}$$

and the result follows.

Exercise 7.1.5. On a probability space $(\Omega, \mathcal{F}, P$, let $(X_n)_{n\geq 0}$ be a sequence of independent and identically distributed random variables, let $\mathcal{F}_n = \sigma(X_1, \dots, X_n)$ and let τ be a stopping time with respect to the filtration $\mathbb{F} = (\mathcal{F}_n)_n$. Prove that $X_{\tau+n}$ is independent of $\mathcal{F}_\tau$ for every integer n, and that $(X_{\tau+n})_{n\geq 0}$ is a sequence of independent and identically distributed random variables with the same distribution as the variables X_n.

Answer. For every set B in $\mathcal{F}_\tau$, the set $B \cap \{\tau = k\}$ belongs to $\mathcal{F}_k$ and it is independent of X_{k+n}, for every integer k. For every A in $\mathcal{F}$, the set $\{X_{\tau+n} \in A\} = \{X_{k+n} \in A\}$ is independent of $\mathcal{F}_k$ if $\tau = k$, therefore

$$\begin{aligned} &P(\{X_{\tau+n} \in A\} \cap B \cap \{\tau = k\}) \\ &= P(\{X_{k+n} \in A\} \cap \{\tau = k\})P(B \cap \{\tau = k\}) \\ &= P(\{X_{\tau+n} \in A\}P(B \cap \{\tau = k\}). \end{aligned}$$

This proves the independence of $X_{\tau+n}$ and $\mathcal{F}_\tau$, for every integer n. For all integers k and n, if $\tau = k$ then the variables $X_{\tau+n} = X_{k+n}$ are mutually independent and they have the same distribution as X_1, by the independence of $X_{\tau+n}$ and $\mathcal{F}_\tau$, it follows that the variables $X_{\tau+n}$ are independent and identically distributed.

Exercise 7.1.6. On a filtered probability space $(\Omega, \mathcal{F}, P, \mathbb{F})$, let $(X_n)_{n\geq 0}$ be a adapted sequence of real random variables such that $E(\max_n |X_n|)$ is finite and let τ be an a.s. finite stopping time. Prove that the random variable $\tau' = \min\{n \geq 1 : X_n \geq E^{\mathcal{F}_n}(X_\tau)\}$ is a stopping time such that $\tau' \leq \tau$ and $X_n < E^{\mathcal{F}_n}(X_{\tau'})$ a.s. on $\{\tau' > n\}$.

Answer. For every integer n, $\tau' = n$ implies

$$E^{\mathcal{F}_n}(X_\tau) = \sum_{k\leq n} X_k 1_{\{\tau=k\}} + \sum_{k>n} E^{\mathcal{F}_n}(X_k 1_{\{\tau=k\}})$$

and the set $\{E^{\mathcal{F}_n}(X_\tau) \leq X_n\}$ is $\mathcal{F}_n$-measurable therefore $\{\tau' = n\}$ is $\mathcal{F}_n$-measurable. If $\tau = k < n$, then $E^{\mathcal{F}_n}(X_k) = X_k$ and the minimum n such that $X_n \geq E^{\mathcal{F}_n}(X_k)$ is smaller or equal to k, hence $\tau' \leq \tau$.

On $\{\tau' > n\}$, we have $X_n < E^{\mathcal{F}_n}(X_\tau)$ and

$$\begin{aligned} E^{\mathcal{F}_n}(X_{\tau'}) &= \sum_{k>n} E^{\mathcal{F}_n}(X_k) 1_{\{\tau'=k\}} \\ &\geq \sum_{k>n} E^{\mathcal{F}_n}\{E^{\mathcal{F}_k}(X_\tau)\} 1_{\{\tau'=k\}} > \sum_{k>n} X_n 1_{\{\tau'=k\}}. \end{aligned}$$

Exercise 7.1.7. On a filtered probability space $(\Omega, \mathcal{F}, P, \mathbb{F})$, let $(X_n)_{n\geq 0}$ be a positive supermartingale and let τ be a stopping time. Prove that the sequence of the variables $Y_n = E^{\mathcal{F}_n}(X_{\tau\vee n})$ is a positive supermartingale a.s. bounded by X_n.

Answer. If $\tau = k$, then

$$Y_n = E^{\mathcal{F}_n}(X_{k\vee n}) = E^{\mathcal{F}_n}(X_k) 1_{\{k>n\}} + X_n 1_{\{k\leq n\}}$$

and $Y_n \leq X_n$. For all integers $m > n$, the events $\{\tau \leq n\}$ and $\{\tau > n\}$ belong to $\mathcal{F}_n$ and

$$\begin{aligned} E^{\mathcal{F}_n}(Y_m) &= E^{\mathcal{F}_n}(X_{\tau\vee m})1_{\{\tau\leq n\}} + E^{\mathcal{F}_n}(X_{\tau\vee m})1_{\{\tau>n\}} \\ &\leq X_n 1_{\{\tau\leq n\}} + E^{\mathcal{F}_n}(X_{\tau\vee m})1_{\{\tau>n\}} \end{aligned}$$

where

$$\begin{aligned} E^{\mathcal{F}_n}(X_{\tau\vee m})1_{\{\tau>n\}} &= E^{\mathcal{F}_n}E^{\mathcal{F}_\tau}(X_m)1_{\{n<\tau\leq m\}} + 1_{\{\tau>m\}}E^{\mathcal{F}_n}(X_\tau) \\ &\leq E^{\mathcal{F}_n}(X_\tau), \end{aligned}$$

hence $E^{\mathcal{F}_n}(Y_m) \leq E^{\mathcal{F}_n}(X_{\tau\vee n}) = Y_n$.

Exercise 7.1.8. On a filtered probability space $(\Omega, \mathcal{F}, P, \mathbb{F})$, let $(X_n)_{n\geq 0}$ be a positive supermartingale and let τ be a stopping time. Prove that the sequence of the variables $(Y_n)_n = (X_{\tau\wedge n})_n$ is a positive supermartingale. Let $\tau_a = \min\{n \in \mathbb{N} : X_n > a\}$ prove that τ_a is a stopping time with respect to $\mathbb{F}$.

Answer. The events $\{\tau_a > n\}$ and $\{X_n \leq a\}$ are equivalent and $\{X_n \leq a\}$ is $\mathcal{F}_n$-measurable which proves that the variable τ_a is a stopping time. Let $n < m$ in $\mathbb{N}$, then $\{n < \tau \leq m\}$ and $\{\tau \leq n\}$ belong to $\mathcal{F}_n$ which implies

$$\begin{aligned} E^{\mathcal{F}_n}(Y_m) &= E^{\mathcal{F}_n}(X_\tau)1_{\{\tau\leq m\}} + E^{\mathcal{F}_n}(X_m)1_{\{\tau>m\}}, \\ E^{\mathcal{F}_n}(X_\tau)1_{\{\tau<m\}} &= X_\tau 1_{\{\tau\leq n\}} + E^{\mathcal{F}_n}(X_\tau)1_{\{n<\tau\leq m\}} \\ &\leq X_\tau 1_{\{\tau\leq n\}} + X_n 1_{\{n<\tau\leq m\}}, \end{aligned}$$

therefore $E^{\mathcal{F}_n}(Y_m) \leq X_{\tau\wedge n}$.

Exercise 7.1.9. On a filtered probability space $(\Omega, \mathcal{F}, P, \mathbb{F})$, let $(X_n)_{n\geq 0}$ be a positive submartingale, let $S_n = \max_{m\leq n} X_m$ and let $\tau_a = \min\{n \in \mathbb{N} : S_n > a\}$, for $a > 0$, prove the inequality $P(S_n > a) \leq a^{-1}E(X_n 1_{\{S_n>a\}})$ and calculate $E(\tau_a)$.

Answer. The equality $\{\tau_a \leq n\} = \{S_n > a\}$ implies $a < \max_{m\leq\tau_a} X_m$ and $X_{\tau_a} > a$, therefore

$$\begin{aligned} aP(S_n > a) &= aP(\tau_a \leq n) \leq E(X_{\tau_a}1_{\{\tau_a\leq n\}}) \\ &\leq E\{E^{\mathcal{F}_{\tau_a}}(X_n)1_{\{\tau_a\leq n\}}\} \\ &= E(X_n 1_{\{\tau_a\leq n\}}). \end{aligned}$$

Exercise 7.1.10. Let $(X_n)_{n\geq 0}$ be a sequence of independent and identically distributed real variables and let g be a measurable function from $\mathbb{R}$ to $\mathbb{R}_+$ such that $Eg(X_0) = 1$. Prove that $Y_n = \prod_{k=1}^n g(X_k)$ is a positive martingale converging a.s. to 1.

Answer. Let $\mathcal{F}_n = \sigma((X_k)_{k\leq n}$, for all integers $m < n$, the variable $Y_m^{-1}Y_n = \prod_{k=m+1}^n g(X_k)$ is independent of $\mathcal{F}_m$, hence

$$Y_m^{-1}E(Y_n \mid \mathcal{F}_n) = \prod_{k=m+1}^{n} E\{g(X_k)\} = 1,$$

equivalently $E(Y_n \mid \mathcal{F}_n) = Y_m$ and $\{g(X_n)\}_{n\geq 0}$ is a positive martingale. The variables Y_n have a finite expectation is $EY_n = \prod_{k=1}^n E\{g(X_k)\} = 1$ therefore it converges a.s. to 1 as n tends to infinity.

Exercise 7.1.11. Let $(Y_n)_{n\geq 0}$ be independent square integrable variables, let $S_n = \sum_{k=1}^n Y_k$ and let τ be an integrable stopping time with respect to a filtration $\mathbb{F}$ such that $\mathcal{F}_n = \sigma(Y_1, \ldots, Y_n)$, prove that $E(S_\tau) = E(\tau)\,E(Y_1)$, $S_\tau - \tau E(Y_1)$ is a martingale and $\operatorname{Var}(S_\tau) = E(\tau)\operatorname{Var}(Y_1)$.

Answer. Let $\mu = E(Y)$, for integers $n < m$ we have

$$E^{\mathcal{F}_n}(S_m - m\mu) = S_n - n\mu + E^{\mathcal{F}_n}\{S_m - S_n - (m-n)\mu\} = S_n - n\mu$$

where $S_m - S_n$ is independent of $\mathcal{F}_n$ hence $(X_n)_n = (S_n - n\mu)_n$ is a martingale with respect to $\mathbb{F}$, and X_τ is a martingale. For every n, $(X_{\tau\wedge n})_n$ is a martingale such that

$$\begin{aligned}E(X_\tau - X_{\tau\wedge n}) &= \sum_{m>n} E(Y_m 1_{\{\tau>n\}}) \leq \sum_{m>n} E(|Y_m| 1_{\{\tau>m\}}) \\ &= E(|Y_1|) \sum_{m>n} P(\tau > m)\end{aligned}$$

where $\{\tau > m\} \in \mathcal{F}_{m-1}$ is independent of Y_m.

As $E(\tau) = \sum_n nP(\tau = n) = \sum_n P(\tau > n)$ is finite, $\sum_{m>n} P(\tau > m)$ tends to zero as n tends to infinity and

$$E(X_\tau) = \lim_{n\to\infty} E(X_{\tau\wedge n}) = \lim_n E(\tau \wedge n)\,E(Y_1) = E(\tau)\,E(Y_1).$$

For every integer n, the variance of X_n is $\operatorname{Var}(X_n) = n\operatorname{Var}(Y_1)$, the variance of $X_{\tau\wedge n}$ is $E(\tau \wedge n)\operatorname{Var}(Y_1)$. As $(Z_n)_n = (S_n^2 - nE(Y_1^2))_n$ is a martingale

(Exercise 7.1.8), the difference of the variances of X_τ and $X_{\tau\wedge n}$ is

$$\begin{aligned} E(X_\tau^2 - X_{\tau\wedge n}^2) &\le E(Z_\tau - Z_{\tau\wedge n}) \\ &= E\Big[1_{\{\tau>n\}}\Big\{\Big(\sum_{k=n+1}^{\tau} Y_k)^2 - (\tau-n)E(Y_1^2)\Big)\Big\}\Big] \\ &\le \sum_{m>n} E\{(Y_m-\mu)^2 1_{\{\tau>n\}}\} \\ &\le \sum_{m>n} E\{(Y_m-\mu)^2 1_{\{\tau>m\}}\} = \mathrm{Var}(Y_1)\sum_{m>n} P(\tau>m), \end{aligned}$$

where $(Y_m-\mu)^2$ is independent of $\{\tau > m\}$ and $\sum_{m>n} P(\tau > m)$ converges to zero as n tends to infinity, this provides the variance of X_τ.

Exercise 7.1.12. On a filtered probability space $(\Omega, \mathcal{F}, P, \mathbb{F})$, let $(X_n)_{n\ge 0}$ be an integrable martingale such that the variables $X_{n+1}-X_n$ are independent, identically distributed and integrable, and let $\tau_a = \min\{n \in \mathbb{N} : X_n \le a\}$, for $a > 0$, prove that τ_a is an integrable stopping time.

Answer. The martingale is the adapted sum $X_n = \sum_{k=1}^n Y_k + X_0$, with independent variables $Y_k = X_k - X_{k-1}$. The equivalence of $\{\tau_a > n\}$ and $\{X_n > a\}$ entails

$$E(\tau_a) = \sum_{n\in\mathbb{N}} nP(\tau_a = n) = \sum_{n\in\mathbb{N}} P(\tau_a > n) = \sum_{n\in\mathbb{N}} P(X_n > a).$$

By the integrability of $|Y_1|$, for every ε in $]0,1[$ there exists $\delta > 0$ such that $P(|Y_1| > \delta) < \varepsilon$, hence

$$P(X_n > n\delta) \le P(|Y_1| > \delta, \dots, |Y_n| > \delta) < \varepsilon^n.$$

Let $n_0 = \delta^{-1}a$, then for every $n \ge n_0$, $P(X_n > n\delta) < \varepsilon^n$. It follows that $\sum_{n\ge n_0} P(X_n > a)$ and therefore $E(\tau_a)$ are finite.

Exercise 7.1.13. On a filtered probability space $(\Omega, \mathcal{F}, P, \mathbb{F})$, let $(X_n)_{n\ge 0}$ be a martingale of L^2 such that $X_0 = 0$, prove that

$$E\Big[\Big\{\sum_{n\ge 1} E^2\Big(X_n \mid F_{n-1}\Big)\Big\}^{\frac{1}{2}}\Big] \le 2E\Big\{\Big(\sum_{n\ge 1} X_n^2\Big)^{\frac{1}{2}}\Big\},$$

deduce a bound for $E\Big(\Big[\sum_{n\ge 1}\{X_n - E(X_n \mid F_{n-1})\}^2\Big]^{\frac{1}{2}}\Big)$.

Answer. By Exercise 1.5.5 the variables $x_n = E(X_n \mid F_{n-1})$ satisfy

$$E\Big\{\Big(\sum_{n\geq 1}|x_n|^2\Big)^{\frac{1}{2}}\Big\} = \sup_{H_n} E\Big\{\sum_{n\geq 1} x_n H_n\Big\}$$

where $(H_n)_n$ has a unit norm $E\Big\{\Big(\sum_{n\geq 1}|H_n|^2\Big)^{\frac{1}{2}}\Big\}$.

Let $h_n = E(H_n \mid F_{n-1})$ and let $Y_n = \sum_{k=1}^n |h_k|^2$ and $Z_n = \sum_{k=1}^n |X_k|^2$, with $Y_0 = Z_0 = 0$, we have

$$\begin{aligned} E\Big\{\Big(\sum_{n\geq 1}|x_n|^2\Big)^{\frac{1}{2}}\Big\} &= \sup_{E\|(H_n)_n\|_{l_2}=1} E\Big(\sum_{n\geq 1} H_n x_n\Big) \\ &= \sup_{E\|(H_n)_n\|_{l_2}=1} E\Big(\sum_{n\geq 1} X_n h_n\Big), \end{aligned}$$

by definition of the conditional mean. By Hölder's inequality

$$E\Big(\sum_{n\geq 1} X_n h_n\Big) \leq E\Big\{\Big(\sum_{n\geq 1}|X_n|^2\Big)^{\frac{1}{2}}\Big\} E\Big\{\Big(\sum_{n\geq 1}|h_n|^2\Big)^{\frac{1}{2}}\Big\},$$

where

$$\begin{aligned} E\Big\{\Big(\sum_{n\geq 1}|X_n|^2\Big)^{\frac{1}{2}}\Big\} &= E\Big[\Big\{\sum_{n\geq 1}(Z_n - Z_{n-1})\Big\}^{\frac{1}{2}}\Big] \\ &\leq 2E(Z_\infty^{\frac{1}{2}}) = 2E\Big\{\Big(\sum_{n\geq 1}|Z_n|^2\Big)^{\frac{1}{2}}\Big\}, \end{aligned}$$

$$\begin{aligned} E\Big\{\Big(\sum_{n\geq 1}|h_n|^2\Big)^{\frac{1}{2}}\Big\} &= E\Big[\Big\{\sum_{n\geq 1}(|Y_n| - |Y_{n-1}|)\Big\}^{\frac{1}{2}}\Big] \\ &\leq 2E(|Y_\infty|^{\frac{1}{2}}) = 2E\Big\{\Big(\sum_{n\geq 1}|h_n|^2\Big)^{\frac{1}{2}}\Big\} \end{aligned}$$

and $E\Big\{\Big(\sum_{n\geq 1}|h_n|^2\Big)^{\frac{1}{2}}\Big\} \leq E\Big[\Big\{\sum_{n\geq 1} E(|H_n|^2 \mid F_{n-1})\Big\}^{\frac{1}{2}}\Big\}\Big] = 1$.

It follows that

$$\begin{aligned} &E\Big(\Big[\sum_{n\geq 1}\{X_n - E(X_n \mid F_{n-1})\}^2\Big]^{\frac{1}{2}}\Big) \\ &\leq \sqrt{2}E\Big(\Big[\sum_{n\geq 1}\{X_n^2 + E^2(X_n \mid F_{n-1})\}\Big]^{\frac{1}{2}}\Big) \\ &\leq 2\sqrt{2}E\Big\{\Big(\sum_{n\geq 1} X_n^2\Big)^{\frac{1}{2}}\Big\}. \end{aligned}$$

7.2 Discrete processes

A random walk is defined as a sequence of sums $S_n = \sum_{i=1}^{n} X_i$ for independent random variables X_i with a common distribution function F, and $S_0 = 0$. Let $\tau_n = \min\{n : S_n \in I\}$ for a real set I, for a set A such that $I \cap A = \emptyset$, let $G_n(I; A) = P(S_k \in A,\, k = 1, \ldots, n-1,\, S_n \in I\}$ and $H_n(I) = \int_A F(I - y)\, G_{n-1}(I; dy)$, the distribution function of S_{τ_1} is

$$H(I) = P(S_{\tau_1} \in I) = \sum_{n \geq 1} H_n(I).$$

The sequence $(\tau_1 + \cdots + \tau_r, S_{\tau_1} + \cdots + S_{\tau_r})$ is a renewal process.

Exercise 7.2.1. Let $(X_n)_{n\geq 1}$ be a sequence of independent and centered variables, let $S_n = \sum_{k=1}^{n} X_k$ and let $T_n = \inf\{k > T_{n-1} : S_k > 0\}$, prove that $Y_n = S_{T_n}$ is a martingale.

Answer. The random variables T_n are stopping times (see Exercise 7.1.11) and for every integer $n \geq 1$ the expectation of S_n is zero, hence $E(S_{T_n}) = 0$. The variable Y_{n-1} is independent of the variables X_k for $k > T_{n-1}$, therefore the conditional expectation

$$\begin{aligned} E(Y_n - Y_{n-1} \mid Y_{n-1}) &= E(S_{T_n} - S_{T_{n-1}} \mid Y_{n-1}) \\ &= E\Big(\sum_{k=T_{n-1}+1}^{T_n} X_k \mid Y_{n-1} \Big) \end{aligned}$$

does not depend on Y_{n-1} and $E(Y_n - Y_{n-1} \mid Y_{n-1}) = 0$.

Exercise 7.2.2. Let $(X_n)_{n\geq 1}$ be a sequence of independent exponential variables on $[0, 1]$, calculate the distribution functions of S_n and $\tau_1 = \min\{n : S_n > x\}$. Calculate $G_n(x; t) = P(S_k \leq t,\, k = 1, \ldots, n-1,\, S_n > x\}$ for $t < x$, $H_n(x)$ and $H(x)$.

Answer. For exponential variables with parameter $\lambda > 0$, the series S_n is increasing with distribution functions $F_1(t) = 1 - e^{-\lambda t}$, $F_2(t) = 1 - e^{-\lambda t}(1 + \lambda t)$, then if we assume that the distribution function of S_{n-1} is $F_{n-1}(t) = 1 - e^{-\lambda t} \sum_{k=0}^{n-2} (k!)^{-1} \lambda^k t^k)$ we obtain the distribution function of S_n

$$F_n(t) = 1 - e^{-\lambda t} \sum_{k=0}^{n-1} \frac{\lambda^k t^k}{k!},$$

is therefore the distribution function of S_n for every $n \geq 1$. The distribution function of τ_1 is

$$P(\tau_1 \leq n) = 1 - P(\tau_1 > n) = 1 - P(S_n \leq x)$$
$$= e^{-\lambda t} \sum_{k=0}^{n-1} \frac{\lambda^k t^k}{k!}.$$

Let $0 < t < x$, $t \leq n-1$, the probabilities $G_n(x;t)$ and $H_n(x)$ are

$$\begin{aligned} G_n(x;t) &= P(S_{n-1} \leq t, S_n > x) = P(S_{n-1} \leq t) P(X_n > x - t) \\ &= F_{n-1}(t) e^{-\lambda(t-x)}, \\ H_n(x) &= \int_0^t F(t-y)\, G_{n-1}(dy;t) = F_{n-1}(t) \int_0^t F(t-y)\, dF(y) \\ &= F_{n-1}(t)\{1 - e^{-\lambda t}(1+\lambda t)\} \end{aligned}$$

and we deduce the distribution of S_{τ_1}

$$H(x) = P(S_{\tau_1} > x) = \{1 - e^{-\lambda t}(1+\lambda t)\} \sum_{n\geq 1} F_{n-1}(t).$$

Exercise 7.2.3. Let $(X_n)_{n\geq 1}$ be an integer sequence of independent variables with expectation μ and variance σ^2, let $Y_0 = S_0$ be an integer, let $Y_1 = X_1 + \cdots + X_{Y_0}$ and for every integer $n \geq 1$ let $S_n = \sum_{k=0}^n Y_k$ and

$$Y_{n+1} = X_{S_{n-1}+1} + \cdots + X_{S_n}.$$

Prove that $(Y_n)_{n\geq 0}$ is a martingale if $\mu = 1$, calculate its expectation and its variance and prove a central limit theorem for Y_n.

Answer. The conditional expectation of $(Y_n)_{n\geq 1}$ is $E(Y_1 \mid Y_0) = Y_0 \mu$ and for every n let $\mathcal{F}_n = \sigma(Y_1, \ldots, Y_n)$, the variables S_{n-2} and S_{n-1} are $\mathcal{F}_{n-1}$ measurable and the variables $X_{S_{n-1}+1}, \cdots, X_{S_n}$ are independent of $\mathcal{F}_{n-1}$ it follows that

$$E(Y_n \mid \mathcal{F}_{n-1}) = Y_{n-1}\mu$$

and $(Y_n)_{n\geq 0}$ is a martingale if $\mu = 1$, it is a submartingale if $\mu > 1$ and a supermartingale if $\mu < 1$. The expectation of Y_1 is $E(Y_1) = E(Y_0)\mu$ and for every n, $E(Y_n) = E(Y_{n-1})\mu$, hence

$$E(Y_n) = E(Y_0)\mu^n.$$

The variance of Y_n conditionally on $\mathcal{F}_{n-1}$ is

$$V_{n,n-1} = Y_{n-1} E(X^2) - Y_{n-1}^2 \mu^2 = \sigma^2 Y_{n-1} - \mu^2 (Y_{n-1}^2 - Y_{n-1})$$

and the variance of Y_n is

$$\begin{aligned} V_n &= E(V_{n,n-1}) + \text{Var}\{E(Y_n \mid \mathcal{F}_{n-1})\} \\ &= \sigma^2 E(Y_{n-1}) - \mu^2 E(Y_{n-1}^2 - Y_{n-1}) + \mu^2 \text{Var}(Y_{n-1}) \\ &= \sigma^2 E(Y_{n-1}) + \mu^2 E(Y_{n-1}) - \mu^2 E^2(Y_{n-1}) \\ &= (\sigma^2 + \mu^2) E(Y_0) \mu^{n-1} - \mu^{2n} E^2(Y_0). \end{aligned}$$

By the central limit theorem and conditionally on $\mathcal{F}_{n-1}$, the variable $V_{n,n-1}^{-\frac{1}{2}}(Y_n - \mu Y_{n-1})$ converges weakly to a normal variable therefore it converges weakly to a normal variable.

Exercise 7.2.4. With the notations of Exercise 7.2.3 let $\mu_n = S_n^{-1} \sum_{k=1}^{S_n} X_k$ and $\sigma_n^2 = \frac{1}{S_n} \sum_{k=1}^{S_n} (X_k - \mu_n)^2$. As $\mu = 1$, prove the a.s. convergence of μ_n to μ and the weak convergence of $S_n^{\frac{1}{2}}(\mu_n - \mu)$ to a normal variable, prove the a.s. convergence of σ_n^2 to σ^2.

Answer. The variables X_k are independent and the expectation of the variable S_n is

$$E(S_n) = E(Y_0) \sum_{k=0}^{n-1} \mu^k,$$

if $\mu = 1$ then $n^{-1} S_n$ converges a.s. to $n^{-1} E(S_n) = E(Y_0)$. Let $m_0 = E(Y_0)$ and let $[nm_0]$ be the integer part of nm_0, then we have

$$\begin{aligned} \mu_n &= \frac{1}{[nm_0]} \sum_{k=1}^{[nm_0]} X_k + \frac{1}{S_n} \sum_{k=[nm_0]}^{S_n} X_k + \Big(\frac{1}{S_n} - \frac{1}{[nm_0]}\Big) \sum_{k=1}^{[nm_0]} X_k \\ &= \Big(\frac{1}{[nm_0]} \sum_{k=1}^{[nm_0]} X_k\Big) \{1 + o_{a.s.}(1)\} \end{aligned}$$

and it converges a.s. to 1 by the strong law of large numbers. By the same arguments, we obtain

$$\sigma_n^2 = \Big(\frac{1}{[nm_0]} \sum_{k=1}^{[nm_0]} (X_k - \mu_n)^2\Big) \{1 + o_{a.s.}(1)\}$$

and it converges a.s. to σ^2.

The variable $S_n^{\frac{1}{2}}(\mu_n - \mu)$ is asymptotically equivalent to

$$[nm_0]^{\frac{1}{2}} \Big(\frac{1}{[nm_0]} \sum_{k=1}^{[nm_0]} X_k - 1\Big)$$

and by the central limit theorem it converges weakly to a normal variable.

7.3 Quadratic variations and predictable compensator

On a filtered probability space $(\Omega, \mathcal{F}, P, \mathbb{F})$, a process $(A_n)_{n\geq 0}$ is predictable if X_n is $\mathcal{F}_{n-1}$-measurable. Let $(X_n)_{n\geq 0}$ be a square integrable submartingale on $(\Omega, \mathcal{F}, P, \mathbb{F})$, there exists a unique increasing predictable process $(A_n)_{n\geq 0}$ such that $M_n = X_n^2 - A_n$ is an integrable martingale. There are defined by $A_0 = 0$ and

$$A_{n+1} - A_n = E^{\mathcal{F}_n}(X_{n+1}^2) - X_n^2 = E^{\mathcal{F}_n}\{(X_{n+1} - X_n)^2\}.$$

A filtration $\mathbb{F} = (\mathcal{F}_t)_{t\in\mathbb{R}_+}$ on a probability space $(\Omega, \mathcal{A}, P)$ is an increasing, complete, right-continuous family of sub-sigma algebras of $\mathcal{A}$ with left-hand limits. A real process $X = (X_t)_{t\in\mathbb{R}_+}$ on a probability space $(\Omega, \mathcal{A}, P, \mathbb{F})$ is a family of measurable applications from $(\Omega, \mathcal{A})$ into $(\mathbb{R}, \mathcal{B}(\mathbb{R}))$, where $\mathcal{B}(\mathbb{R})$ is the Borel space of $\mathbb{R}$. It is
- measurable or adapted to $\mathbb{F}$ if for every t, X_t is $\mathcal{F}_t$-measurable,
- predictable if X_t is $\mathcal{F}_{t^-}$-measurable,
- integrable (respectively square integrable) if for every t, X_t is integrable (respectively square integrable),
- X is a martingale if X is $\mathbb{F}$-adapted, integrable and such that for every $0 < s < t$, $E(X_t \mid \mathcal{F}_s) = X_s$.

Theorem 7.3. *For a martingale $(X_n)_{n\geq 0}$ with an increasing process A_n such that $n^{-1}A_n$ converges in probability to a limit σ^2 and satisfying the Lindeberg condition*

$$n^{-1}\sum_{k=0}^{n} E\{(X_n - X_{n-1})^2 1_{|X_n - X_{n-1}| > n^{-\frac{1}{2}}\varepsilon} \mid \mathcal{F}_n\} \to 0$$

for every $\varepsilon > 0$, then $n^{-1}X_n$ converges a.s. to zero and $n^{-\frac{1}{2}}X_n$ converges weakly to a Gaussian variable $\mathcal{N}(0, \sigma^2)$.

Let $(X_n)_{n\geq 0}$ be a square integrale martingale on a filtered probability space $(\Omega, \mathcal{F}, P, \mathbb{F})$, its quadratic variations are an increasing sequence

$$V_n(X) = X_0^2 + \sum_{i=1}^{n}(X_i - X_{i-1})^2.$$

For every $n \geq 0$

$$\begin{aligned} EX_n^2 &= \sum_{i=1}^{n} E(X_n^2 - X_{n-1}^2) + EX_0^2 \\ &= E\{\sum_{i=1}^{n}(X_i - X_{i-1})^2\} + EX_0^2 = EV_n(X). \end{aligned}$$

The conditional expectation of V_{n+1} is

$$E\{V_{n+1}(X)|\mathcal{F}_n\} = V_n(X) + E\{(X_{n+1} - X_n)^2|\mathcal{F}_n\} \geq V_n(X),$$

$(V_n(X))_{n\geq 1}$ is a submartingale and it converges a.s. to a limit $V(X)$ in $L^1(P)$.

Let $(X_n)_{n\geq 1}$ be a submartingale on a filtered probability space $(\Omega, \mathcal{F}, (\mathcal{F}_n)_{n\geq 0}, P)$, with $X_0 = 0$, and let

$$\widetilde{X}_n = \sum_{i=1} E(X_n \mid F_{n-1}).$$

For all $m < n$, $E\{X_n - \widetilde{X}_n \mid F_m\} = X_m - \widetilde{X}_m$ then $(\widetilde{X}_n)_{n\geq 1}$ is the unique predictable process such that $(X_n - \widetilde{X}_n)_{n\geq 1}$ is a $(\mathcal{F}_n)_n$-martingale.

Let $(\widetilde{V}_n(X))_{n\geq 1}$ be the sequence of the predictable quadratic variations $\widetilde{V}_n(X) = E\{V_{n+1}(X)|\mathcal{F}_n\}$. Then

$$\begin{aligned}E(X_n^2 - X_{n-1}^2|\mathcal{F}_{n-1}) &= E\{(X_n - X_{n-1})^2|\mathcal{F}_{n-1}\}\\ &= E(V_n - V_{n-1}|\mathcal{F}_{n-1}) = E(\widetilde{V}_n|\mathcal{F}_{n-1}) - \widetilde{V}_{n-1},\end{aligned}$$

$(X_n^2 - V_n(X))_{n\geq 1}$ and $(X_n^2 - \widetilde{V}_n(X))_{n\geq 1}$ are $(\mathcal{F}_n)_n$-martingales.

Exercise 7.3.1. Let $(Y_n)_{n\geq 0}$ be independent square integrable and centered variables, prove that $Z_n = (\sum_{k=1}^n Y_k)^2 - nE(Y_1^2)$ is a martingale with respect to the filtration $\mathbb{F}$ generated by the variables Y_n, $\mathcal{F}_n = \sigma(Y_1, \ldots, Y_n)$.

Answer. Let $S_n = \sum_{k=1}^n Y_k$, for $n < m$

$$\begin{aligned}E^{\mathcal{F}_n}(Z_m) &= S_n^2 + E^{\mathcal{F}_n}\{2S_n(S_m - S_n) + (S_m - S_n)^2\} - mE(Y_1^2)\\ &= Z_n + E\{(S_m - S_n)^2\} - (m - n)E(Y_1^2) = Z_n.\end{aligned}$$

Exercise 7.3.2. Let $(X_n)_{n\geq 1}$ and $(A_n)_{n\geq 1}$ sequences of $L^1(P)$ random variables such that X_n is a $(\mathcal{F}_n)$-martingale and A_n is $\mathcal{F}_{n-1}$-measurable, and let $Y_0 = X_0$ and for $n \geq 1$, let $Y_{n+1} = Y_n + A_n(X_{n+1} - X_n)$. Prove that $(Y_n)_{n\geq 1}$ is a $(\mathcal{F}_n)$-martingale and that the process $V_n(Y) = X_0^2 + \sum_{i=1}^{n-1} A_i^2\{V_{i+1}(X) - V_i(X)\}$ is a convergent submartingale.

Answer. The expectation of Y_{n+1} conditionally on $\mathcal{F}_n$ is

$$E\{Y_{n+1}|\mathcal{F}_n\} = Y_n + A_nE(X_{n+1} - X_n|\mathcal{F}_n),$$

so Y_n is a $(\mathcal{F}_n)$-martingale. If X_n is a $\mathcal{F}_n$-submartingale (respectively supermartingale), then Y_n is a $\mathcal{F}_n$-submartingale (respectively supermartingale). The quadratic variations $V_n(Y)$ of $(Y_n)_{n\geq 1}$ satisfy

$$E\{V_{n+1}(Y)|\mathcal{F}_n\} - V_n(Y) = A_n^2E\{(X_{n+1} - X_n)^2|\mathcal{F}_n\} \geq 0,$$

therefore the process $V_n(Y) = X_0^2 + \sum_{i=1}^{n-1} A_i^2\{V_{i+1}(X) - V_i(X)\}$ is a convergent submartingale.

Exercise 7.3.3. Let $M = (M_n)_{n\geq 0}$ be a square integrable local martingale with respect to a filtration $\mathbb{F}$, find its conditional variance and the conditional variance of $M_{m\wedge\tau}$, for a stopping time τ.

Answer. The variance of a square integrable martingale M_m conditionally on $\mathcal{F}_n$ such that $n < m$ is $E\{(M_m - M_n)^2|\mathcal{F}_n\} = E(M_m^2|\mathcal{F}_n) - M_n^2$. If M is a local martingale, for every stopping time τ and for every integer n

$$\mathrm{Var}(M_{n\wedge\tau} \mid \mathcal{F}_{n-1}) = E\{M_{n\wedge\tau}^2 \mid \mathcal{F}_n) - M_{(n-1)\wedge\tau}^2,$$

where

$$\begin{aligned}E(M_{n\wedge\tau}^2 \mid \mathcal{F}_{n-1}) &= E(M_n^2 1_{\{\tau\geq n\}} + M_\tau^2 1_{\{\tau<n\}} \mid \mathcal{F}_{n-1})\\ &= E(M_n^2 \mid \mathcal{F}_{n-1})1_{\{\tau>n-1\}} + M_\tau^2 1_{\{\tau\leq n-1\}},\end{aligned}$$

$\{\tau \leq n-1\}$, $\{\tau \geq n\}$ and $M_\tau^2 1_{\{\tau<n\}}$ are $\mathcal{F}_{n-1}$-measurable, therefore $\mathrm{Var}(M_{n\wedge\tau} \mid \mathcal{F}_{n-1}) = \mathrm{Var}(M_n \mid \mathcal{F}_{n-1})1_{\{\tau>n-1\}}$. For an increasing sequence $(\tau_n)_n$ of stopping times tending a.s. to infinity it follows that $\mathrm{Var}(M_{n\wedge\tau_n} \mid \mathcal{F}_{n-1})$ converges a.s. to $\mathrm{Var}(M_n \mid \mathcal{F}_{n-1})$. In the same way, for $n < m$

$$\mathrm{Var}(M_{m\wedge\tau}1_{\{\tau\leq m\}} \mid \mathcal{F}_n) = \mathrm{Var}(M_\tau \mid \mathcal{F}_n)1_{\{n<\tau\leq m\}},$$

and

$$\mathrm{Var}(M_{m\wedge\tau} \mid \mathcal{F}_n) = \mathrm{Var}(M_m \mid \mathcal{F}_n)1_{\{\tau>m\}} + \mathrm{Var}(M_\tau)1_{\{\tau\leq m\}}$$

and for an increasing sequence $(\tau_n)_n$ of stopping times tending a.s. to infinity, $\mathrm{Var}(M_{m\wedge\tau_n} \mid \mathcal{F}_n)$ converges a.s. to $\mathrm{Var}(M_m \mid \mathcal{F}_n)$.

Exercise 7.3.4. Let $(X_n)_n$ be a martingale with increasing predictable process $(A_n)_n$, for $\eta > 0$ and $\lambda > 0$ prove the inequality

$$P(|X_n| > \lambda) \leq \frac{\eta}{\lambda^2} + P(A_n > \eta).$$

Answer. For all $\eta > 0$ and $\lambda > 0$

$$\begin{aligned}P(|X_n| > \lambda) &= P(|X_n| > \lambda, A_n \leq \eta) + P(|X_n| > \lambda, A_n > \eta)\\ &\leq \frac{E(X_n^2 1_{\{A_n\leq\eta\}})}{\lambda^2} + P(A_n > \eta)\\ &\leq \frac{\eta}{\lambda^2} + P(A_n > \eta),\end{aligned}$$

by the equality $E(X_n^2 1_{\{A_n \leq \eta\}}) = E(A_n 1_{\{A_n \leq \eta\}}) \leq \eta$ for the predictable process A_n.

Exercise 7.3.5. For a martingale $(X_n)_n$ with respect to a filtration $\mathbb{F}$, with increasing predictable process $(A_n)_n$, prove the inequality

$$P(\max_{k \leq n} |X_k| \geq \lambda) \leq \lambda^{-2} E(A_n).$$

Answer. For every integer n, let $\tau_n = \min\{k : X_k = \max_{i \leq n} X_i\}$, then τ_n is a stopping time: for every $m \leq n$, $\tau_n \leq m$ is equivalent to $X_m^\star =: \max_{i \leq m} X_i \geq X_k$ for every $k \leq n$ and $X_m^\star$ is $\mathcal{F}_m$-measurable, it follows that $X_n^\star$ is $\mathcal{F}_m$-measurable hence $\{\tau_n \leq m\}$ is $\mathcal{F}_m$-measurable.

For every stopping time τ, we have $E(X_\tau^2) = E(A_\tau)$ and

$$P(\max_{k \leq n} X_k > \lambda) \leq \frac{E(X_{\tau_n}^2)}{\lambda^2} = \frac{E(A_{\tau_n})}{\lambda^2} \leq \frac{E(A_n)}{\lambda^2}.$$

Exercise 7.3.6. Let $(X_n)_{n \geq 0}$ be a martingale of L^{2k}, $k \geq 1$, with respect to a filtration $\mathbb{F} = (\mathcal{F}_n)_{n \geq 0}$, with increasing predictable process $(A_n)_n$ such that $A_0 = 0$. Prove the inequality $E(X_n^{2k}) \geq c_k E(A_n^k)$, for a strictly positive constant c_k.

Answer. For all integers n and $k \geq 1$, $M_n = X_n^2 - A_n$ is a martingale with expectation zero, then by convexity we have $E(M_n^k) \geq 0$, and A_n is $\mathcal{F}_{n-1}$ measurable. In particular

$$E(X_n^2 A_n) = E\{E(X_n^2 A_n \mid \mathcal{F}_{n-1})\} = E(A_n^2),$$

it follows that $0 \leq E(M_n^4) = E(X_n^4) - E(A_n^2)$ hence $E(X_n^4) \geq E(A_n^2)$, by the same arguments

$$\begin{aligned} 0 &\leq E(M_n^3) = E(X_n^6 - 3X_n^4 A_n + 3X_n^2 A_n^2 - A_n^3) \\ &= E(X_n^6 - 3X_n^4 A_n + 2A_n^3) \\ &\leq E(X_n^6 - A_n^3). \end{aligned}$$

Assuming that for every $j \leq k$, $E(X_n^{2j}) \geq c_j E(A_n^j)$ then we have

$$E(X_n^{2j} A_n^{k+1-j}) = E\{E(X_n^{2j} \mid \mathcal{F}_{n-1}) A_n^{k+1-j}\} \geq c_j E(A_n^{k+1})$$

and

$$
\begin{aligned}
E(M_n^{k+1}) &= E\Big\{X_n^{2(k+1)} + \sum_{j=1}^{k} \binom{k+1}{j} X_n^{2(k+1-j)}(-A_n)^j\Big\} + E(-A_n)^{k+1} \\
&\le E\Big\{X_n^{2(k+1)} + \sum_{j\,even, j=1}^{k} \binom{k+1}{j} X_n^{2(k+1-j)} A_n^j \\
&\quad - \sum_{j\,odd, j=1}^{k} \binom{k+1}{j} c_{k+1-j} A_n^{k+1}\Big\} + E(-A_n)^{k+1}
\end{aligned}
$$

$$
\begin{aligned}
E\{X_n^{2(k+1)}\} &\ge \Big\{(-1)^{k+1} + \sum_{j\,odd, j=1}^{k} c_{k+1-j} \binom{k+1}{j}\Big\} E(A_n^{k+1}) \\
&\quad - \sum_{j\,even, j=2}^{k-1} \binom{k+1}{j} E\{X_n^{2(k+1-j)} A_n^j\},
\end{aligned}
$$

this proves the inequality

$$
E(X_n^{2(k+1)}) \ge c_{k+1} E(A_n^{k+1}).
$$

Exercise 7.3.7. Let $(X_n)_n$ be a martingale with respect to a filtration $\mathbb{F}$ and let $(b_n)_n$ be a strictly positive decreasing sequence, for $\lambda > 0$ prove the inequality $P(\max_{m \le k \le n} b_k |X_k| \ge \lambda) \le \lambda^{-2} b_m^2 \sum_{m \le k \le n} E\{(X_k - X_{k-1})^2\}$.

Answer. Let $Y_k = X_k - X_{k-1}$, by the martingale property of $(X_n)_n$ the variables Y_k are centered and the variance of X_k is

$$
E(X_k^2) = E[\{\sum_{i=1}^{k} (X_i - X_{i-1})\}^2] = \sum_{i \le k} \sigma_i^2
$$

where $X_0 = 0$, $\sigma_i^2 = E\{(X_i - X_{i-1})^2\}$ and $E\{(X_i - X_{i-1})(X_j - X_{j-1})\} = 0$ for all distinct i and j, then

$$
\begin{aligned}
P(\max_{m \le k \le n} b_k |X_k| > \lambda) &\le \sum_{k=m}^{n} P(b_k |X_k| > \lambda) \\
&\le \lambda^{-2} \sum_{k=m}^{n} b_k^2 E(X_k^2) \\
&\le \lambda^{-2} b_m^2 \sum_{k=m}^{n} \sum_{i=1}^{k} \sigma_i^2.
\end{aligned}
$$

Problem 7.2.1. (a). Let $(X_n)_n$ a positive submartingale on a filtered probability space $(\Omega, \mathcal{F}, (\mathcal{F}_n)_{n\geq 0}, P)$ and let $(c_n)_n$ be a decreasing sequence of $\mathbb{R}_+$. Prove that for every $n \geq 1$

$$Z_n = c_0 X_0 + \sum_{k=1}^{n} c_k (X_k - X_{k-1}) \geq c_n X_n$$

and for every $a > 0$

$$P(\max_{k\leq n} c_k X_k > a) \leq a^{-1} E(Z_n).$$

(b). Let $(Y_n)_n$ a positive martingale of L^2 on a filtered probability space $(\Omega, \mathcal{F}, (\mathcal{F}_n)_{n\geq 0}, P)$ and let $(c_n)_n$ be a decreasing sequence of $\mathbb{R}_+$. Prove that for every $a > 0$

$$a^2 P(\max_n c_n |Y_n| > a) \leq c_0^2 E(Y_0^2) + \sum_{k\geq 1}^{n} c_k^2 E\{(Y_k - Y_{k-1})^2\},$$

$P(\limsup_n c_n |Y_n| = 0) = 1$ if $\sum_{k\geq 1}^{n} c_k^2 E\{(Y_k - Y_{k-1})^2\}$ is finite and $\lim_n c_n = 0$.

Answer. (a). For every $n \geq 1$, the assumptions imply

$$Z_n = (c_0 - c_1)X_0 + \sum_{k=1}^{n-1} (c_{k-1} - c_k)X_k - X_{k-1}) + c_n X_n \geq c_n X_n.$$

Then for every $a > 0$

$$P(\max_{k\leq n} c_k X_k > a) \leq P(Z_n > a) \leq a^{-1} E(Z_n).$$

(b). The process $Z_n = c_0 Y_0 + \sum_{k=1}^{n} c_k (Y_k - Y_{k-1})$ is a martingale, therefore

$$E(Z_n^2) = c_0^2 E(Y_0^2) + \sum_{k\geq 1}^{n} c_k^2 E\{(Y_k - Y_{k-1})^2\} \geq E^2(Z_n),$$

it follows that for every $a > 0$

$$P(\max_{k\leq n} c_k |Y_k| > a) \leq P(Z_n^2 > a^2) \leq a^{-2} E(Z_n^2)$$

where the right term is finite therefore

$$\lim_{a\to 0} P(\limsup_n c_n |Y_n| > a) = \lim_{a\to 0} \lim_n P(\max_{k\leq n} c_n |Y_n| > a) = 0,$$

reversely $P(\limsup_n c_n |Y_n| = 0) = 1$.

7.4 Exponential martingales, sub and supermartingales

On a probability space $(\Omega, \mathcal{F}, (\mathcal{F}_n)_{n\geq 0}, P)$, let $(X_n)_{n\leq 1}$ be a local martingale such that $X_0 = 0$ and for $\lambda \neq 0$ let

$$\phi_n(\lambda) = \log E \prod_{k=1}^{n} E\{e^{\lambda(X_k - X_{k-1})} \mid F_{k-1}\},$$

the functions ϕ_n are concave. For every λ such that $\phi_n(\lambda)$ is finite, let

$$Z_n(\lambda) = \exp\{\lambda X_n - \phi_n(\lambda)\}.$$

Exercise 7.4.1. Prove that $\phi_n(\lambda) \geq 0$ for every integer n.

Answer. For every n, by convexity we have

$$E\{e^{\lambda(X_n - X_{n-1})} \mid F_{t_{n-1}}\} \geq \exp\{\lambda E(X_n - X_{n-1} \mid F_{n-1})\} = 1,$$

as ϕ_n is the logarithm of their product, it is positive.

Exercise 7.4.2. Let X be a local martingale with independent and integrable increments and with $X_0 = 0$, prove that for every $\lambda \neq 0$ such that $\phi_n(\lambda)$ is finite, the sequence $(Z_n(\lambda))_{n\leq 0}$ is a local martingale with mean $EZ_n(\lambda) = 1$ and a strictly positive variance $\exp\{\phi_n(2\lambda) - 2\phi_n(\lambda)\} - 1$.

Answer. For a process X with independent increments and $X_0 = 0$, the equality $X_n = \sum_{k=1}^{n}(X_k - X_{k-1})$ entails

$$E(e^{\lambda X_n}) = E\Big\{\prod_{k=1}^{n} e^{\lambda(X_k - X_{k-1})}\Big\} = \prod_{k=1}^{n} E\Big\{e^{\lambda(X_k - X_{k-1})}\Big\}$$

and

$$\begin{aligned}
\phi_n(\lambda) &= \log \prod_{k=1}^{n} E\Big\{e^{\lambda(X_k - X_{k-1})}\Big\} \\
&= \sum_{k=1}^{n} \log E\Big\{e^{\lambda(X_k - X_{k-1})}\Big\}, \\
e^{\phi_n(\lambda)} &= \prod_{k=1}^{n} E\Big\{e^{\lambda(X_k - X_{k-1})}\Big\} = E(e^{\lambda X_n}),
\end{aligned}$$

hence $E\{Z_n(\lambda)\} = 1$. For every integrable stopping time τ

$$\begin{aligned}
E\{Z_{(n+1)\wedge\tau}(\lambda) \mid F_n\} &= Z_{n\wedge\tau}(\lambda) E\Big[\frac{\exp\{\lambda(X_{(n+1)\wedge\tau} - X_{n\wedge\tau})\}}{e^{\phi_{n+1}(\lambda) - \phi_n(\lambda)}} \mid F_n\Big] \\
&= \frac{E[\exp\{\lambda(X_{n+1} - X_n)\}]}{e^{\phi_{n+1}(\lambda) - \phi_n(\lambda)}} 1_{\{n+1\leq\tau\}} \\
&\quad + 1_{\{n\geq\tau\}},
\end{aligned}$$

and by convexity

$$E[\exp\{\lambda(X_{n+1} - X_n)\} \mid F_n] \geq \exp[E\{\lambda(X_{n+1} - X_n)\}] = 1.$$

Therefore $E\{Z_{(n+1)\wedge T}(\lambda) \mid F_n\} \geq Z_{n\wedge T}(\lambda)$ and $(Z_n(\lambda))_{n\leq 0}$ is a local submartingale. Since $E\{Z_n(\lambda)\} = 1$, $(Z_n(\lambda))_{n\leq 0}$ is a local martingale. Moreover

$$\begin{aligned} EZ_n^2(\lambda) &= E\exp\{2\lambda X_n - 2\phi_n(\lambda)\} \\ &= E\{Z_n(2\lambda)\}\exp\{\phi_n(2\lambda) - 2\phi_n(\lambda)\} \\ &= \exp\{\phi_n(2\lambda) - 2\phi_n(\lambda)\}, \end{aligned}$$

by the concavity of the logarithm, the variance $\exp\{\phi_n(2\lambda) - 2\phi_n(\lambda)\} - 1$ of Z_n is strictly positive.

Exercise 7.4.3. On a probability space $(\Omega, \mathcal{F}, (\mathcal{F}_n)_{n\geq 0}, P)$, let $(X_n)_{n\leq 1}$ be a sequence of $\mathcal{M}_0^{2loc}$ and let $(A_n)_{n\leq 1}$ be the predictable increasing process of X_n such that $Z_n = X_n^2 - A_n$ is a local martingale. Prove that for every $\lambda > 0$

$$M_n(\lambda) = \exp\Big(\lambda X_n - \frac{\lambda^2}{2}A_n\Big)$$

is a $(\mathcal{F}_n)_{n\geq 1}$ submartingale.

Answer. For every $\lambda > 0$, $Z_n(\lambda) = \lambda X_n - \frac{\lambda^2}{2}A_n$ is a local martingale such that

$$\begin{aligned} E(M_n \mid \mathcal{F}_{n-1}) &\geq \exp\Big\{E\Big(\lambda X_n - \frac{\lambda^2 A_n}{2} \mid \mathcal{F}_{n-1}\Big)\Big\} \\ &= \exp\Big(\lambda X_{n-1} - \frac{\lambda^2 A_{n-1}}{2}\Big) \end{aligned}$$

and $(M_n(\lambda))_{n\leq 1}$ is a submartingale with respect to $(\mathcal{F}_n)_{n\geq 1}$.

7.5 Random series

On a probability space $(\Omega, \mathcal{A}, P)$, a sequence $(Y_n)_{n\in\mathcal{Z}}$ of L^2 variables is second order stationary if for all integers n and m, the expectation of Y_n is a constant μ and the covariance of Y_n and Y_m satisfies

$$\gamma_{n,m} = E\{(Y_n - \mu)(Y_m - \mu)\} = \gamma_{0,n-m}.$$

A random walk $(Y_n)_{n\geq 0}$ is defined by a sequence $(\varepsilon_i)_{n\geq 0}$ of independent centered variables with finite variance σ^2, as $Y_n = \sum_{i=1}^n \varepsilon_i = Y_{n-1} + \varepsilon_n$.

The covariance of Y_n and Y_m is $\gamma_{n,m} = (n \wedge m)\sigma^2$. A sequence $(Y_n)_{n\in\mathcal{Z}}$ such that

$$Y_n = \alpha Y_{n-1} + \varepsilon_n, n \geq 1 \tag{7.1}$$

is an autoregressive series of first order AR(1), with ε_n a centered variable independent of Y_{n-1} and with finite variance σ^2.

Exercise 7.5.1. Let $X_k = \sum_{i=1}^n R_i \sin(2kf\pi+\phi_i)$, k in $\mathcal{Z}$, with independent variables $R_1,\dots,R_n,\phi_1,\dots,\phi_n$ such that the variables ϕ_k are uniform on $[0,2\pi]$. Calculate the expectations, the variances and the covariances of the series $(X_k)_{k\in\mathcal{Z}}$ and prove its second order stationarity.

Answer. For every k and for i in $\mathbb{N}$, the expectation of $\sin(2kf\pi+\phi_i)$ is

$$\begin{aligned} m_k &= \frac{1}{2\pi}\int_0^{2\pi} \sin(2kf\pi+\phi_i)\,d\phi_i \\ &= -\frac{1}{2\pi}\{\cos(2kf\pi+2\pi) - \cos(2kf\pi)\} = 0, \end{aligned}$$

and $E(X_k) = 0$, the covariance of $\sin(2kf\pi+\phi_i)$ and $\sin(2kf\pi+\phi_j)$ is therefore zero if $i \neq j$ and the variance of $\sin(2kf\pi+\phi_i)$ is

$$\begin{aligned} s_k &= \frac{1}{2\pi}\int_0^{2\pi} \sin^2(2kf\pi+\phi_i)\,d\phi_i \\ &= \frac{1}{2} - \frac{1}{4\pi}\int_0^{2\pi} \cos(4kf\pi+2\phi_i)\,d\phi_i \\ &= \frac{1}{2} - \frac{1}{8\pi}\{\sin(4kf\pi+4\pi) - \sin(4kf\pi)\} = \frac{1}{2}. \end{aligned}$$

The covariance of X_k and X_n is therefore

$$E(X_k X_n) = \sum_{i=1}^n \sum_{j=1}^n E(R_i R_j) E\{\sin(2kf\pi+\phi_i)\sin(2nf\pi+\phi_j)\},$$

where the value of $c_{k,n} = E\{\sin(2kf\pi+\phi_i)\sin(2nf\pi+\phi_i)\}$ is

$$\begin{aligned} c_{k,n} &= \frac{1}{2\pi}\int_0^{2\pi} \sin(2kf\pi+\phi)\sin(2nf\pi+\phi)\,d\phi \\ &= \frac{1}{4\pi}\int_0^{2\pi} \{\cos(2(k-n)f\pi) - \cos(2(k+n)f\pi+2\phi_i)\}\,d\phi \\ &= \frac{1}{2}\cos(2(k-n)f\pi) \end{aligned}$$

and $E\{\sin(2kf\pi + \phi_i)\sin(2nf\pi + \phi_j)\} = 0$ if $i \neq j$, hence

$$\mathrm{Cov}(X_k, X_n) = \frac{1}{2}\sum_{i=1}^{n} E(R_i^2)\cos(2|k-n|f\pi)$$

and the series is stationary.

Exercise 7.5.2. An AR(2) series $Y_n = \alpha_1 Y_{n-1} + \alpha_2 Y_{n-2} + \varepsilon_n$, for $n \geq 1$, is defined by $Y_1 = \alpha_1 Y_0 + \varepsilon_1$ and for every $n \geq 2$ by a centered variables ε_n of L^2, independent of Y_{n-1}. Calculate the variances and the covariances of the series under the conditions and $|\alpha_2| < 1$, $\alpha_1 + \alpha_2 < 1$ and $\alpha_2 - \alpha_1 < 1$.

Answer. The variance of Y_1 is $\sigma_1^2 = \alpha_1^2\sigma_0^2 + \sigma^2$ and for every integer $n \geq 2$, the variance of Y_n is

$$\begin{aligned}\sigma_n^2 &= \alpha_1^2\sigma_{n-1}^2 + 2\alpha_1\alpha_2\gamma_{n-1,n-2} + \alpha_2^2\sigma_{n-2}^2 + \sigma^2\\ &= \alpha_2^2\sigma_{n-2}^2 - \alpha_1^2\sigma_{n-1}^2 + 2\alpha_1\gamma_{n,n-1} + \sigma^2\end{aligned}$$

where $\alpha_2\gamma_{n-1,n-2} = \gamma_{n,n-1} - \alpha_1\sigma_{n-1}^2$.

For all integers m and $n \geq m + k$,

$$\begin{aligned}\gamma_{n,m} &= \alpha_1\gamma_{n,m-1} + \alpha_2\gamma_{n,m-2}\\ &= (\alpha_1^2 + \alpha_2)\gamma_{n,m-2} + \alpha_1\alpha_2\gamma_{n,m-3}\\ &= \alpha_1(\alpha_1^2 + 2\alpha_2)\gamma_{n,m-3} + \alpha_2(\alpha_1^2 + \alpha_2)\gamma_{n,m-4}.\end{aligned}$$

Problem 7.5.1. Calculate the variances and the covariances of an AR(1) series (7.1) under the conditions $E(\varepsilon_n) = 0$ and $|\alpha| < 1$ and prove the convergence of Y_n.

Answer. For all integers n and m, the equation $E(Y_nY_m) = \alpha E(Y_nY_{m-1}) + E(Y_n\varepsilon_m)$ implies $\gamma_{n,m} = \alpha\gamma_{n,m-1}$. The series is written as

$$Y_n = \alpha^{n-m}Y_m + \alpha^{n-m-1}\varepsilon_{m+1} + \cdots + \alpha\varepsilon_{n-1} + \varepsilon$$

the covariance of Y_n and Y_m is $\gamma_{n,m} = \alpha^{n-m}E(Y_m^2)$. The variance of Y_n is $\sigma_n^2 = \alpha^2\sigma_{n-1}^2 + \sigma^2$ where σ^2 is the variance of the variables ε_n, therefore

$$\sigma_n^2 = \alpha^{2n}\sigma_0^2 + \frac{1-\alpha^{2n}}{1-\alpha^2}\sigma^2,$$

where σ_0^2 is the variance of Y_0, it converges to

$$\sigma_Y^2 = (1-\alpha^2)^{-1}\sigma^2$$

as n tends to infinity and

$$\gamma_{n,m} = \alpha^{n-m}\Big\{\alpha^{2n}\sigma_0^2 + \frac{1-\alpha^{2n}}{1-\alpha^2}\sigma^2\Big\},$$

$(Y_n)_n$ is not stationary but asymptotically stationary for large n as it is asymptotically equivalent to α^{n-m}. The series $(Y_n)_n$ is a martingale of L^2 and $E(Y_n^2)$ converges, it follows that $\sup_{n\in\mathbb{N}} E(Y_n^2)$ is finite and Y_n converges a.s. to a variable Y_∞ of L^2.

Problem 7.5.2. In the AR(1) model $X_n = \mu + \alpha X_{n-1} + \varepsilon_n$, with ε_n a centered variable independent of X_{n-1} and with finite variance, $|\alpha| < 1$, let $\bar{X}_n = n^{-1}\sum_{k=1}^n X_k$. Prove the a.s convergence of $\bar{X}_n$ and of the empirical correlation of X_k and X_{k-1}

$$r_n = \frac{\sum_{k=1}^n (X_{k-1} - \bar{X}_n)(X_k - \bar{X}_n)}{\sum_{k=1}^n (X_{k-1} - \bar{X}_n)^2}.$$

Calculate the variance of X_k and prove that its covariance with r_n converges to zero.

Answer. The model for X_n is equivalent to (7.1) with $Y_n = X_n - m$ where $\mu = (1-\alpha)m$ and the expectation of $\bar{X}_n$ is $m = E\bar{X}_n = \mu + \alpha m$. The a.s. convergence of Y_n implies it satisfies the Cauchy criterion and $(\bar{X}_n - \bar{X}_m)_{m,n\geq 1}$ converges a.s. to zero as m and n tend to infinity, therefore $\bar{X}_n$ converges a.s. to m.

The variance of $\bar{X}_n$ is $v_n = n^{-1}\bar{\sigma}_n^2 + n^{-1}\bar{\gamma}_{n,n-1}$ where

$$\bar{\sigma}_n^2 = n^{-1}\sum_{k=1}^n E(Y_k^2) = n^{-1}\sum_{k=1}^n \sigma_k^2,$$

$$\bar{\gamma}_{n,n-1} = n^{-1}\sum_{k=1}^n E(Y_{k-1}Y_k) = n^{-1}\sum_{k=1}^n \gamma_{k,k-1}.$$

The variance σ_n^2 of Y_n converges to σ_Y^2 (Exercise 7.5.1) then for every $\varepsilon > 0$ there exists n_0 such that $|\sigma_n^2 - \sigma_Y^2| < \varepsilon$ for every $n \geq n_0$ which implies

$$|\bar{\sigma}_n^2 - \sigma_Y^2| \leq |n^{-1}\sum_{k=1}^{n_0} \sigma_k^2| + n^{-1}\sum_{k=n_0+1}^{n} |\sigma_k^2 - \sigma_Y^2| < 2\varepsilon.$$

As n tends to infinity, $\gamma_{n,n-1}$ converges to $\alpha\sigma_Y^2$ and $\bar{\gamma}_{n,n-1}$ converges to $\alpha\sigma_Y^2$, by the same arguments as the convergence of $\bar{\sigma}_n^2$. It follows that nv_n converges to $(1+\alpha)\sigma_Y^2$.

The vector Z with components $X_i - \bar{X}_n$ is written like in (5.2) as a linear combination $Z = A_n\mathbb{X}_n$ of the components of $\mathbb{X}_n = (X_1, \ldots, X_n)^T$. Its expectation is zero and its variance is $\mathrm{Var}(Z) = A_n\Sigma_n A_n$ where the matrix $\Sigma_n = \mathrm{Var}(\mathbb{X}_n)$ has the components $E\{(X_i - m)(X_j - m)\} = \gamma_{i,j}$, it is asymptotically equivalent to the diagonal matrix D of the variances σ_i^2. By the a.s. convergence of $\bar{X}_n$ to m, the series $E\{(X_k - \bar{X}_n)(X_{k-1} - \bar{X}_n)$ converges to $\gamma_{k,k-1}$, for $k = 1, \ldots, n$, and r_n converges a.s. to the correlation α of X_k and X_{k-1}.

For $k = 1, \ldots, n$, $E\{(X_k - \bar{X}_n)(\bar{X}_n - m)\}$ converges to zero by the a.s. convergence of $\bar{X}_n$ to m it follows that the covariance of r_n and $\bar{X}_n$ converges to zero.

Problem 7.5.3. In the AR(1) model (7.1) with $|\alpha| < 1$, let $S_{0,\alpha} = 0$ and for $k \geq 1$ let $S_{k,\alpha} = \sum_{j=1}^k \alpha^{k-j}\varepsilon_j$, let $\widetilde{S}_{k,\alpha} = \sum_{j=1}^k \varepsilon_j \sum_{l=0}^{j-1} \alpha^l \varepsilon_{j-l}$ and let $\bar{S}_{k,\alpha} = \sum_{j=1}^k S_{j,\alpha}$. Prove the weak convergence of the process

$$(S_{[ns],\alpha}, n^{-1}s^{1/2}\bar{S}_{[ns],\alpha}, n^{-1/2}s^{1/2}\widetilde{S}_{[ns],\alpha})_{s\in[0,1]}$$

to $\frac{\sigma}{(1-\alpha^2)^{1/2}}(W, \sqrt{3}\bar{W}, \sigma\sqrt{2}\int_0^s W\,dW)_{s\in[0,1]})$ where W is the standard Brownian motion and $\bar{W} = \int_0^{\cdot} W(t)\,dt$. Prove the weak convergence of $n^{-1}s\sum_{k=1}^{[ns]} S_{k,\alpha}^2$ to $2\sigma^2(1-\alpha^2)^{-1}\int_0^s W_x^2\,dx$.

Answer. By the martingale property of $S_{k,\alpha}$ and the expression of the variance, we have

$$\mathrm{Var}S_k = \sigma^2 \sum_{l=0}^{k-1} \alpha^{2l} = \sigma^2(1-\alpha^{2k})(1-\alpha^2)^{-1}$$

and the variance of $\bar{S}_{t,\alpha}$ is

$$\begin{aligned}
\mathrm{Var}\bar{S}_{t,\alpha} &= \sigma^2 2\sum_{l=1}^{t}(t-l)\sum_{k=0}^{l-1}\alpha^{2k} + o(1)\\
&= \sigma^2 2\sum_{l=1}^{t}(t-l)\frac{\alpha^{2l}-1}{\alpha^2-1} + o(1)\\
&= \frac{\sigma^2}{1-\alpha^2}t^2 + o(t^2).
\end{aligned}$$

As the variance of $\bar{W}_s$ is $\frac{s^3}{3}$, $n^{-2}\mathrm{Var}\bar{S}_{[ns],\alpha}$ converges to $\frac{3}{s}\sigma^2(1-\alpha^2)^{-1}\mathrm{Var}\bar{W}_s$. The convergence of the weighted sums of the $S_{k,\alpha}$ is obtained by similar arguments, the variances are asymptotically equivalent to $t\sigma^2\alpha^2(1-\alpha^2)^{-1}$ if $|\alpha| < 1$, with $\mathrm{Var}\int_0^x \alpha^s\,d\bar{W}_s = \int_0^x s^2\alpha^{2s}\,ds$.

The convergence of $\widetilde{S}_{[ns],\alpha}$ is deduced from the behaviour of $S_{[ns],\alpha}$ and the weak convergence of $n^{-1/2}\sum_{k=}^{[ns]}\varepsilon_k$ to σW. The variance of $\int_0^s W\,dW = \frac{1}{2}(W_s^2 - s)$ is the limiting variance of $\sigma^{-2}\widetilde{S}_{[ns],1}$ which equals $t^2/2$. The variance of $\widetilde{S}_{[ns],\alpha}$ is $\sigma^4 t(1-\alpha^2)^{-1}$. The expectation of $t^{-1}\alpha^{-2t}\sum_{k=1}^t kS_{k,\alpha}^2$ is $\sigma^2(\alpha^2-1)^{-1}t^{-1}\alpha^{-2t}\sum_{k=1}^t k\alpha^{2k}$ if $|\alpha|>1$ and it converges to $\sigma^2\alpha^2(\alpha^2-1)^{-2}$.

Problem 7.5.4. In the AR(1) model $X_n = \mu + \alpha X_{n-1} + \varepsilon_n$ with $|\alpha|<1$, prove the weak convergence of the process $(X_{[ns]} - m, \bar{X}_{[ns]} - m)_{s\in[0,1]}$ to a Brownian motion $\sigma(1-\alpha^2)^{-1/2}(W_s, s^{-1/2}\sqrt{3}\bar{W}_s)_s$. Prove the weak convergence of the process $n^{1/2}(\widehat{\alpha}_{[ns]} - \alpha))_{s\in[0,1]}$.

Answer. The variable $X_k - m$ develops as the sum of the independent and centred variables

$$X_k - m = \alpha^k(X_0 - m) + S_{k,\alpha} = S_{k,\alpha} + o_{L_2}(1)$$

which converges weakly to a Brownian motion $\sigma(1-\alpha^2)^{-1/2}W$ and $\bar{X}_{[ns]} - m = [ns]^{-1}\bar{S}_{[ns],\alpha} + o_p(1)$ converges to the limit defined in Problem 7.5.3. By definition of the estimator $\widehat{\alpha}_n$ we have

$$\begin{aligned}
\widehat{\alpha}_t - \alpha &= \frac{\sum_{k=1}^t (X_{k-1} - \bar{X}_t)(X_k - \bar{X}_t - \alpha(X_{k-1} - \bar{X}_t))}{\sum_{k=1}^t (X_{k-1} - \bar{X}_t)^2} \\
&= \frac{\sum_{k=1}^t (X_{k-1} - \bar{X}_t)((1-\alpha)(m - \bar{X}_t) + \varepsilon_k)}{\sum_{k=1}^t (X_{k-1} - \bar{X}_t)^2} \\
&= \frac{-(X_t - X_0)(1-\alpha)t^{-1}\bar{S}_{t,\alpha} + \sum_{k=1}^t \varepsilon_k X_{k-1} - t\bar{X}_t\bar{\varepsilon}_t)}{\sum_{k=1}^t (X_{k-1} - \bar{X}_t)^2},
\end{aligned}$$

with $\bar{\varepsilon}_t = t^{-1}\sum_{k=1}^t \varepsilon_k$. When $|\alpha|<1$ and t tends to infinity, X_t, $\bar{X}_t$ and $t^{-1}\bar{S}_{t,\alpha}$ are $O_p(1)$ and the other terms of the numerator are $O_p(t^{1/2})$. It follows

$$t^{1/2}(\widehat{\alpha}_t - \alpha) = \frac{t^{-1/2}\sum_{k=1}^t \varepsilon_k(X_{k-1} - m) - t^{1/2}\bar{\varepsilon}_t(\bar{X}_t - m)}{t^{-1}\sum_{k=1}^t (X_{k-1} - \bar{X}_t)^2} + o_p(1).$$

For the denominator of $\widehat{\alpha}_{[ns]} - \alpha$,

$$\begin{aligned}
\sum_{k=1}^{[ns]} (X_{k-1} - \bar{X}_t)^2 &= \sum_{k=1}^{[ns]-1} \{S_{k,\alpha} - [ns]^{-1}\bar{S}_{[ns],\alpha} + o_p(1)\}^2 \\
&= \sum_{k=1}^{[ns]-1} S_{k,\alpha}^2 - [ns]^{-1}\bar{S}_{[ns],\alpha}^2 + o_p(1),
\end{aligned}$$

hence

$$n^{-1}\sum_{k=1}^{[ns]}(X_{k-1}-\bar{X}_{[ns]})^2 \xrightarrow{\mathcal{D}} \frac{\sigma^2}{1-\alpha^2}s^{-1}\Big(\int_0^s (2W_x^2-3s^{-2}\bar{W}_s^2)\,dx\Big).$$

For the numerator of $\widehat{\alpha}_{[ns]}-\alpha$, the process

$$n^{-1/2}\sum_{k=1}^{[ns]}(X_{k-1}-m)\varepsilon_k = n^{-1/2}\widetilde{S}_{[ns]-1,\alpha}+o_p(1)$$

converges weakly to $\sigma^2\sqrt{2}s^{-1/2}(1-\alpha^2)^{-1/2}\int_0^s W\,dW$ and $n^{1/2}\bar{\varepsilon}_{[ns]}(\bar{X}_{[ns]}-m)$ converges weakly to $\sigma^2\sqrt{3}s^{-1/2}(1-\alpha^2)^{-1/2}W\bar{W}$.

7.6 Markov chains

On a probability space $(\Omega,\mathcal{A},P)$, a Markov chain $(X_k)_{k\geq 0}$ with values in a measurable state space $(E,\mathcal{E})$ is defined by the probability ν of the initial state X_0 on $(E,\mathcal{E})$ and the transition probability Π on $(E\times E,\mathcal{E}\otimes\mathcal{E})$ such that the distribution of X_{n+1} conditionally on $X_0,\ldots,X_n$ is determined by $\Pi(X_n,\cdot)$ on $(E,\mathcal{E})$, for every set of $\mathcal{E}$

$$P(X_{n+1}\in A\mid X_0,\ldots,X_n)=\Pi(X_n,A).$$

The entrance time T_A in a set A is $T_A=\inf\{n\in\mathbb{N}: X_n\in A\}$, it is a stopping time, and the recurrence set of A is

$$R_A=\{\omega:\sum_{n\in\mathbb{N}}1_{\{X_n(\omega)\in A\}}=\infty\}.$$

A state x in E is recurrent if

$$P(T_x<\infty)=1,\ P(R_x)=1$$

and it is transient if

$$P(T_x<\infty)<1,\ P(R_x)=0,$$

a set A is absorbing if $\Pi(A,x)=0$ for every x which does not belong to A. The image of a real function f of $C_b(E)$ by Π is

$$\Pi f(x)=\int_E f(x)\Pi(x,dy)$$

and the image of a real function f of $C_b(E^n)$ by $\nu\otimes\Pi^{\otimes n}$ is

$$\nu\otimes\Pi^{\otimes n}(f)=\int_{E^n} f(x_0,\ldots,x_n)\,\nu(dx_0)\,\Pi(x_0,dx_1)\cdots\Pi(x_{n-1},dx_n),$$

$$\Pi^{\otimes n}(x,\,dy)=\int_E \Pi_{n-1}(s,dy)\,\Pi(x,ds)=\int_E \Pi_{n-1}(x,ds)\,\Pi(s,dy).$$

A Markov chain is ergodic if $\nu \otimes \Pi^{\otimes n}$ converges to a probability measure $\mu > 0$ which does not depend on the initial probability ν. If a Markov chain has a recurrent state a, then it is ergodic and the measure

$$\mu(\Gamma) = E_a\Big\{\sum_{k=1}^{n} 1_\Gamma(X_k)\Big\}$$

is σ-finite on the set $\{x : P_x(R_a = 1)\}$, such that $\mu\Pi = \mu$, it defines the invariant probability

$$\mu(\Gamma) = \frac{1}{E_a(T_a)} E_a\Big\{\sum_{k=1}^{n} 1_\Gamma(X_k)\Big\}.$$

Exercise 7.6.1. The empirical probability of transition of a Markov chain $(X_n)_{n\geq 1}$ is defined for every recurrent set Γ of $\mathcal{E}^{\otimes 2}$ as

$$\Pi_n(\Gamma) = n^{-1} \sum_{k=0}^{n} 1_\Gamma(X_k, X_{k+1}).$$

(a). Prove the a.s. convergence of $\Pi_n(\Gamma)$, for every measurable set Γ.
(b). Prove the a.s. convergence of $F_n(\Gamma) = n^{-1} \sum_{k=0}^{n} 1_\Gamma(X_k)$, for every measurable set Γ.

Answer. (a). Let $\mathcal{F}_n$ the sigma-algebra generated by $(X_0, \ldots, X_n)$, for every measurable set Γ the conditional probability

$$P\{(X_n, X_{n+1}) \in \Gamma \mid \mathcal{F}_n\} = \int_\Gamma \Pi(X_n, dx)$$

is positive and

$$E\{\Pi_n(\Gamma) - \Pi_{n-1}(\Gamma) \mid \mathcal{F}_n\} = -\frac{1}{n(n-1)} P\{(X_n, X_{n+1}) \in \Gamma \mid \mathcal{F}_n\} \leq 0,$$

$\Pi_n(\Gamma)$ is therefore a supermartingale and it converges a.s. to an integrable variable as n tends to infinity.

(b). For every measurable set Γ, the variable $F_n(\Gamma)$ is $\mathcal{F}_n$-measurable and the conditional probability $P(X_n \in \Gamma \mid \mathcal{F}_{n-1}$ is positive then

$$E\{F_n(\Gamma) - F_{n-1}(\Gamma) \mid \mathcal{F}_{n-1}\} = -\frac{1}{n(n-1)} P(X_n \in \Gamma \mid \mathcal{F}_{n-1} \leq 0$$

and $F_n(\Gamma)$ is a supermartingale, therefore it converges a.s. to an integrable variable as n tends to infinity.

Exercise 7.6.2. Let $(X_n)_{n\geq 1}$ be a Markov chain on a recurrent space E and let f be a measurable function of $L^2(\mu\Pi)$, with the invariant probability μ of the chain, prove that the variable

$$U_n = n^{-\frac{1}{2}} \sum_{k=0}^{n} \{f(X_k, X_{k+1}) - \Pi f(X_k)\}$$

converges weakly to a centered Gaussian variable with finite variance $\sigma^2 = \lim_{n\to\infty} \mathrm{Var} U_n$.

Answer. The variables $A_n = f(X_n, X_{n+1}) - \Pi f(X_n)$ are $\mathcal{F}_{n+1}$-measurables and the sequence of variables $(A_n)_{n\geq 0}$ is a martingale with respect to the filtration $(\mathcal{F}_{n+1})_{n\geq 0}$, as its conditional expectations are

$$E\{f(X_n, X_{n+1}) - \Pi f(X_n) \mid \mathcal{F}_n\} = 0$$

and $E\{f(X_n, X_{n+1})\} = \mu\Pi f$ is finite for every n. By convexity, $(A_n^2)_{n\geq 0}$ is a positive submartingale and

$$E(A_n^2 \mid \mathcal{F}_n) = \Pi f^2(X_n) - \{\Pi f(X_n)\}^2,$$
$$E(A_n^2) = \mu\Pi f^2 - \int_E \{\Pi f\}^2 \, d\mu$$

where $\int_E \{\Pi f\}^2 \, d\mu \leq \mu\Pi f^2$ therefore A_n^2 is integrable with and $(A_n^2)_{n\geq 0}$ converges a.s. to an integrable limit, then it satisfies the Lindeberg condition

$$n^{-1} \sum_{k=0}^{n} E\{A_n^2 1_{\{A_n > n^{\frac{1}{2}}\varepsilon\}} \mid \mathcal{F}_n\} \to 0$$

for every $\varepsilon > 0$ and $n^{-1} \sum_{k=1}^{n} E(A_n^2)$ converges to a finite limit, the weak convergence of U_n to a centered Gaussian variable follows.

Exercise 7.6.3. Let $(X_n)_{n\geq 1}$ be a Markov chain on a recurrent space E, let f be a measurable function of $L^1(\mu\Pi)$ and let $\Pi_n f = n^{-1} \sum_{k=0}^{n} f(X_k, X_{k+1})$, give an exponential bound for $P(\Pi_n f > a)$, for every $a > 0$.

Answer. For all $a > 0$ and $t > 0$ we have

$$\begin{aligned} P(\Pi_n f > a) &= \inf_{t>0} P(e^{nt\Pi_n f} > e^{nat}) \leq \inf_{t>0} e^{-nat} E(e^{nt\Pi_n f}) \\ &= \inf_{t>0} \exp\{n \log \varphi_f(t) - nat\} \end{aligned}$$

where

$$\varphi_f(t) = \int_E e^{tf(x,y)}\,\mu(dx)\,\Pi(x,dy) = \mu\Pi u_t f$$

with the function $u_t f(x,y) = e^{tf(x,y)}$.

The minimum of this exponential bound is reached at t_0 such that

$$\frac{\mu\Pi\, f u_0 f}{\mu\Pi\, u_0 f} = a,$$

denoting $u_0 = u_{t_0}$. It follows that

$$P(\Pi_n f > a) \le \exp\{-nat_0 + n\log\mu\Pi u_0 f\}.$$

Exercise 7.6.4. Find sequences of variables satisfying
(a). $Y_{n+1} = aY_n$,
(b). $Y_{n+2} = a^2 Y_n$,
(c). $Y_{n+2} + a^2 Y_n = 0$, with $|a| < 1$.

Answer. (a). The differential equation $y'_x = ay_x$ has the solution $y_x = y_0 e^{ax}$, by discretization on integers we obtain the Markov chain $Y_n = Y_0 a^n$.

(b). The differential equation $y''_x = a^2 y_x$ has the solution $y_x = y_0\{e^{ax} \pm e^{-ax}\}$, a solution of the discrete equation is the process

$$Y_n = \frac{Y_0}{2}\{a^n \pm (-a)^n\}.$$

(c). The equation $y''_x + a^2 y_x = 0$ has the solution $y_x = y_0\{\sin(ax) + \cos(ax)\}$, a discrete solution is the process $Y_n = Y_0\{(ai)^n \pm (-ai)^n\}$.

Exercise 7.6.5. Let $(X_k)_{k\ge1}$ be a Markov on $\{0,1\}$ with transition probabilities $\pi_1 = P(X_{k+1} = 1 \mid X_k = 1)$ and $\pi_0 = P(X_{k+1} = 1 \mid X_k = 0)$ and with $p = P(X_k = 1)$ for every k, write the distribution of (X_{k+1}, X_k).

Answer. Let $q = P(X_k = 0)$ and let $P_{11} = P(X_{k+1} = 1, X_k = 1)$ and $P_{10} = P(X_{k+1} = 1, X_k = 0)$ then

$$p = P_{11} + P_{10} = \pi_1 p + \pi_0(1-p) = \frac{\pi_0}{1-\pi_1+\pi_0},$$

$$q = \frac{1-\pi_1}{1-\pi_1+\pi_0},$$

$$P_{11} = \pi_1 p = \frac{\pi_0\pi_1}{1-\pi_1+\pi_0},$$

$$P_{10} = \pi_0 q = \frac{\pi_0(1-\pi_1)}{1-\pi_1+\pi_0}.$$

Let $\rho_1 = P(X_{k+1} = 0 \mid X_k = 1) =$ and $\rho_0 = P(X_{k+1} = 0 \mid X_k = 0)$, then $\rho_1 = 1 - \pi_1$ and $\rho_0 = 1 - \pi_0$, the probabilities P_{00} and P_{01} are deduced by the same arguments

$$P_{00} = \rho_0 q = \frac{(1 - \pi_1)(1 - \pi_0)}{1 - \pi_1 + \pi_0},$$

$$P_{01} = \rho_1 p = \frac{\pi_0(1 - \pi_1)}{1 - \pi_1 + \pi_0} = P_{10}.$$

Reversely, these probabilities and the expressions of p and q provide the conditional probabilities $\pi_1 = P(X_{k+1} = 1 \mid X_k = 1)$ and $\pi_0 = P(X_{k+1} = 1 \mid X_k = 0)$.

Chapter 8

Time-continuous martingales

8.1 Martingales and predictable compensator

A filtration $\mathbb{F} = (\mathcal{F}_t)_{t\in\mathbb{R}_+}$ on a probability space $(\Omega, \mathcal{A}, P)$ is an increasing, complete, right-continuous family of sub-sigma algebras of $\mathcal{A}$ with left-hand limits, i.e., $\mathcal{F}_s \subset \mathcal{F}_t$ for every $s < t$, $\mathcal{F}_t = \cap_{s>t}\mathcal{F}_s$ and $\mathcal{F} = \mathcal{F}_\infty$. A real process $X = (X_t)_{t\in\mathbb{R}_+}$ on a probability space $(\Omega, \mathcal{A}, P, \mathbb{F})$ is a family of measurable applications from $(\Omega, \mathcal{A})$ into $(\mathbb{R}, \mathcal{B}(\mathbb{R}))$, where $\mathcal{B}(\mathbb{R})$ is the Borel space of $\mathbb{R}$. It is
- measurable or adapted to $\mathbb{F}$ if for every t, X_t is $\mathcal{F}_t$-measurable,
- predictable if X_t is $\mathcal{F}_{t^-}$-measurable,
- integrable (respectively square integrable) if for every t, X_t is integrable (respectively square integrable),
- X is a martingale if X is $\mathbb{F}$-adapted, integrable and such that for every $0 < s < t$, $E(X_t \mid \mathcal{F}_s) = X_s$.

A real process X may also be considered as a measurable application from $(\Omega \times \mathbb{R}_+, \mathcal{A} \bigotimes \mathcal{B}(\mathbb{R}_+))$ into $(\mathbb{R}, \mathcal{B}(\mathbb{R}))$. The space of the predictable processes on $(\Omega, \mathcal{A}, P, \mathbb{F})$ is generated by the set $\mathcal{E}$ of the left-continuous processes with right-hand limits in the form

$$Y = \sum_{i\leq n} Y_i 1_{[t_i, t_{i+1}[}$$

where Y_i is a $\mathcal{F}_{t_i}$-measurable real variable and $(t_i)_i$ is an increasing partition on $\mathbb{R}_+$, with $t_0 = 0 < t_1 < \cdots < t_n$.

A real random variable T on $(\Omega, \mathcal{A}, P, \mathbb{F})$ is a stopping time on $\mathbb{R}_+$ if $\{T \leq t\}$ is $\mathcal{F}_t$-measurable for every $t \geq 0$, T is a predictable stopping time if there exists an increasing sequence of stopping times $(T_n)_{n\in\mathbb{N}}$ such that $T_n < T$ for every n and T_n converges to T. For a real process $(X_t)_{t\geq 0}$

on $(\Omega, \mathcal{A}, P, \mathbb{F})$ and for every real interval I, the variable $T_I = \inf\{t \geq 0 : X_t \in I\}$ is a $\mathbb{F}$-stopping time.

A time-continuous martingale $X = (X_t)_{t\geq 0}$ on $(\Omega, \mathcal{F}, (\mathcal{F}_t)_{t\geq 0}, P)$ is defined with respect to a right-continuous and increasing filtration $(\mathcal{F}_t)_{t\geq 0}$ of $\mathcal{F}$ by the property

$$E(X_t \mid \mathcal{F}_s) = X_s, \ s < t.$$

For the natural filtration $(\mathcal{F}_t)_{t\geq 0}$ of a process $(X_t)_{t\geq 0}$, $\mathcal{F}_t$ is the sigma-algebra generated by $\{X_s; s \leq t\}$ and for all $s < t$, a martingale satisfies $E(X_t \mid X_s) = X_s$ for $s < t$.

A martingale X is uniformly integrable if

$$\lim_{C\to\infty} \sup_{t\geq 0} E(|X_t| 1_{\{|X_t|>C\}}) = 0.$$

Then $\lim_{t\to\infty} E|X_t - X_\infty| = 0$ and $X_t = E(X_\infty|\mathcal{F}_t)$ for every t, with the natural filtration $(\mathcal{F}_t)_{t\geq 0}$ of X.

Theorem 8.1. *Let A be an increasing process on a space $(\Omega, \mathcal{A}, P, \mathbb{F})$, then there exists an unique increasing predictable process $\widetilde{A}$, called the "predictable compensator" of A on $(\Omega, \mathcal{A}, P, \mathbb{F})$ and such that*

$$\widetilde{A}(t) = \lim_{n\to\infty} \sum_{0\leq t_i \leq t} E\{A(t_i) - A(t_{i-1} \mid \mathcal{F}_{t_i}\},$$

for every partition $(t_i)_{i\leq n}$ of $[0, t]$ such that $0 = t_0 < t_1 < \ldots < t_n = t$ and $t_{i+1} - t_i$ converges to zero as n tends to infinity.

Then the process $A - \widetilde{A}$ is a local martingale. For a square-integrable local martingale M, there exists a process of predictable variation denoted $< M >$ such that $M_t^2 - < M >_t$ is a local martingale, it follows that $EM_t^2 = E < M >_t$ and for $0 < s < t$

$$E(M_t - M_s)^2 = EM_t^2 - EM_s^2 = E(< M >_t - < M >_s).$$

On a filtered probability space $(\Omega, \mathcal{A}, P, \mathbb{F})$, metric spaces of martingales are defined as

(1) $\mathcal{M}_0^2$, the space of the right-continuous square integrable martingales with value zero at $t = 0$, with the norm $\|M\|_2 = \sup_t (EM_t^2)^{1/2}$,
(2) $L^2(M)$, the metric space of the predictable processes X, with finite integral $\|X\|_M^2 = E\int_0^\infty X_s^2 \, d < M >_s$.

Let M be a martingale of $\mathcal{M}_0^2$, then M_t converge a.s. to a limit M_∞ as t tends to infinity and $\|M\|_2 = (EM_\infty^2)^{1/2}$ is finite.

For M_1 and M_2 in $\mathcal{M}_0^2$, a scalar product is defined as

$$< M_1, M_2 > = \frac{1}{2}(< M_1 + M_2, M_1 + M_2 > \\ - < M_1, M_1 > - < M_2, M_2 >),$$

two martingales M_1 and M_2 of $\mathcal{M}_0^2$ are orthogonal if $< M_1, M_2 >= 0$.

On a filtered probability space $(\Omega, \mathcal{F}, P, \mathbb{F})$, let $(M_t)_{t\geq 0}$ be in $\mathcal{M}_{loc}^2$, sum of a continuous process M^c and a jump process $M^d = \sum_{0<s\leq t} \Delta M_s$, with jumps $\Delta M_s = M_s - M_{s^-}$. The increasing predictable process of its quadratic variations $[M]_t =< M^c >_t + \sum_{0<s\leq t}(\Delta M_s)^2$ has the predictable projection $\widetilde{[M]}_t - < M >_t$ such that $[M]_t =< M >_t$ belongs to $\mathcal{M}_{0,loc}$.

Every measurable increasing right-continuous function with left-hand limits V on $\mathbb{R}_+$ defines a σ-finite measure μ_V on $(\mathbb{R}_+, \mathcal{B}(\mathbb{R}_+))$ as $\mu_V(]s,t]) = V(t) - V(s)$ for $s < t$.

Exercise 8.1.1. Let M_1 and M_2 be martingales of $\mathcal{M}_0^2$, prove that M_1 and M_2 are orthogonal if M_1M_2 is a martingale.

Answer. In the development

$$(M_1 + M_2)^2 - < M_1 + M_2, M_1 + M_2 > -(M_1^2 - < M_1, M_1 >) \\ -(M_2^2 - < M_2, M_2 >) = 2(M_1M_2 - < M_1, M_2 >)$$

the three first terms of the left member are martingales. If M_1M_2 is a martingale, it follows that for every $t > 0$

$$E\{M_1(t)M_2(t)\} = E\{M_1(0)M_2(0)\} = 0$$

and M_1 and M_2 are orthogonal.

Exercise 8.1.2. Let A and $\widetilde{A}$ be positive increasing process such that $\widetilde{A}$ is predictable and $\int X\,d\mu_A = \int X\,d\mu_{\widetilde{A}}$ for every predictable process X, prove that $A - \widetilde{A}$ is a martingale.

Answer. For all $0 < s < t$ and B in $\mathcal{F}_s$, the process $X = 1_{[s,t[\times B}$ is predictable and the equality $\mu_A(X) = \mu_{\widetilde{A}}(X)$ is equivalent to

$$\mu_A(]s,t] \times B) = \mu_{\widetilde{A}}(]s,t] \times B), \\ E[1_B\{A(t) - A(s)\}] = E[1_B\{\widetilde{A}(t) - \widetilde{A}(s)\}], \\ E\{A(t) - A(s) \mid \mathcal{F}_s\} = E\{\widetilde{A}(t) - \widetilde{A}(s) \mid \mathcal{F}_s\}$$

this implies that $A - \widetilde{A}$ is a martingale. Reciprocally, the sets $]s,t] \times B$, B in $\mathcal{F}_s$, generate all predictable processes so the martingale property of $A - \widetilde{A}$ entails $\mu_A(X) = \mu_{\widetilde{A}}(X)$ for every predictable process X.

Exercise 8.1.3. On a filtered probability space $(\Omega, \mathcal{F}, P, \mathbb{F})$, let $(X_t)_{t\geq 0}$ be a positive supermartingale and let τ be a stopping time. Prove that the sequence of the variables $(Y_t)_t = (X_{\tau\vee t})_t$ is a positive supermartingale a.s. bounded by X_t.

Answer. For every $t > 0$, the sets $\{t < \tau\}$ and $\{\tau \leq t\}$ belong to $\mathcal{F}_t$ and

$$\begin{aligned} Y_t &= E^{\mathcal{F}_t}(X_t 1_{\{\tau\leq t\}}) + E^{\mathcal{F}_t}(X_\tau 1_{\{\tau>t\}}) \\ &= X_t 1_{\{\tau\leq t\}} + E^{\mathcal{F}_t}(X_\tau) 1_{\{\tau>t\}} \leq X_t. \end{aligned}$$

For all $0 < s < t$, we have

$$E^{\mathcal{F}_s}(Y_t) = E^{\mathcal{F}_s}(X_{\tau\vee t} 1_{\{\tau\leq s\}}) + E^{\mathcal{F}_s}(X_{\tau\vee t} 1_{\{\tau>s\}}),$$

where $E^{\mathcal{F}_s}(X_{\tau\vee t} 1_{\{\tau\leq s\}}) = E^{\mathcal{F}_s}(X_t) 1_{\{\tau\leq s\}} \leq X_s 1_{\{\tau\leq s\}}$ and

$$\begin{aligned} E^{\mathcal{F}_s}(X_{\tau\vee t}) 1_{\{\tau>s\}} &= E^{\mathcal{F}_s} E^{\mathcal{F}_\tau}(X_t) 1_{\{s<\tau\leq t\}} + E^{\mathcal{F}_s}(X_\tau) 1_{\{\tau>t\}} \\ &\leq E^{\mathcal{F}_s}(X_\tau) 1_{\{\tau>s\}}, \end{aligned}$$

therefore $E^{\mathcal{F}_s}(Y_t) \leq X_{\tau\vee s} = Y_s$.

Exercise 8.1.4. Let $(M_t)_{t\geq 0}$ be a continuous local martingale with respect to the filtration $\mathbb{F} = (\mathcal{F}_t)_{t\geq 0}$ such that $\mathcal{F}_t = \sigma(M_s, s \leq t)$, prove that for every $\mathbb{F}$ stopping time τ, $(M_{t+\tau})_{t\geq 0}$ is a $\mathbb{F}$ martingale. Prove that the process $Z_t = \exp(M_t^2 - < M >_t)$ is a local submartingale.

Answer. For all $s > 0$

$$E\{M_{t+\tau\wedge s} - M_t \mid \mathcal{F}_t\} \leq E\{M_{t+s} - M_t \mid \mathcal{F}_t\} = 0$$

hence $(M_{t+\tau})_{t\geq 0}$ is a supermartingale such that $E(M_{t+\tau\wedge s}) = E(M_t)$, therefore it is a martingale. For all $0 < s < t$, we have

$$\begin{aligned} E\Big(\frac{Z_{t\wedge\tau}}{Z_{s\wedge\tau}} \mid \mathcal{F}_s\Big) &= E[\exp\{M_{t\wedge\tau}^2 - M_{s\wedge\tau}^2 \\ &\quad -(< M >_{t\wedge\tau} - < M >_{s\wedge\tau})\} \mid \mathcal{F}_s], \end{aligned}$$

by convexity and by the martingale property this implies

$$E\Big(\frac{Z_{t\wedge\tau}}{Z_{s\wedge\tau}} \mid \mathcal{F}_s\Big) \geq 1,$$

as $Z_t > 0$ this is equivalent to $E(Z_{t\wedge T} \mid \mathcal{F}_s) \geq Z_{s\wedge T}$ and Z_t is a local submartingale.

Exercise 8.1.5. Let $(X_t)_{t\geq 0}$ be a local martingale of L^p, $p \geq 1$, and let $X_t^* = \sup_{s\geq t} X_s$, prove the inequalities

$$P(X_t^* \geq x) \leq x^{-1} E|X_t|$$

and $\|X_t^*\|_p \leq \frac{p}{p-1}\|X_t\|_p$, for all $x > 0$ and $t > 0$.

Answer. Let $(t_i)_i$ be an increasing partition on $[0, t]$, with $t_0 = 0$ and $t_n = t$, and let $X_i = X(t_i)$ and $X_n^* = \max_{k=0,\dots,n} X_k$. As n tends to infinity, the variable X_n^* converges to X_t^* and it satisfies

$$P(X_n^* \geq x) \leq x^{-1} E|X_n|$$

for every n and the first inequality for X_t^* follows by limits as n tends to infinity. For $p \geq 1$, the inequality $\|X_n^*\|_p \leq \frac{p}{p-1}\|X_n\|_p$ of Exercise 7.1.3, for every integer n, entails the second inequality as n tends to infinity.

Exercise 8.1.6. Let $(X_t)_{t\geq 0}$ be a submartingale such that $\sup_{t\geq 0} E(X_t^2)$ is finite, prove its L^2 convergence to a limit X_∞ such that $E(X_\infty|\mathcal{F}_t) \geq X_t$ for every integer $t \geq 0$.

Answer. The inequality $E(X_t \mid \mathcal{F}_s) \geq X_s$, for $0 < s < t$, and the integrability condition implies

$$E(X_s) \leq E(X_t) \leq \sup_{t\geq 0} E|X_t| < \infty,$$

the sequence $E(X_s)_{s\geq 0}$ is increasing and bounded, therefore it converges a.s. to a limit X_∞. For every integer $t \geq 0$, $E(X_\infty|\mathcal{F}_t) \geq X_t$ by the submartingale property. As s and t tend to infinity, $s < t$, the convergence of $E\{(X_t - X_s)^2\}$ to zero entails the convergence of X_t to X_∞ in L^2.

Exercise 8.1.7. On a filtered probability space $(\Omega, \mathcal{F}, P, \mathbb{F})$, let $(X_t)_{t\geq 0}$ be a positive supermartingale and let $\tau = \inf\{t \in \mathbb{R}_+ : X_t > a\}$ prove that τ is a stopping time with respect to $\mathbb{F}$ and that $Y_t = X_{\tau\wedge t}$ is a supermartingale such that $Y_t \leq a$ for every t if τ is infinite.

Answer. The events $\{\tau > s\}$ and $\{X_t \leq a\}$ are equivalent and $\{X_t \leq a\}$ is $\mathcal{F}_t$-measurable which proves that the variable τ is a stopping time. Let

$Y_t = X_\tau 1_{\{\tau<t\}} + X_t 1_{\{\tau\geq t\}}$, and for all $0 < s < t$, the sets $\{s < \tau\}$ and $\{\tau \leq s\}$ belong to $\mathcal{F}_s$, hence

$$\begin{aligned} E^{\mathcal{F}_s}(Y_t) &= E^{\mathcal{F}_s}(X_\tau 1_{\{\tau\leq s\}}) + E^{\mathcal{F}_s}(X_{\tau\wedge t} 1_{\{s<\tau\}}), \\ E^{\mathcal{F}_s}(X_\tau 1_{\{\tau<t\}}) &= X_\tau 1_{\{\tau\leq s\}} + E^{\mathcal{F}_s}(X_\tau) 1_{\{s<\tau\leq t\}} + E^{\mathcal{F}_s}(X_t) 1_{\{t<\tau\}} \\ &\leq X_\tau 1_{\{\tau<s\}} + X_s 1_{\{s<\tau\}}, \end{aligned}$$

therefore $E^{\mathcal{F}_s}(Y_t) \leq X_{\tau\wedge s} = Y_s$. If τ is infinite, $Y_t = X_t$ for every finite t and it is lower than a.

Exercise 8.1.8. On a space $(\Omega, \mathcal{F}, P, \mathbb{F})$, let $(X_t)_{t\geq 0}$ be a positive submartingale, let $S_t = \sup_{s\leq t} X_s$ and for $a > 0$ let $\tau_a = \inf\{t \geq 0 : S_t > \ a\}$, prove the inequality $P(S_t > a) \leq a^{-1}E(X_t 1_{\{S_t>a\}})$ and calculate $E(\tau_a)$.

Answer. The equality $\{\tau_a \leq t\} = \{S_t > a\}$ implies that τ_a is a stopping time such that $a < \sup_{s<\tau_a} X_s = S_{\tau_a}$ and on $[0, \tau_a[$, $S_t \leq a$ therefore $X_t \leq a$, it follows that $X_{\tau_a} > a$ and

$$\begin{aligned} aP(S_t > a) = aP(\tau_a \leq t) &\leq E(X_{\tau_a} 1_{\{\tau_a\leq t\}}) \\ &\leq E\{E^{\mathcal{F}_{\tau_a}}(X_t) 1_{\{\tau_a\leq t\}}\} \\ &= E(X_t 1_{\{\tau_a\leq t\}}). \end{aligned}$$

The expectation of τ_a is

$$\begin{aligned} E(\tau_a) &= \int_0^\infty t\, dP(\tau_a \leq t) \\ &= a^{-1} \int_0^\infty t\, dP(S_t > a) \\ &\leq a^{-1} \int_0^\infty t\, dP(X_t > a), \end{aligned}$$

by the inequality of Exercise 8.1.5.

Exercise 8.1.9. Let $(X_t)_{t\geq 0}$ be in $\mathcal{M}_0^{2,loc}$, with an increasing process of predictable variations $(A_t)_{t\geq 0}$, prove that for all $\varepsilon > 0$ and $\eta > 0$

$$P(X_t^* > \varepsilon) \leq P(A_t > \eta) + \frac{\eta}{\varepsilon^2}.$$

Answer. Let $\varepsilon > 0$, the process $X_t^2 - A_t$ belongs to $\mathcal{M}_0^{2,loc}$ and it is centered, furthermore $X_t^* \geq X_t$, let $\tau = \inf\{t : X_t > \varepsilon\}$ then $X_\tau > \varepsilon$ and $X_t > \varepsilon$ is equivalent to $\tau \leq t$. The process A is increasing then for every t, $A_t \geq A_\tau$

if $X_t > \varepsilon$. The Bienaymé–Chebychev inequality implies

$$P(X_{t\wedge\tau}^2 > \varepsilon, A_t \leq \eta) \leq \frac{E(X_{t\wedge\tau}^2 1_{\{A_t\leq\eta\}})}{\varepsilon^2} = \frac{E(A_{t\wedge\tau} 1_{\{A_t\leq\eta\}})}{\varepsilon^2} \leq \frac{\eta}{\varepsilon^2},$$

the result follows from the inequality

$$P(X_t^* > \varepsilon) = P(X_t^* > \varepsilon, A_t > \eta) + P(X_t^* > \varepsilon, A_t \leq \eta) \leq P(A_t > \eta) + P(X_{t\wedge\tau}^2 > \varepsilon, A_t \leq \eta).$$

Exercise 8.1.10. Let $(M_t)_{t\geq 0}$ be in $\mathcal{M}_0^{2,loc}$, with an increasing process of predictable variations $(A_t)_{t\geq 0}$ prove that $A_t = < M^c, M^c >_t + \sum_{s\leq t} \Delta M_s^2$. Deduce that for a counting process N with predictable compensator $\widetilde{N}$, the increasing process of predictable variations of $N - \widetilde{N}$ is $\widetilde{N}$.

Answer. Let $M_t = M^c + M^d$ where M^c is continuous in $\mathcal{M}_0^{2,loc}$ and $M^d = \sum_{s\leq t} \Delta M_s$ is a jump process, the increasing process of predictable variations of M^c is $< M^c, M^c >$ and $M^d = M_+^d - M_-^d$ where $M_+^d = \sum_{s\leq t} \Delta M_s 1_{\{\Delta M_s > 0\}}$ and $M_-^d = \sum_{s\leq t} |\Delta M_s| 1_{\{\Delta M_s > 0\}}$ are increasing. Their predictable compensators $\widetilde{M_+^d}$ and $\widetilde{M_-^d}$ define the predictable compensator $\widetilde{M^d} = \widetilde{M_+^d} - \widetilde{M_-^d}$ of M^d. The processes $M_t^{c2} - < M^c, M^c >_t$ is a centered local martingale hence $E(M_t^{c2}) = E(< M^c, M^c >_t)$ and the martingale property implies $E(\Delta M_s^d \Delta M_t^d) = 0$ for all distinct s and t, hence $E(M_t^{d2}) = \sum_{s\leq t} \Delta M_s^2$, and therefore

$$E(M_t^2) = E(M_t^{c2}) + E(M_t^{d2}) = A_t.$$

By the same arguments, for $s < t$ we have

$$\begin{aligned} E(M_t^2 - M_s^2 \mid \mathcal{F}_s) &= E\{(M_t - M_s)^2 \mid \mathcal{F}_s\} \\ &= E\{(M_t^c - M_s^c)^2 \mid \mathcal{F}_s\} + E\{(M_t^d - M_s^d)^2 \mid \mathcal{F}_s\} \\ &= E\{< M^c, M^c >_t - < M^c, M^c >_s \mid \mathcal{F}_s\} \\ &\quad + E\{\sum_{s<x\leq t} \Delta M_x^2 \mid \mathcal{F}_s\} = E(A_t - A_s \mid \mathcal{F}_s). \end{aligned}$$

For a counting process N with a continuous predictable compensator, let $M = N - \widetilde{N}$ in $\mathcal{M}_0^{2,loc}$ then the process $\sum_{s\leq t} \Delta M_s^2$ reduces to N_t and $A_t = \widetilde{N}$.

8.2 Stochastic integral

On a filtered probability space $(\Omega, \mathcal{A}, P, \mathbb{F})$, every measurable increasing right-continuous process with left-hand limits V on $\mathbb{R}_+$ defines a σ-finite random measure μ_V on $(\mathbb{R}_+, \mathcal{B}(\mathbb{R}_+))$ as $\mu_V(]s,t]) = V(t) - V(s)$ for $s < t$. The integral of a predictable process B on $(\Omega, \mathcal{A}, P, \mathbb{F})$ with respect to an increasing integrable process A on $(\Omega, \mathcal{A}, P, \mathbb{F})$ is defined as a Stieljes integral with respect to $\mu_{A(\omega)}$ for each sample-path, when $E \int_0^\infty |B_t| \, d\mu_A(t)$ is finite. Let $0 = t_0 < t_1 < \ldots < t_n = t$ be a partition of $[0,t]$ such that $t_{i+1} - t_i$ converges to zero as n tends to infinity, the integral of a predictable process B with respect to an increasing process A is defined for every $\omega \in \Omega$ as

$$\int_0^t B(\omega, s) \, d\mu_{A(\omega)}(s) = \lim_{n\to\infty} \sum_{1\le i\le n} B(\omega, t_i)\mu_{A(\omega)}(]t_i, t_{i+1}]).$$

The stochastic integral with respect to M in $\mathcal{M}_0^2$ is defined as the extension of the linear application defined on the subspace $\mathcal{E}$ of $L^2(M)$. Let $t > 0$, M in $\mathcal{M}_0^2$ and X in $L^2(M) \cap \mathcal{E}$ such that $X(t) = \sum_{i\le n} U_i 1_{]t_{i-1},t_i]}(t)$ where U_i is a $\mathcal{F}_{t_{i-1}}$-measurable random variable and $0 = t_0 < t_1 < \ldots < t_n = t$, $k = 1, 2$. The stochastic integral $X.M = \int_0^t X(s) dM(s)$ is defined as the continuous linear application

$$X.M_t = \sum_{i\le n} U_i(M_{t_i} - M_{t_{i-1}}).$$

It is a square integrable martingale

$$\begin{aligned} E(X.M_t)^2 &= E\sum_{i\le n} U_i^2(< M >_{t_i} - < M >_{t_{i-1}}) \\ &= E\int_0^t X_s^2 \, d < M >_s \end{aligned}$$

with an integral defined as a Stieljes integral of X^2 with respect to the increasing process $< M >$. Moreover, $((X.M_t)^2 - \int_0^t X_s^2 \, d < M >_s)_{t>0}$ is still a martingale. These results extend by continuation to predictable processes of $L^2(M)$.

Theorem 8.2. *Let M in $\mathcal{M}_0^2$ and X in $L^2(M)$, then X belongs to $L^1(< M, M' >)$ for every martingale M' of $\mathcal{M}_0^2$ and there exists an unique process in $\mathcal{M}_0^2$ denoted $X.M$ such that for every $t > 0$*

$$< X.M, M' >_t = \int_0^t X \, d < M, M' > .$$

These definitions and properties extend to the space of local martingales $\mathcal{M}_0^{2,loc}$, where there exists an increasing sequence of stopping times $(S_n)_n$ tending to infinity and such that $(M(t \wedge S_n))_t$ belongs to $\mathcal{M}_0^2$. Similarly, it extends to the space $L^{2,loc}(M)$ of the predictable processes X for which there exists an increasing sequence of stopping times $(S_n)_n$ such that S_n tends to infinity and for all integer n and $t > 0$, $E \int_0^{t \wedge S_n} X_s^2 \, d < M >_s$ is finite.

Let M in $\mathcal{M}_0^{2,loc} \cap \mathcal{C}$ and X in $L^2(M, loc)$, then the stochastic integral $X.M$ is identical to the Stieljes integral $\int X \, dM$.

A special case is the integral of a process X with respect to the martingale $M = N - \widetilde{N}$ defined for a point process N having unpredictable jumps and a predictable compensator $\widetilde{N}$. If $\widetilde{N}$ is continuous, then $< M >= \widetilde{N}$.

If $\widetilde{N}$ is not continuous, it splits into the sum of a point process and a continuous part $\widetilde{N}(t) = \sum_{s \leq t} \Delta \widetilde{N} + \widetilde{N}^c$, and $M = M^c + M^d$. The above results are generalized with

$$\widetilde{M^2}(t) = \int_0^t (1 - \Delta \widetilde{N}(s)) \, d\widetilde{N}(s),$$

$$\widetilde{M_1 M_2}(t) = \int_0^t \Delta \widetilde{N}_1(s) \, d\widetilde{N}_2(s).$$

Exercise 8.2.1. Let M_1 and M_2 in $\mathcal{M}_0^2$ and let X_1 and X_2 be predictable processes of $L^2(M)$, prove that for every $t > 0$

$$< X_1.M_1, X_2.M_2 >_t = \{(X_1 X_2). < M_1, M_2 >\}_t,$$
$$E\{(X_1.M_1)_t (X_2.M_2)_t\} = E\{(X_1 X_2). < M_1, M_2 >\}_t.$$

Answer. By Theorem 8.2 we have

$$< X_1.M_1, X_2.M_2 >_t = \int_0^t X_1(s) \, d < M_1, X_2.M_2 >_s,$$
$$< M_1, X_2.M_2 >_s = \int_0^s X_2 \, d < M_1, M_2 >,$$

it follows that

$$\begin{aligned} < X_1.M_1, X_2.M_2 >_t &= \int_0^t X_1(s) X_2(s) \, d < M_1, M_2 >_s \\ &= \{(X_1 X_2). < M_1, M_2 >\}_t. \end{aligned}$$

The processes $Y_k = (X_k.M_k)^2 - \int_0^{\cdot} X_k^2 \, d < M_k >$ are martingales with expectation zero, therefore $E(X_k.M_k)^2(t) = E \int_0^t X_k^2 \, d < M_k >$ for every

$t \geq 0$. The second equality for $E\{(X_1X_2)_t. < M_1, M_2 >_t\}$ is deduced from the expectation

$$E\{X_1.M_1(t) + X_2.M_2(t)\}^2 = E\{X_1.M_1(t)\}^2 + E\{X_2.M_2(t)\}^2$$
$$+2E\{X_1.M_1(t)\, X_2.M_2(t)\}$$
$$= E\int_0^t X_1^2\, d < M_1 >_s + E\int_0^t X_2^2\, d < M_2 >$$
$$+2E\Big\{\int_0^t X_1\, dM_1 \int_0^t X_2\, dM_2\Big\}$$

and the martingale property of he processes Y_k entails

$$E\{X_1.M_1(t)\, X_2.M_2(t)\} = E\Big\{\int_0^t X_1\, dM_1 \int_0^t X_2\, dM_2\Big\}$$
$$= E\int_0^t X_1X_2\, d < M_1, M_2 >$$
$$= E\{(X_1X_2). < M_1, M_2 >\}_t.$$

Exercise 8.2.2. On a filtered probability space $(\Omega, \mathcal{A}, P, \mathbb{F})$, let N be a non-homogeneous Poisson process with a continuous compensator Λ and let Y be a predictable process, calculate the characteristic function of $\int_0^t Y_s\, dN_s$.

Answer. The process Y is the limit of a sequence of stepwise processes $Y_{nt} = \sum_{i=1}^n Y_i 1_{[t_{i-1},t_i[}$ on a partition of $[0,t]$ such that $t_0 = 0$ and $t_n = t$, and the variables Y_i are $\mathcal{F}_{t_{i-1}}$ measurable. The stochastic integral $I = \int_0^t Y_s\, dN_s$ is the limit of $I_n = \sum_{i=1}^n Y_i \Delta N_i$ where the variables $\Delta N_i = N(t_i) - N(t_{i-1})$ are independent with Poisson distributions

$$P(\Delta N_i = k) = e^{-\Delta\Lambda_i} \frac{(\Delta\Lambda_i)^k}{k!}.$$

The characteristic function ϕ of I is the limit of $\phi_n(x) = E(e^{ixI_n})$

$$\phi_n(x) = \sum_{k\geq 0} E(e^{ix\sum_{j=1}^n Y_j k}) P(\Delta N_j = k)$$
$$= \prod_{j=1}^n e^{-\Delta\Lambda_j} \sum_{k\geq 0}^n E\frac{\{(\Delta\Lambda_j)^k e^{ixkY_j}\}}{k!}$$
$$= \prod_{j=1}^n E[\exp\{\Delta\Lambda_j(e^{ixY_j} - 1)\}]$$
$$= E\Big[\exp\Big\{\sum_{j=1}^n \Delta\Lambda_j(e^{ixY_j} - 1)\Big\}\Big]$$

and for every x, $\phi_n(x)$ converges to $\phi(x) = E\exp\{\int_0^t (e^{ixY_s} - 1)\}\, d\Lambda_s\}$.

Exercise 8.2.3. Let B be a the standard Brownian motion and let Y be a predictable process, calculate the characteristic function of $\int_0^t Y_s\,dB_s$.

Answer. Like above, the integral $I = \int_0^t Y_s\,dB_s$ is the limit of the integral $I_n = \sum_{i=1}^n Y_i \Delta B_i$ of a stepwise process $Y_{nt} = \sum_{i=1}^n Y_i 1_{[t_{i-1},t_i[}$ on a partition $(t_i)_{i=0,\dots,n}$ of $[0,t]$, where the variables Y_i are $\mathcal{F}_{t_{i-1}}$ measurable and the variables $\Delta B_i = B(t_i) - B(t_{i-1})$ are independent centered $\mathcal{F}_{t_i}$ measurable Gaussian variables with variance $\mathrm{Var}(\Delta B_i) = t_i - t_{i-1}$. The characteristic function $\phi(x) = E(e^{ixI}$ of I is the limit of the characteristic function ϕ_n of the stepwise process Y_n

$$\begin{aligned}
\phi_n(x) &= \prod_{j=1}^n E\{E(e^{ixY_j\Delta B_j}) \mid \mathcal{F}_{t_{j-1}}\} \\
&= \prod_{j=1}^n E\exp\Big\{-\frac{x^2 Y_j^2 (t_j - t_{j-1})}{2}\Big\} \\
&= E\exp\Big\{-\frac{x^2 \int_0^t Y_{ns}^2\,ds}{2}\Big\}
\end{aligned}$$

and $\phi(x) = E\exp\{-\frac{x^2 \int_0^t Y_s^2\,ds}{2}\}$.

Exercise 8.2.4. Let N be a counting process with a continuous predictable compensator $\widetilde{N}$, let X be a predictable process and let T be a stopping time such that $\widetilde{N}_T$ and $\int_0^T X^2\,d\widetilde{N}$ belong to $L^2(P)$, prove that $\int_0^T X\,dN$ belongs to $L^2(P)$.

Answer. The local martingale $M = N - \widetilde{N}$ is such that $Z = M^2 - \widetilde{N}$ is a local martingale, and the stochastic integral $\int_0^t X\,dM$ is a local martingale. Let $Y_t = \int_0^t X\,dN$ and $\widetilde{Y}_t = \int_0^t X\,d\widetilde{N}$, as $\widetilde{Y}$ is predictable, it follows that

$$\begin{aligned}
E\Big\{\Big(\int_0^T X\,dM\Big)^2\Big\} &= E\Big(\int_0^T X^2\,d\widetilde{N}\Big) \\
&= E(Y_T^2) - 2E(Y_T\widetilde{Y}_T) + E(\widetilde{Y}_T^2) \\
&= E(Y_T^2) - E(\widetilde{Y}_T^2)
\end{aligned}$$

where $E(Y_T\widetilde{Y}_T) = E\{E(Y_T\widetilde{Y}_T \mid \mathcal{F}_{T^-})\} = E(\widetilde{Y}_T^2)$.

Under the conditions, the process $\int_0^{\cdot} X\,dM$ belongs to $L^2(P)$ and by the Cauchy–Schwarz inequality

$$E(\widetilde{Y}_T^2) \le E\Big(\widetilde{N}_T \int_0^T X^2\,d\widetilde{N}\Big) \le \|\widetilde{N}_T\|_{L^2(P)}\|X^2.\widetilde{N}_T\|_{L^2(P)},$$

this entails that $\widetilde{Y}$ and therefore Y belong to $L^2(P)$.

Exercise 8.2.5. Let N be a counting process with predictable compensator $\widetilde{N}$ and let X be a predictable process in $L^2(\widetilde{N}, loc)$, calculate the variance of $X.N$.

Answer. Let $N_t = \sum_{s\le t} 1_{\{T_i\le t\}}$ and let $M = N - \widetilde{N}$, by Exercise 8.1.10 the process of $M^2 - \widetilde{N}$ belongs to $\mathcal{M}_0^{2,loc}$. The stochastic integral $X.N_t = \sum_{s\le t} X_{T_i} 1_{\{T_i\le t\}}$ is a jump process and $X.M = X.N - X\widetilde{N}$. The variance of $X.N$ is therefore $E\{(X.M)^2\} = E(X^2.\widetilde{N})$.

Exercise 8.2.6. Let M in $\mathcal{M}_0^2$ and let X be a predictable process in $L^2(M)$, prove that for all $t > 0$ and $a > 0$

$$P(\sup_{s\le t}(X.M)_s > a) \le \frac{1}{a^2} E\int_0^t X^2\,d < M > .$$

Answer. By the Bienaymé–Chebychev inequality we obtain

$$P(\sup_{s\le t}(X.M)_s > a) \le \frac{1}{a^2} E\Big[\Big\{\sup_{s\le t}(X.M)_s\Big\}^2\Big],$$

let $\tau \le t$ be the random time such that $(X.M)_\tau = \sup_{s\le t}(X.M)_s$, τ is a stopping time and

$$E[\{(X.M)_\tau\}^2] = E\int_0^\tau X^2\,d < M >,$$

as the process $\int_0^{\cdot} X^2\,d < M >$ is increasing, the result follows.

Exercise 8.2.7. Let $(M_t)_t$ be a martingale with increasing predictable process $(A_t)_t$, and $(X_t)_t$ be a predictable process. Prove that for all $\eta > 0$ and $\lambda > 0$

$$P(\sup_{s\le t} X.M_s > \lambda) \le \frac{\eta}{\lambda^2} + P(X.A_t > \eta).$$

Answer. Under the conditions, the process $X.M - X^2.A$ is a martingale with expectation zero. Let $\tau \le t$ be the random time such that

$X.M_\tau = \sup_{s\leq t} X.M_s$, for all $\eta > 0$ and $\lambda > 0$

$$P(\sup_{s\leq t} X.M_s > \lambda) \leq P(\sup_{s\leq t} X.M_s > \lambda, X^2.A_\tau \leq \eta) + P(X^2.A_\tau > \eta)$$
$$\leq \frac{E(M_\tau^2 1_{\{X^2.A_\tau\leq\eta\}})}{\lambda^2} + P(X^2.A_\tau > \eta)$$

where the process $X^2.A$ is predictable hence

$$E(M_\tau^2 1_{\{X^2.A_\tau\leq\eta\}}) = E\{E(M_\tau^2 \mid \mathcal{F}_{\tau^-})1_{\{X^2.A_\tau\leq\eta\}})$$
$$= E(X^2.A_\tau 1_{\{X^2.A_t\leq\eta\}}) \leq \eta,$$

this proves the result.

8.3 Exponential submartingales

Let $\mathbb{F} = (F_t)_{t\leq 0}$ and let $M = (M_t)_{t\geq 0}$ be continuous in $\mathcal{M}_0^{2,loc}(\mathbb{F})$, for $\lambda > 0$ it defines the process

$$X_t(\lambda) = \exp\Big(\lambda M_t - \frac{\lambda^2}{2} < M >_t\Big).$$

Exercise 8.3.1. Prove that $X_t(\lambda)$ is a submartingale.

Answer. Let $(\mathcal{F}_t)_{t\geq 0}$ be the filtration generated by the process M. As $M_0 = 0$ and $< M >$ is increasing, then $< M >_0= 0$ and $< M >_t\geq 0$ for every $t \geq 0$. By convexity of the exponential it follows that for $< t$

$$E\Big\{\frac{X_t(\lambda)}{X_{t-}(\lambda)} \mid \mathcal{F}_{t-}\} = E\Big\{e^{\lambda(M_t-M_{t-})-\frac{\lambda^2}{2}(<M>_t-<M>_{t-})} \mid \mathcal{F}_{t-}\Big\}$$
$$\geq \exp\Big[E\Big\{\lambda(M_t - M_{t-})$$
$$-\frac{\lambda^2}{2}(< M >_t - < M >_{t-})\Big\} \mid \mathcal{F}_{t-}\Big]$$
$$= 1,$$

hence $E\{X_t(\lambda) \mid \mathcal{F}_{t-}\} \geq X_{t-}(\lambda)$ and $X_t(\lambda)$ is a submartingale.

Exercise 8.3.2. Let $(X_t)_{t\geq 0}$ be a martingale with independent increments such $EX_0 = 0$. Prove that for all λ and t such that the Laplace transform $L_{X_t}(\lambda)$ of X_t is finite, the process $Y_t(\lambda) = \exp\{\lambda X_t - \log L_{X_t}(\lambda)\}$ is a submartingale if $L_{X_s}L_{X_{t-s}} \geq L_{X_t}$ for $0 < s < t$ and a supermartingale if $L_{X_s}L_{X_{t-s}} \leq L_{X_t}$ for every $0 < s < t$.

Answer. Let $(\mathcal{F}_t)_{t\geq 0}$ be the filtration generated by $(X_t)_{t\geq 0}$, for all $\lambda > 0$ and $t > s > 0$, $L_{X_t}(\lambda) - L_{X_s}(\lambda) = E(e^{\lambda X_t} - e^{\lambda X_s})$ and

$$\begin{aligned}
E\{Y_t(\lambda)|\mathcal{F}_s\} &= Y_s E\Big\{\frac{Y_t}{Y_s}(\lambda)|\mathcal{F}_s\Big\} \\
&= Y_s \frac{L_{X_s}}{L_{X_t}}(\lambda) E[\exp\{\lambda(X_t - X_s)|\mathcal{F}_s\}] \\
&= Y_s \frac{L_{X_s}}{L_{X_t}}(\lambda) E[\exp\{\lambda(X_{t-s})\}] \\
&= Y_s \frac{L_{X_s} L_{X_{t-s}}}{L_{X_t}}(\lambda),
\end{aligned}$$

by the independence of the increments of the process X. Under the conditions, the result follows.

Exercise 8.3.3. Let $(N_t)_{t\geq 0}$ be a counting process with predictable compensator $(A_t)_{t\geq 0}$ and let Y be a predictable process with respect to the filtration $\mathbb{F}$ generated by N. Prove that

$$Z_t = \exp\Big\{\int_0^t \log Y_s \, dN_s + \int_0^t (Y_s - 1)\, dA_s\Big\}$$

is a local submartingale.

Answer. Let $N_t = \sum_{i\geq 0} 1_{T_i \leq t}$, the process $M_t = N_t - A_t$ belongs to $\mathcal{M}_0^{2loc}$ and the processes N and A are increasing then by convexity, for $0 < s < t$ we have

$$E(Z_t \mid \mathcal{F}_s) \geq Z_s \exp\Big[E\Big\{\int_s^t \log Y \, dN + \int_s^t (Y-1)\, dA \mid \mathcal{F}_s\Big\}\Big]$$

and the inequality $y - 1 \geq \log(y)$ entails

$$\begin{aligned}
E\Big\{\int_s^t \log Y \, dN + \int_s^t (Y-1)\, dA \mid \mathcal{F}_s\Big\} &\geq E\Big\{\int_s^t \log Y \, dM \mid \mathcal{F}_s\Big\} \\
&= E\Big\{\int_s^t \log Y \, dM\Big\},
\end{aligned}$$

where $\int_0^{\cdot} \log Y \, dM$ belongs to $\mathcal{M}_0^{loc}$ which implies that for $0 < s < t$, $E\Big\{\int_s^t \log Y \, dM\Big\} = 0$ and $E(Z_t \mid \mathcal{F}_s) \geq Z_s$.

8.4 Inequalities for martingales indexed by $\mathbb{R}_+$

Exercise 8.4.1. Prove that for $p > 1$ we have

$$E(M_t^p) \leq 2^{p-1} E(< M^c >_t^{\frac{p}{2}}) + 2^{p-1} E(< M^d >_t^{\frac{p}{2}}).$$

Answer. For $p > 1$ and every real x, the function $h(x) = x^p$ is convex, hence

$$\Big(\frac{1+x}{2}\Big)^p \leq \frac{1+x^p}{2},$$

it follows that for all a and b in $\mathbb{R}$ we have $(a+b)^p \leq 2^{p-1}(a^p+b^p)$, writing the inequality with $x = a^{-1}b$ or $x = b^{-1}a$, and

$$E(M_t^p) \leq 2^{p-1}E(M_t^{d\,p}) + 2^{p-1}E(M_t^{c\,p}).$$

By Hölder's inequality

$$E(M_t^{d\,p}) = E[\{E(M_t^{d\,2} \mid \mathcal{F}_{t-})\}^{\frac{p}{2}}] = E(< M^d >_t^{\frac{p}{2}}),$$
$$E(M_t^{c\,p}) = E[\{E(M_t^{c\,2} \mid \mathcal{F}_{t-})\}^{\frac{p}{2}}] = E(< M^c >_t^{\frac{p}{2}}),$$

this yields the result.

Exercise 8.4.2. Let $M = (M_t)_{t\geq 0}$ in $\mathcal{M}^2_{0,loc}$ and let $M_t^* = \sup_{s\leq t} M_s$, prove that for every $t > 0$

$$E(M_t^{*2}) \leq 4E([M]_t).$$

Answer. From Exercise 8.1.5 and integrating by parts, for $\tau \leq t$ we have

$$\begin{aligned} E(M_t^{*2}) &= \int_0^\infty \lambda^2\, dP(M_t^* \leq \lambda) \leq 2\int_0^\infty \lambda P(M_t^* \geq \lambda)\, d\lambda \\ &\leq 2E\{M_\tau \int_0^\infty 1_{\{M_t^*\geq\lambda\}}\, d\lambda\} \\ &= 2E\{M_\tau M_t^*\} \leq 2\{E(M_\tau^2)\}^{\frac{1}{2}}\, \{E(M_t^{*2})\}^{\frac{1}{2}}. \end{aligned}$$

By definition, $E(M_\tau^2) = E([M]_\tau) \leq E([M]_t)$ for the increasing process $[M]_t$, which implies the inequality $\{E(M_t^{*2})\}^{\frac{1}{2}} \leq 2\{E([M]_t)\}^{\frac{1}{2}}$.

Exercise 8.4.3. Let $p \geq 2$, M in $\mathcal{M}^2_{0,loc}$ and $M_t^* = \sup_{s\leq t} M_s$, prove that

$$EM_t^{*p} \leq p^{\frac{p}{p-1}} E([M]_t^{\frac{p}{2}}).$$

Answer. Let $\tau = \inf\{t : |M_t| \geq \lambda\}$, $\lambda > 0$, τ is a stopping time with respect to $\mathbb{F}$. The events $\{M_t^* \geq \lambda\}$ and $\{t \geq \tau\}$ are equivalent and for every $\lambda > 0$

$$\lambda P(M_t^* \geq \lambda) \leq E(|M_\tau| 1_{\{M_t^*\geq\lambda\}}),$$

this inequality generalizes to $\mathcal{M}^p_{0,loc}$

$$\lambda^p P(M_t^* \geq \lambda) \leq E(|M_\tau|^p 1_{\{M_t^*\geq\lambda\}}).$$

By the same arguments

$$\begin{aligned}
E(M_t^{*p}) &= \int_0^\infty \lambda^p \, dP(M_t^* \le \lambda) \\
&\le p \int_0^\infty \lambda^{p-1} P(M_t^* \ge \lambda) \, d\lambda \\
&\le pE\{M_\tau^{p-1} \int_0^\infty 1_{\{M_t^* \ge \lambda\}} \, d\lambda\} \\
&= pE\{M_\tau^{p-1} M_t^*\}
\end{aligned}$$

where $t \ge \tau$ and by Hölder's inequality with $p' = (p-1)^{-1}p$

$$E\{M_\tau^{p-1} M_t^*\} \le \{E(M_\tau^p)\}^{\frac{p-1}{p}} \{E(M_t^{*p})\}^{\frac{1}{p}},$$

therefore $E(M_t^{*p}) \le p^{\frac{p}{p-1}} E(M_\tau^p)$. Then for every $t > 0$, the equality $E(M_t^p) = E[\{E(M_t^2 \mid \mathcal{F}_{t-})\}^{\frac{p}{2}}]$ implies

$$E(M_t^{*p}) \le p^{\frac{p}{p-1}} E\Big([M]_\tau^{\frac{p}{2}}\Big) \le p^{\frac{p}{p-1}} E\Big([M]_t^{\frac{p}{2}}\Big)$$

for every $t \ge \tau$.

Exercise 8.4.4. Let M in $\mathcal{M}_{0,loc}^2$ and A be a predictable process such that $\int_0^t A_s^2 \, d < M >_s$ is locally finite. Let $Y_t = \int_0^t A_s \, dM_s$ prove that

$$EY_t^{*p} \le p^{\frac{p}{p-1}} E\Big\{\Big(\int_0^t A_s^2 \, d < M >_s\Big)^{\frac{p}{2}}\Big\}.$$

Answer. Like in Exercise 8.4.3 we have

$$EY_t^{*p} \le p^{\frac{p}{p-1}} E\Big([Y]_t^{\frac{p}{2}}\Big)$$

where

$$E\Big([Y]_t^{\frac{p}{2}}\Big) = E\Big\{E\Big([Y]_t^{\frac{p}{2}} \mid \mathcal{F}_{t-}\Big)\Big\} = E\Big(< Y >_t^{\frac{p}{2}}\Big)$$

and $< Y >_t = \int_0^t A_s^2 \, d < M >_s$.

Chapter 9

Jump processes

9.1 Renewal processes

A renewal process is a sequence of random variables $(T_k)_{k\geq 0}$ such that the durations $X_k = T_k - T_{k-1} > 0$, $k \geq 1$, are independent and identically distributed. The distribution of the counting process

$$N(t) = \sum_{n\geq 1} \{T_n \leq t\}$$

on $\mathbb{R}_+$, with jump times $(T_k)_{k\geq 0}$, is determined by the equality

$$\{N(t) \geq n\} = \{T_n \leq t\},$$

for $n \geq 1$. If the distribution function F of the durations X_k has a density f the probability of the event $\{T_1 \leq t\}$ is the convolution

$$f^{\star 1}(t) = \int_0^\infty P(T_0 \leq t - s)\, dP(X_1 \leq s) = \int_0^\infty F_0(t-s) f(s)\, ds$$

and the probability of $\{T_n \leq t\}$ is the nth convolution

$$\begin{aligned} P(T_n \leq t) &= \int_0^\infty P(T_{n-1} \leq t - s)\, dF(s) \\ &= \int_0^\infty F_0(t-s) f^{\star n}(s)\, ds := F^{\star n}(t). \end{aligned}$$

Let $\mu = EX_n$, then

$$E(N_t) = \frac{E(T_{N_t})}{\mu}.$$

The current waiting time C_t and the residual waiting time R_t of the process $(T_n)_n$ at time t are defined as

$$R_t = T_{N_t+1} - t, \quad C_t = t - T_{N_t},$$

so $X_{N_t+1} = R_t + C_t$, their expectations are

$$EC_t = \int_0^t \bar{F}_{T_{N_t}}(s)\, ds,$$

$$ER_t = \int_t^\infty \bar{F}_{T_{N_t}}(s)\, ds.$$

Exercise 9.1.1. Define the distribution function of T_n as F is the uniform distribution on $[0,1]$.

Answer. For every t in $[0,1]$ we have $P(T_1 \le t) = t$ and

$$P(T_n \le t) = \int_0^t P(T_{n-1} \le t-s)\, ds.$$

Assuming that $P(T_{n-1} \le t) = \frac{t^{n-1}}{(n-1)!}$, it follows that

$$\begin{aligned} P(T_n \le t) &= \int_0^t \frac{(t-s)^{n-1}}{(n-1)!}\, ds \\ &= \int_0^t \frac{x^{n-1}}{(n-1)!}\, dx = \frac{t^n}{(n)!}. \end{aligned}$$

Exercise 9.1.2. Write the expression of expectation $M(t)$ of $N(t)$ and prove that for every $s > 0$

$$\int_0^\infty e^{-sx}\, dM(x) = \frac{L_{F_0}(s)}{1 - L_f(s)}.$$

Answer. The expectation of the $N(t)$ is

$$\begin{aligned} M(t) = EN(t) &= \sum_{n\ge1} nP(N(t) = n) \\ &= \sum_{n\ge1} P(N(t) \ge n) = \sum_{n\ge1} F^{\star n}(t). \end{aligned}$$

For every $s > 0$, the Laplace transform of T_1 is

$$L_{T_1}(s) = \int_0^\infty e^{-st} f^{\star 1}(t)\, dt = L_{F_0}(s) L_f(s),$$

an integration by parts yields

$$\begin{aligned} L_{F_0}(s) &= \frac{1}{\mu} \int_0^\infty e^{-sx} \bar{F}(t)\, dt \\ &= \frac{1}{s\mu}\{1 - L_f(s)\}, \end{aligned}$$

and

$$L_f(s) = \int_0^\infty e^{-sx} f(t)\, dt,$$

the Laplace transform of T_n is

$$L_{T_n}(s) = L_{F_0}(s) L_X^n(s) = L_{F_0}(s) L_f^n(s).$$

It follows that

$$\begin{aligned}\int_0^\infty e^{-sx}\, dM(x) &= \sum_{n\geq 0} L_{F^{\star n}}(s) \\ &= L_{F_0}(s) \sum_{n\geq 0} L_f^n(s) \\ &= \frac{L_{F_0}(s)}{1 - L_f(s)}.\end{aligned}$$

Exercise 9.1.3. Let $X_1, \ldots, X_n, \ldots$ be a renewal process and let N be an independent geometric variable with probabilities $p_k = p(1-p)^{k-1}$, write the Laplace transform of $T_N = X_1 + \cdots + X_N$ and the expectation of T_N.

Answer. Let F the distribution function of the variables X_i, the Laplace transform of T_N is

$$\begin{aligned}L(t) = E(e^{tT_N}) &= \sum_{k\geq 1} P(N=k) \sum_{i=1}^{k} E(e^{tX_i}) \\ &= \sum_{k\geq 1} p(1-p)^{k-1} L_X^k(t) \\ &= \frac{pL_X(t)}{1-(1-p)L_X(t)},\end{aligned}$$

the expectation of T_N is

$$E(T_N) = L'(0) = E(X) + \frac{(1-p)E(X)}{p} = \frac{E(X)}{p}.$$

Exercise 9.1.4. Prove that a renewal process $(T_k)_{k\geq 0}$ is stationary and that its increments are independent.

Answer. As the variables $X_k = T_k - T_{k-1}$ are independent and identically distributed, the distribution of $T_{k+n} - T_k = \sum_{i=k+1}^{k+n} X_i$ is identical to the distribution of $T_n - T_0 = \sum_{i=1}^{n} X_i$ and the process is stationary. For all n

and k, the variables $T_{k+n} - T_k$ and $T_k = \sum_{i=1}^k X_i$ are independent so the increments of the process are independent.

Exercise 9.1.5. Let $(T_n)_{n\geq 1}$ be a renewal process and let $(Y_n)_{n\geq 1}$ be an independent sequence of independent and identically distributed variables such that $E(T_{n+1} - T_n)$ and $E(Y_n)$ are finite. Let $Y_t = \sum_{n\geq 1} Y_n 1_{\{T_n \leq t\}}$, determine $\lim_{t\to\infty} t^{-1} Y_t$.

Answer. Let $N_t = \sum_{n\geq 1} 1_{\{T_n\leq t\}}$ and $Y_t = \sum_{n\geq N_t} Y_n$, the expectation of Y_t is

$$\begin{aligned} E(Y_t) &= E(N_t)E(Y_1) \\ \varphi_t(s) &= E(e_{it^{-1}sY_t}) = E\{\varphi_Y^{Y_t}(t^{-1}s)\}, \\ \log \varphi_t(s) &\leq E(t^{-1}Y_t) \log \varphi_Y(s) \end{aligned}$$

where $\lim_{t\to\infty} t^{-1}E(N_t) = \sum_{n\geq 1} F^{\times n}(t)$ is finite: for all positive integers n and $m < n$

$$\begin{aligned} F^{\times n}(t) &= \int_0^t F^{\times n-m}(t-s)\, dF^{\times m}(s) \\ &\leq F^{\times n-m}(t)\, F^{\times m}(t) \end{aligned}$$

let k such that $F^{\times k}(t) < 1$ and let $n = kI + J$, then

$$F^{\times n}(t) \leq \{F^{\times k}(t)\}^I,$$

it converges to zero and the series $\sum_{n\geq 1} F^{\times n}(t)$ converges as n tends to infinity.

Exercise 9.1.6. Let $X_1, \ldots, X_n, \ldots$ be a renewal process and let N be an independent Binomial variable $Bin(n,p)$, write the Laplace transform of $T_N = X_1 + \cdots + X_N$ if the variables X_i have an exponential distribution.

Answer. Let L_X be the Laplace transform of the variables X_i, the Laplace transform of T_N is

$$\begin{aligned} L(t) &= \sum_{k=1}^n P(N=k) \prod_{i=1}^k E(e^{tX_i}) \\ &= \sum_{k=1}^n \binom{n}{k} p^k (1-p)^{n-k} L_X^k(t) \\ &= \{1 - p + pL_X(t)\}^n - 1, \end{aligned}$$

under an exponential distribution $\mathcal{E}_\lambda$ it is

$$L(t) = \left\{1 - p + \frac{\lambda p}{\lambda + t}\right\}^n - 1 = \left\{1 - \frac{tp}{\lambda + t}\right\}^n - 1.$$

9.2 Poisson process

A Poisson process $N = (N_t)_{t\geq 0}$ with intensity λ is a counting process $N_t = \sum_{n\geq 0} 1_{\{T_n \leq t\}}$ with $T_0 = 0$, with independent increments and such that for all s and $t \geq 0$

$$P(N_t = n) = P(T_n \leq t) - P(T_{n-1} \leq t)$$

and

$$P(N_{t+s} - N_s = n) = e^{-\lambda t}\frac{(\lambda t)^n}{n!}.$$

Equivalently, a Poisson process N is a renewal process with exponentially distributed variables $X_n = T_n - T_{n-1}$ (see problem 4.3.1). The expectation of a Poisson process N is

$$E(N_t) = \sum_{n\geq 0} nP(N_t = n) = \lambda t$$

and the conditional expectation with respect to the sigma-algebra $\mathcal{F}_s$ generated by $(N_s, s \leq t)$ is

$$E(N_t - \lambda t \mid \mathcal{F}_s) = N_s - \lambda s,$$

the process $N_t - \lambda t$ is a martingale with respect to the filtration $\mathcal{F}_t)_{t\geq 0}$.

Exercise 9.2.1. Prove that the random times of a Poisson process have a Gamma distribution and write the Laplace transform of $P(N_t = n)$.

Answer. The variable $T_n = \sum_{k=1}^n X_k$ has the distribution function $F^{\times n}$ and its the Laplace transform is

$$L_X^n(s) = \frac{\lambda^n}{(\lambda + s)^n}$$

where L_X is the Laplace transform of the variables $X_k \sim \mathcal{E}_\lambda$, this is the Laplace transform of a $\Gamma_{n,\lambda}$ distribution. The Laplace transform of the distribution $P(N_t = n) = F^{\times n}(t) - F^{\times n+1}(t)$ is therefore

$$\begin{aligned} L_n(s) &= \frac{\lambda^n}{(\lambda + s)^n} - \frac{\lambda^{n+1}}{(\lambda + s)^{n+1}} \\ &= L_X^n(s)\frac{s}{\lambda + s} = L_X^{n+1}(s)\frac{s}{\lambda}. \end{aligned}$$

Exercise 9.2.2. Let N be a Poisson process with intensity λ, calculate the probability $P(N_s = k \mid N_t = n)$ for integers $k \leq n$ and positive $s < t$.

Answer. For all integer n and positive real t we have

$$P(N_t = n) = e^{-\lambda t}\frac{(\lambda t)^n}{n!}$$

and by the independence of the increments of the Poisson process

$$\begin{aligned}
P(N_s = k \mid N_t = n) &= \frac{P(N_s = k, N_t - N_s = n-k)}{P(N_t = n)} \\
&= \frac{P(N_s = k)P(N_{t-s} = n-k)}{P(N_t = n)} \\
&= \frac{(\lambda s)^k\{\lambda(t-s)\}^{n-k}}{(\lambda t)^n}\frac{n!}{k!(n-k)!} \\
&= \binom{n}{k}\left(\frac{s}{t}\right)^k\left(1-\frac{s}{t}\right)^{n-k},
\end{aligned}$$

therefore $P(N_s = k \mid N_t = n)$ is the binomial distribution with probability $t^{-1}s$.

Exercise 9.2.3. Calculate the covariance of N_t and N_s for all positive reals s and t.

Answer. For $s < t$, the covariance is $\text{Cov}(N_s, N_t) = E(N_s N_t) - \lambda^2 st$ and by the independence and stationarity of the process

$$\begin{aligned}
E(N_s N_t) &= E(N_s^2) + E\{N_s(N_t - N_s)\} \\
&= (\lambda s)^2 + \lambda s + \lambda^2 s(t-s) \\
&= \lambda s\{1 + \lambda t\},
\end{aligned}$$

hence $\text{Cov}(N_s, N_t) = \lambda(s \wedge t)$.

Exercise 9.2.4. Write the distribution of the sum of two independent Poisson processes and their marginal distributions conditionally on their sum. Generalize to the sum of n independent Poisson processes.

Answer. Let N_k be a Poisson process with intensity $\lambda_k > 0$, $k = 1, 2$. For all positive s and t and for every integer n, we have

$$\begin{aligned}
P(N_{1t} + N_{2s} = n) &= \frac{e^{-(\lambda_1 t + \lambda_2 s)}}{n!}\sum_{k=0}^{n}\binom{n}{k}(\lambda_1 t)^k(\lambda_2 s)^{n-k} \\
&= \frac{e^{-(\lambda_1 t + \lambda_2 s)}}{n!}(\lambda_1 t + \lambda_2 s)^n,
\end{aligned}$$

$N_{s,t} = N_{1t} + N_{2s}$ has the distribution of a Poisson variable with intensity $\lambda_1 t + \lambda_2 s$. The joint distribution of $(N_{1t}, N_{s,t})$ is defined by the probabilities

$$\begin{aligned} P(N_{1t} = k, N_{s,t} = n) &= P(N_{1t} = k)P(N_{2s} = n - k) \\ &= e^{-(\lambda_1 t + \lambda_2 s)} \frac{(\lambda_1 t)^k (\lambda_2 s)^{n-k}}{k!(n-k)!}, \end{aligned}$$

and the conditional distribution of N_{1t} is

$$P(N_{1t} = k \mid N_{s,t} = n) = \binom{n}{k} \frac{(\lambda_1 t)^k (\lambda_2 s)^{n-k}}{(\lambda_1 t + \lambda_2 s)^n},$$

it is a binomial distribution with parameters n and the probability

$$p_{s,t} = \frac{\lambda_1 t}{\lambda_1 t + \lambda_2 s}.$$

For n independent Poisson processes, let $t = (t_1, \ldots, t_n)$ in $\mathbb{R}_+^n$ and let integers $k > i$, the variables $S_{n,t} = N_{1t_1} + \cdots + N_{nt_n}$ and $S_{n,t} - N_{1t_1}$ are independent and they have the distributions of Poisson variables with parameters $\mu_{n,t} = \lambda_1 t_1 + \cdots + \lambda_n t_n$ and respectively $\mu_{n,t} - \lambda_1 t_1$. The probabilities $p_{i,k} = P(N_{1t_1} = i, S_{n,t} = k)$ and $p_{i|k} = P(N_{1t_1} = i \mid S_{n,t} = k)$ are

$$\begin{aligned} p_{i,k} &= P(N_{1t_1} = i, S_{n,t} - N_{1t_1} = k - i) \\ &= e^{-\mu_{n,t}} \frac{(\lambda_1 t_1)^i (\mu_{n,t} - \lambda_1 t_1)^{k-i}}{i!(k-i)!}, \end{aligned}$$

$$p_{i|k} = \binom{k}{i} p_{n,t}^i (1 - p_{n,t})^{k-i}$$

with the probability

$$p_{n,t} = \frac{\lambda_1 t_1}{\mu_{n,t}}.$$

The distribution of N_{jt_j} conditionally on $S_{n,t} = k$ is binomial with parameters k and probabilities $p_{in,t} = \lambda_i t_i \mu_{n,t}^{-1}$, for every $j = 1, \ldots, n$.

Exercise 9.2.5. Let $(X_i)_{i \leq n}$ be a sequence of independent variables with distribution function F and partial sums $S_n = \sum_{i=1}^n X_i$, and let $(N_t)_{t \geq 0}$ be an independent Poisson process, write the distribution function of S_{N_t}.

Answer. If N is a Poisson process with intensity λ, S_{N_t} has the distribution function

$$\begin{aligned} P(S_{N_t} \leq x) &= \sum_{k=1}^{\infty} P(S_k \leq x,)P(N_t = k) \\ &= e^{-\lambda t} \sum_{k=1}^{\infty} \frac{(\lambda t)^k}{k!} F^{\star k}(x) \end{aligned}$$

where $F^{\star k}$ is the kth convolution of F.

Problem 9.2.1. Let Λ an increasing function on $\mathbb{R}_+$ and let $(T_k)_{k\geq 0}$ be the jump times of a renewal process such that $P(T_k - T_{k-1} \geq x) = e^{-\Lambda(x)}$, prove that the variables $S_k = \Lambda(T_k)$ are the jump times of a Poisson process with intensity 1 if and only if Λ is stationary.

Answer. Let $T_k = \Lambda^{-1}(S_k)$ such that $P(T_k - T_{k-1} \geq x) = e^{-\Lambda(x)}$, then

$$P(\Lambda^{-1}(S_k) - \Lambda^{-1}(S_{k-1}) \geq x) = e^{-\Lambda(x)},$$

if Λ is stationary it follows that for $y = \Lambda(x)$

$$P(\Lambda^{-1}(S_k - S_{k-1}) \geq x) = P(S_k - S_{k-1} \geq y) = e^{-y}$$

and $(S_k)_{k\geq 0}$ is a Poisson process with intensity 1.

Reversely if $(S_k)_{k\geq 0}$ is a Poisson process with intensity 1, its distribution is determined by the probabilities $P(S_k - S_{k-1} \geq y) = e^{-y}$, then

$$\begin{aligned} P(T_k - T_{k-1} \geq x) &= P(\Lambda^{-1}(S_k) - \Lambda^{-1}(S_{k-1}) \geq x) = e^{-\Lambda(x)} \\ &= P(\Lambda^{-1}(S_k - S_{k-1}) \geq x) \end{aligned}$$

and Λ is stationary.

Problem 9.2.2. Let Λ be an increasing left-continuous process with right limits and let $(T_k)_{k\geq 0}$ be the jump times of a renewal process such that $P(T_k - T_{k-1} \geq x \mid \Lambda) = e^{-\Lambda(x)}$.

(a). Prove that the variables $S_k = \Lambda(T_k)$ are the jump times of a Poisson process with intensity 1 if and only if Λ is a stationary process.

(b). Find the Laplace transform of the process $(T_k)_{k\geq 0}$ such that $T_k = \Lambda^{-1}(S_k)$, with a stationary process Λ.

Answer. (a). The proof is similar to the proof of Exercise 3 with the conditional probabilities $P_\Lambda(T_k - T_{k-1} \geq x) = e^{-\Lambda(x)}$, for $k \geq 1$.

(b). If Λ is a stationary process, for all integers k and n and for every $t > 0$ we have

$$\begin{aligned} \Lambda(kt) &= \Lambda(t) + \Lambda((k-1)t) = k\Lambda(t), \\ n\Lambda\Big(\frac{kt}{n}\Big) &= nk\Lambda\Big(\frac{t}{n}\Big) = k\Lambda(t), \end{aligned}$$

by limit it follows that for all $s < t$ in $\mathbb{R}_+$

$$\Lambda(t-s) = (t-s)\Lambda(1).$$

Let $\theta = \Lambda(1)$, the inverse process is $\Lambda^{-1}(t) = \theta^{-1}t$. Conditional on Λ, the variables $X_k = T_k - T_{k-1}$, $k \geq 1$, have a distribution function $F_\Lambda = 1 - e^{-\Lambda}$ such that $F(0) = 0$ hence $\Lambda(0) = 0$ a.s. and $\lim_{x\to\infty} \Lambda(x)$ is a.s. infinite. The variables $S_k - S_{k-1}$ and S_0 have the exponential distribution with parameter 1, T_0 has therefore the conditional distribution function F_Λ like the variables X_k. For $k \geq 1$, the conditional probabilities of the variables X_k are

$$P_\Lambda(T_k - T_{k-1} \geq x) = P(T_k - T_{k-1} \geq x \mid \Lambda) = e^{-\Lambda(x)},$$

they determine the Laplace transform of the variables X_k

$$\begin{aligned}
L_{X_k}(u_k) &= E[e^{-u_k\{\Lambda^{-1}(S_k)-\Lambda^{-1}(S_{k-1})\}}] \\
&= E[e^{-u_k(S_k-S_{k-1})\theta^{-1}}] = E[\{L_{\theta^{-1}}(u_k)\}^{S_k-S_{k-1}}] \\
&= \int_{\mathbb{R}_+} \{L_{\theta^{-1}}(u_k)\}^s e^{-s}\, ds \\
&= \int_{\mathbb{R}_+} e^{-s\{1-\log L_{\theta^{-1}}(u_k)\}}\, ds \\
&= \frac{1}{1-\log L_{\theta^{-1}}(u_k)}.
\end{aligned}$$

The Laplace transform of the variable T_0 is also L_{X_k} and the vector $(T_k)_{k=1,\ldots,n}$ has the Laplace transform

$$\begin{aligned}
L_T(u) &= E(e^{-u_0T_0}) \prod_{k=1}^n E(e^{-u_k(T_k-T_{k-1})}) \\
&= \prod_{k=0}^n \frac{1}{1-\log L_{\theta^{-1}}(u_k)}.
\end{aligned}$$

Problem 9.2.3. Find the distributions of the current time C_t and the residual waiting time R_t of a Poisson process with intensity λ, and their joint distribution.

Answer. The residual waiting time $R_t = T_{N_t+1} - t$ has the distribution function

$$\begin{aligned}
F_{R_t}(x) &= 1 - P(T_{N_t+1} > x + t) \\
&= 1 - P(N_{x+t} - N_t = 0) \\
&= 1 - P(N_x = 0) = e^{-\lambda x},
\end{aligned}$$

for every $x < t$, the current time $C_t = t - T_{N_t}$ has the distribution function

$$\begin{aligned} F_{C_t}(x) &= 1 - P(T_{N_t} \leq t - x) \\ &= 1 - P(N_t - N_{t-x} = 0) \\ &= 1 - P(N_x = 0) = 1 - e^{-\lambda x} \end{aligned}$$

and for $x > t$

$$F_{C_t}(x) = 1 - P(T_{N_t} \leq -|t - x|) = 1.$$

For every $y < t$, the joint distribution of (R_t, C_t) is

$$\begin{aligned} F_t(x, y) &= 1 - P(T_{N_t+1} > x + t, T_{N_t} \leq t - y) \\ &= 1 - P(N_{t+x} - N_{t-y} = 0) \\ &= 1 - P(N_{x+y} = 0) = 1 - e^{-\lambda(x+y)} \end{aligned}$$

and for $y > t$

$$F_t(x, y) = 1 - P(T_{N_t+1} > x + t, T_{N_t} \leq -|t - y|)) = 1,$$

therefore $P(R_t > x, C_t > y) = P(R_t > x)P(C_t > y)$, this proves the independence of R_t and C_t.

Problem 9.2.4. Let N be a randomized Poisson process observed at random times $(S_i)_{i \leq n}$ such that $S_i - S_{i-1}$ are independent with the distribution F on a real interval $[0, \tau]$.

(a). Find the distribution of $\{N(S_i)\}_{i \leq n}$ as F is uniform on a finite interval $[0, \tau]$, and as F is a $\Gamma_{a,b}$ distribution on $\mathbb{R}_+$.

(b). Find the distribution of $\{N(S_1), N(S_2)\}$ conditionally on S_2 as $S_1 = XS_2$ and $S_2 - S_1 = (1 - X)S_2$ where X has a $B_{a,b}$ distribution on $[0, 1]$.

(c). Find the distribution of $\{N(S_i)\}_{i \leq n}$ conditionally on S_n as $S_i - S_{i-1} = X_i S_n$ for $i \leq n$, with $S_0 = 0$, where $(X_i)_{i \leq n}$ has a Dirichlet distribution D_a.

Answer. (a). Let $S_0 = 0$, the probabilities of $\{N(S_i)\}_{i \leq n}$ are the product of the marginal probabilities

$$P(N(S_i) - N(S_{i-1}) = k_i) = \frac{1}{k_i!} \int e^{-\lambda u_i} (\lambda u_i)^{k_i} \, dF(u_i),$$

with a uniform distribution, they are

$$P(N(S_i) = k) = \frac{1}{\lambda k_i!} \int_0^{\lambda^{-1}\tau} e^{-s} s^{k_i} \, ds.$$

If $S_i - S_{i-1} \sim \Gamma_{a,b}$, they are

$$P(N(S_i) - N(S_{i-1}) = k_i) = \frac{\lambda^{k_i}}{k_i!\Gamma_a} \int_0^\infty e^{-(\lambda+b)t} t^{k_i+a-1} \, dt$$
$$= \binom{a+k-1}{k_i} \frac{\lambda^{k_i}}{(b+\lambda)^{k_i+a}},$$

they have binomial distributions.

(b). Conditionally on S_2, the joint probabilities of the events $\{N(S_1) = k\}$ and $\{N(S_2) - N(S_1) = n-k\}$ is

$$P_{k,n}(S_2) = e^{-\lambda S_2} \frac{B_{a+k,b+n-k}}{k!(n-k)!B_{a,b}} S_2^{n+a+b-2}.$$

(c). Let $\bar{a}_n = \sum_{i=1}^n a_i$ the sum of the parameters of the Dirichlet distribution D_a. Conditionally on S_n, the joint probabilities of the events $\{N(S_i) = k_i\}$, for $i = 1, \ldots, n$ is the product of the probabilities of the events $\{N(S_i) - N(S_{i-1}) = k_i - k_{i-1}\}$, with $k_0 = 0$

$$P_{k,n}(S_n) = \lambda^{k_n} e^{-\lambda S_n} S_n^{k_n+\bar{a}_n-n} \frac{\Gamma_{\bar{a}_n}}{\prod_{i=1}^n \Gamma_{a_i} \prod_{i=1}^n (k_i - k_{i-1})!}$$
$$\cdot \int \prod_{i=1}^{n-1} x_i^{a_i-1+k_i-k_{i-1}} \Big(1 - \sum_{j=1}^{n-1} x_j\Big)^{a_n-1+k_n-k_{n-1}}$$
$$= \frac{\lambda^{k_n} e^{-\lambda S_n} S_n^{k_n+\bar{a}_n-n} \Gamma_{\bar{a}_n}}{\prod_{i=1}^n \Gamma_{a_i} \prod_{i=1}^n (k_i - k_{i-1})!} \frac{\prod_{i=1}^n \Gamma_{a_i-1+k_i-k_{i-1}}}{\Gamma_{\bar{a}_n+k_n-n}}$$
$$= \frac{\lambda^{k_n} e^{-\lambda S_n} S_n^{k_n+\bar{a}_n-n} B_c}{\prod_{i=1}^n (k_i - k_{i-1})! B_a}$$

with $c = (a_i - 1 + k_i - k_{i-1})_{i=1,\ldots,n}$.

Problem 9.2.5. Let P_1 and P_2 be the probability distributions of Poisson processes with parameters λ_1 and λ_2, find the Hellinger distance $h(P_{1t}, P_{2t})$ and the Kullback–Leibler information $K(P_{1t}, P_{2t})$.

Answer. The affinity of P_{1t} and P_{2t} is

$$\rho(P_{1t}, P_{2t}) = e^{-\frac{1}{2}(\lambda_1 t + \lambda_2 t)} \sum_{k \geq 0} \frac{(\lambda_1 \lambda_2)^{\frac{k}{2}} t^k}{k!}$$
$$= \exp\Big\{-\frac{1}{2}(\lambda_1 t + \lambda_2 t) + (\lambda_1 \lambda_2 t^2)^{\frac{1}{2}}\Big\}$$
$$= \exp\Big[-\frac{1}{2}\{(\lambda_1 t)^{\frac{1}{2}} - (\lambda_2 t)^{\frac{1}{2}}\}^2\Big]$$

and their Hellinger distance is deduced. The information $K(P_1, P_2)$ is defined as

$$\begin{aligned} K(P_{1t}, P_{2t}) &= e^{-\lambda_1 t} \sum_{k\geq 0} \frac{\lambda_1^k t^k}{k!} \log\Big\{ e^{-(\lambda_1 t - \lambda_2 t)} \frac{\lambda_1^k}{\lambda_2^k} \Big\} \\ &= (\lambda_2 - \lambda_1)t + e^{-\lambda_1 t} \log \frac{\lambda_1}{\lambda_2} \sum_{k\geq 1} \frac{\lambda_1^k t^k}{(k-1)!} \\ &= (\lambda_2 - \lambda_1)t + \lambda_1 t \log \frac{\lambda_1}{\lambda_2}. \end{aligned}$$

Problem 9.2.6. Let $(N_t)_{t\geq 0}$ be a Poisson process with intensity λ and with jump $\Delta N_t = N_t - N_t^-$ at t, calculate $E(e^{-uN_t} - e^{-uN_s} \mid \mathcal{F}_s)$ for $s < t$.

Answer. Let $\Delta N_t = N_t - N_t^-$ denote the jump of N at t, for $s < t$

$$\begin{aligned} e^{-uN_t} - e^{-uN_s} &= \sum_{s<x\leq t} (e^{-uN_x} - e^{-uN_x^-}) \\ &= \sum_{s<x\leq t} e^{-uN_x^-} (e^{-u} - 1)\Delta N_x \\ &= (e^{-u} - 1) \int_s^t e^{-uN_x^-} \, dN_x, \end{aligned}$$

and for $s \leq x \leq t$, $E(N_x \mid \mathcal{F}_s) = N_s + \lambda(x - s)$, therefore

$$E(e^{-uN_t} - e^{-uN_s} \mid \mathcal{F}_s) = (e^{-u} - 1)\lambda \int_s^t E(e^{-uN_x^-} \mid \mathcal{F}_s) \, dx.$$

9.3 Markov process

On a filtered probability space $(\Omega, \mathcal{A}, P, \mathbb{F})$, a process $(X_t)_{t\geq 0}$ is a Markov process with values in a measurable space $(E, \mathcal{E})$ if for all $A, \ldots, A_k$ of $\mathcal{E}$ and for all $0 \leq t_1 < t_2 < \cdots < t_k$

$$P(X_{t_1} \in A_1, \ldots, X_{t_k} \in A_k \mid \mathcal{F}_{t_{k-1}}) = P(X_{t_k} \in A_k \mid \mathcal{F}_{t_{k-1}}).$$

The density of probability $P_{xy}(s, t + s) = P(X_{t+s} = y \mid X_s = x)$ of X_{t+s} conditionally on X_s is homogeneous if it does not depend on s, for all positive s and t. A Markov jump process with finite and irreducible countable state space E is defined by a Markov chain of state variables $J = (J_k)_{k\geq 0}$, with initial state J_0, and by the sequence of transition times $T = (T_k)_{k\geq 0}$, where T_0 is the initial time of the process at $t_0 = 0$ and T_k is the arrival time in state J_k. The sojourn duration in state J_{k-1} is $\tau_k = T_{k+1} - T_k$, for

$k \geq 1$. The Markov chain of state variables is defined by the probabilities of the initial state and the probabilities of direct transitions

$$\rho_j = P(J_0 = j),$$
$$p_{ij} = P(J_k = j \mid J_{k-1} = i),$$

with $\sum_{j \neq i} \pi_{ij} = 1$. The sojourn durations are independent variables with the same distribution conditionally on the initial state j of the transition, with exponential conditional distribution functions

$$F_j(x) = P(\tau_k \leq x \mid J_{k-1} = j) = 1 - e^{\mu_j x}, \ x \geq 0.$$

The process $X_t = \sum_{k \geq 1} J_k 1_{\{T_k \leq t < T_{k+1}\}}$ is a Markov jump process with a homogeneous transition probability if $p_{ij} = P(J_{a+k} = j | J_{a+k-1} = i)$ for every integer a and

$$P_{ij}(x) = P(\tau_k \leq x, J_k = j \mid T_{k-1} \in]s, s + ds], J_{k-1} = i)$$

does not depend on s. Let $\lambda_{ij} = \mu_j p_{ij}$, for every $i \neq j$, $P_{ij}(h) = \lambda_{ij} h + o(h)$ and $\mu_j = \sum_{j \neq i} \lambda_{ij}$, the infinitesimal generator of the homogeneous process is the matrix

$$\Lambda = \begin{pmatrix} -\mu_0 & \lambda_{01} & \lambda_{02} & \cdots \\ \lambda_{10} & -\mu_1 & \lambda_{12} & \cdots \\ \lambda_{20} & \lambda_{21} & -\mu_2 & \cdots \\ \vdots & \vdots & \vdots & \vdots \end{pmatrix}.$$

Kolmogorov's equation for the vector $p(t) = \{P(X_t = i)\}_{i \in E}$ is $p'(t) = p(t)\Lambda$.

Exercise 9.3.1. For a homogeneous Markov process on a countable space E, prove that $P_{xy}(s+t) = \sum_{z \in E} P_{xz}(s)\, P_{z,y}(t)$.

Answer. The Markov property implies

$$\begin{aligned} P_{xy}(s+t) &= P(X_{s+t} = y \mid X_0 = x) \\ &= \sum_{z \in E} P(X_{s+t} = y, X_s = z \mid X_0 = x) \\ &= \sum_{z \in E} P(X_{s+t} = y \mid X_s = z, X_0 = x) P(X_s = z \mid X_0 = x) \\ &= \sum_{z \in E} P(X_{s+t} = y \mid X_s = z) P(X_s = z \mid X_0 = x) \\ &= \sum_{z \in E} P(X_t = y \mid X_0 = z) P(X_s = z \mid X_0 = x). \end{aligned}$$

Exercise 9.3.2. For a homogeneous Markov process on a countable space E, let $p_i(t) = P(X_t = i)$ and let $P_{ij}(t) = p_{ij}\{1 - e^{-\lambda_i t}\}$ with $\sum_{j \neq i \in E} p_{ij} = 1$, write the derivative of $p_i(t)$.

Answer. For $s > 0$ and $t > 0$, i in E, we have

$$P(X_{s+t} = i) = \sum_{k \in E} P(X_{s+t} = i \mid X_t = k) p_k(t)$$
$$= \sum_{k \in E} P(X_s = i \mid X_0 = k) p_k(t),$$

it follows that

$$p_i(s+t) = \sum_{k \in E} P_{ik}(s) p_k(t),$$
$$\frac{p_i(s+t) - p_i(t)}{s} = \sum_{k \neq i \in E} \frac{P_{ik}(s)}{s} p_k(t) + \frac{P_{ii}(s) - 1}{s} p_i(t).$$

As s tends to zero $P_{ik}(s)$ is equivalent to $s\lambda_i p_{ik} P_{ik}(0) = s\lambda_i p_{ik}$ and $1 - P_{ii}(s)$ is equivalent to $s\lambda_i P_{ii}(0) = s\lambda_i$, then

$$\lim_{s \to 0} \frac{p_i(s+t) - p_i(t)}{s} = \sum_{k \neq i \in E} \lambda_i p_{ik} p_k(t) - \lambda_i p_i(t).$$

Exercise 9.3.3. Let T_1 and T_2 be independent exponential variables and let $X_t = 1$ if $T_1 < t$ and $T_2 < t$, $X_t = 2$ if $T_1 \geq t$ and $T_2 < t$, and $X_t = 3$ if $T_1 \geq t$ and $T_2 \geq t$. Calculate the probabilities $p_{kt} = P(X_t = k)$, the infinitesimal generator of X_t and solve Kolmogorov's equations for X_t.

Answer. Let $T_1 \sim \mathcal{E}_\lambda$ and $T_2 \sim \mathcal{E}_\mu$, the probabilities are

$$P(T_1 < t)P(T_2 < t) = (1 - e^{-\lambda t})(1 - e^{-\mu t}),$$
$$P(T_1 \geq t)P(T_2 < t) = e^{-\lambda t}(1 - e^{-\mu t}),$$
$$P(T_1 \geq t)P(T_2 \geq t) = e^{-(\lambda+\mu)t}),$$

the generator of X_t is

$$\Lambda = \begin{pmatrix} -(\lambda+\mu) & \lambda & \mu \\ 0 & -\mu & \mu \\ 0 & 0 & 0 \end{pmatrix}.$$

The process X_t is a Markov process and the equation $p_t' = p_t \Lambda$ is equivalent to $p_{1t} = e^{-(\lambda+\mu)t})$ and $p_{2t}' = -\mu p_{2t} + \lambda p_{1t}$, its solution is

$p_{2t} = e^{-\mu t} - e^{-(\lambda+\mu)t}$) then $p_{1t} + p_{2t} = e^{-\mu t}$ is the probability of $\{T_1 \geq t, T_2 \geq t\}$.

Exercise 9.3.4. Let X be a birth and death process with birth rate λ and death rate μ, write the generator of X and solve Kolmogorov's equations with $\mu = 0$.

Answer. The generator of X_t is

$$\Lambda = \begin{pmatrix} -\lambda & \lambda & 0 & \cdots & \\ \mu & -(\lambda+\mu) & \lambda & 0 & \cdots \\ 0 & \mu & -(\lambda+\mu) & \lambda & \cdots \\ 0 & 0 & \mu & \cdots & 0 \\ \vdots & \vdots & \vdots & \vdots & \vdots \end{pmatrix}$$

and Kolmogorov's equations are

$$p_0' = -\lambda p_0 + \mu p_1,$$
$$p_i' = \lambda p_{i-1} - (\lambda+\mu)p_i + \mu p_{i+1},\ i \geq 1,$$

its solutions have the form $p_{0t} = c_t e^{-\lambda t}$ and $p_{it} = k_{it} e^{-(\lambda+\mu)t}$ where

$$c_t' = e^{\lambda t}\mu p_{1t},$$
$$k_{it}' = e^{(\lambda+\mu)t}(\lambda p_{i-1} + \mu p_{i+1}).$$

If $\mu = 0$, $p_{it} = e^{-\lambda t}$ for $i \geq 0$ and $k_{it}' = e^{\lambda t}\lambda p_{i-1}$ hence $k_{1t} = \lambda t$, $k_{2t} = \frac{1}{2}\lambda^2 t^2$, and by induction p_{it} is the probability of the value i for a Poisson process with intensity λ.

Exercise 9.3.5. Let T_1 and T_2 be independent and identically distributed exponential variables, let $\pi = \frac{1}{2}$ be the probability of an independent failure for T_1 or T_2, and let T_3 be an exponential duration variable to repair after a failure. Let $X_t = 1$ if $T_1, T_2 < t$, $X_t = 2$ or 3 after 1 or 2 failures before t and it returns to the first state after being repared. Write the generator of the process X_t and solve Kolmogorov's equations for X_t under the condition $\lambda = \mu$.

Answer. Let T_1 and $T_2 \sim \mathcal{E}_\lambda$ and let $T_3 \sim \mathcal{E}_\mu$, the process X_t has the probabilities

$$P(X_t = 1) = P(T_1 < t)P(T_2 < t) = (1 - e^{-\lambda t})^2,$$
$$P(X_t = 2) = P(T_1 + T_3 \leq t)P(T_2 \leq t) + P(T_2 + T_3 \leq t)P(T_3 \leq t),$$
$$P(X_t = 3) = P(T_1 + T_3 \leq t, T_2 + T_3 \leq t).$$

The generator of the Markov process X_t is

$$\Lambda = \begin{pmatrix} -2\lambda & \lambda & \lambda \\ \mu & -(\lambda+\mu) & \lambda \\ \mu & \mu & -2\mu \end{pmatrix}$$

such that $|\Lambda| = 0$. If $\mu = \lambda$, $\Lambda = \lambda A$ with the matrix

$$A = \begin{pmatrix} -2 & 1 & 1 \\ 1 & -2 & 1 \\ 1 & 1 & -2 \end{pmatrix},$$

it has the eigenvalues 0 with eigenvector $v_1 = (1,0,0)^T$ and -3 with eigenvectors $v_2 = (1,0,-1)^T$ and $v_3 = (1,-1,0)^T$. Let

$$P = \begin{pmatrix} 1 & 1 & 1 \\ 0 & 0 & -1 \\ 0 & -1 & 0 \end{pmatrix},$$

$|P| = 1$ and P has an inverse P^{-1}. Let $A = P^{-1}\Lambda P$ be the diagonal matrix with values $(0,3,-3)^T$ on the diagonal, the vector $\pi = pP$ satisfies the equation $\pi' = \pi A$ therefore $\pi_{1t} = c$, $\pi_{2t} = e^{3\lambda t}$ and $\pi_{3t} = e^{-3\lambda t}$, and a solution of Kolmogorov's equation $p' = p\Lambda$ is $p = \pi P^{-1}$.

9.4 Survival process

A strictly positive real variable X with density f and distribution function F has a hazard rate λ defined as

$$\lambda(t)\,dt = P(X \in]t, t+dt] \mid X \geq t) = \frac{f(t)\,dt}{\bar{F}(t)}.$$

The cumulative hazard function $\Lambda(t) = \int_0^t \lambda(u)\,du$ uniquely defines the survival function of X

$$\bar{F}(t) = 1 - F(t) = \exp\{-\Lambda(t)\}.$$

A point process $N(t)$ is defined by an indicator with value 0 on $[0, X[$ and 1 after X, $N(t) = 1\{X \leq t\}$ is a right-continuous function with left-limits. Its distribution is described by the hazard function λ

$$P\{dN(t) = 1 \mid N(t-) = 0\} = P(X \in [t, t+dt] \mid X \geq t) = \lambda(t)\,dt$$
$$P\{dN(t) = 1 \mid N(t-) = 1\} = 0$$

then

$$P\{dN(t) = 1 \mid \mathcal{F}_{t-}\} = 1_{\{X \geq t\}}\lambda(t)\,dt.$$

A variable X with an arbitrary distribution function $F(t) = P(X \leq t)$ and survival function $\bar{F}(t) = P(X > t)$ has a cumulative hazard function

$$\Lambda(t) = \int_0^t \frac{dF(s)}{1 - F(s^-)}.$$

Writting the distribution function as the sum of its jumps and of its continuous part $F(t) = F^c(t) + \sum_{s \leq t} \Delta F(s)$, the cumulative hazard function is the sum

$$\Lambda(t) = \int_0^t \frac{dF^c(s)}{1 - F(s)} + \sum_{s \leq t} \frac{\Delta F(s)}{1 - F(s^-)} = \Lambda^c(t) + \sum_{s \leq t} \Delta\Lambda(s).$$

Reversely there exists a unique solution of the equation

$$\bar{F}(t) = P(X \geq t) = \exp\{-\Lambda^c(t)\} \prod_{s \leq t} \{1 - \Delta\Lambda(s)\}. \tag{9.1}$$

Considering the filtration $\mathbb{F} = (\mathcal{F}_t)_{t \geq 0}$ generated by the events of the process X previous t, such that $\mathcal{F}_t = \sigma(\{X \leq s\}, \{X \geq s\}, s \leq t)$, the predictable compensator

$$\widetilde{N}(t) = \int_0^t 1_{\{X \geq s\}} \, d\Lambda(s)$$

of $N(t)$ is the unique increasing, left-continuous process with right-limits such that $M = N - \widetilde{N}$ is a local martingale with respect to $\mathbb{F}$. A counting process

$$N_n(t) = \sum_{i=1}^n 1_{\{X_i \leq t\}} \tag{9.2}$$

is defined for a sample of n independent variables X_i having the same distribution function as X. Its predictable compensator is

$$\widetilde{N}_n(t) = \int_0^t \sum_{i=1}^n 1_{\{X_i \geq s\}} \, d\Lambda(s).$$

Exercise 9.4.1. Let $N_t = \sum_{i \geq 1} 1_{\{T_i \leq t\}}$ be a counting process with independent increments and such that $N_t - N_s$ has a Poisson distribution with parameter $\Lambda_t - \Lambda_s$ for $0 > s > t$. Calculate $P(T_1 > t)$, $P(T_2 > t \geq T_1 > s)$ and $P(T_1 > s, T_2 > t)$ for $0 > s > t$, find the density of (T_1, T_2) and the density of $(T_1, T_2 - T_1)$.

Answer. The probabilities are

$$\begin{aligned}
P(T_1 > t) &= P(N_t = 0) = e^{-\Lambda_t},\\
P(T_2 > t \geq T_1 > s) &= P(N_s = 0, N_t - N_s = 1)\\
&= P(N_s = 0)P(N_t - N_s = 1)\\
&= e^{-\Lambda_s} e^{-(\Lambda_t - \Lambda_s)}(\Lambda_t - \Lambda_s) = e^{-\Lambda_t}(\Lambda_t - \Lambda_s),\\
P(T_1 > s, T_2 > t) &= P(N_s = 0, N_t - N_s \leq 1)\\
&= e^{-\Lambda_s} e^{-(\Lambda_t - \Lambda_s)}(1 + \Lambda_t - \Lambda_s)\\
&= e^{-\Lambda_t}(1 + \Lambda_t - \Lambda_s),
\end{aligned}$$

the variable T_1 has the density $f_1(t) = \lambda_t e^{-\Lambda_t}$ and the density of (T_1, T_2) is

$$f(s,t) = \lambda_s \lambda_t e^{-\Lambda_t},$$

the density of $(T_1, T_2 - T_1)$ is deduced

$$g(s,x) = \lambda_s \lambda_{s+x} e^{-\Lambda_{s+x}}.$$

Exercise 9.4.2. Let N_n be a point process (9.2) with intensity function λ, calculate the probability $P(N_n(t) = m \mid N_n(s) = k)$ for integers $k \leq n$ and for positive real $s < t$.

Answer. The conditional probability is deduced from the distribution function $F = 1 - \bar{F}$ defined by (9.1)

$$P(N_n(t) = m \mid N_n(s) = k) = \frac{P(N_n(t) - N_n(s) = m - k)}{P(N_n(s) = k)}$$

with

$$\begin{aligned}
P(N_n(s) = k) &= P\Big\{\sum_{i=1}^n 1_{\{X_i \leq s\}} = k\Big\}\\
&= \binom{n}{k} F^k(s) \bar{F}^{n-k}(s),\\
P(N_n(t) - N_n(s) = m - k) &= P\Big\{\sum_{i=1}^n 1_{\{s < X_i \leq t\}} = m - k\Big\}\\
&= \binom{n}{m-k} \{F(t) - F(s)\}^{m-k}\\
&\quad .[1 - \{F(t) - F(s)\}]^{n-m+k}.
\end{aligned}$$

Exercise 9.4.3. Let $Y_n(t) = \sum_{i=1}^n 1_{\{X_i \geq t\}}$, and let $W_n = n^{-\frac{1}{2}}(N_n - nF)$ and $U_n(t) = n^{-\frac{1}{2}}\{Y_n(t) - \bar{F}(t^-)\}$, prove the weak convergence of the process (W_n, U_n) to a centered Gaussian process.

Answer. The process W_n is an empirical process and it converges weakly to a transformed Brownian bridge $W = B^0 \circ F$, where B^0 is the standard Brownian bridge. As $Y_n(t) = n - N_n(t^-)$, the weak convergence of (W_n, U_n) to a centered Gaussian process (W, U) follows. The covariance of $N_n(t)$ and $N_n(s)$ is

$$\mathrm{Cov}\{N_n(t), N_n(s)\} = \mathrm{E}\{N_n(t), N_n(s)\} - n^2 F(s)F(t)$$

where

$$\begin{aligned}
\mathrm{E}\{N_n(t), N_n(s)\} &= \mathrm{E}\Big\{\sum_{i=1}^n 1_{\{X_i \leq s \wedge t\}} + \sum_{i \neq j = 1}^n 1_{\{X_i \leq s\}} 1_{\{X_j \leq t\}}\Big\} \\
&= nF(s \wedge t) + n(n-1)F(s)F(t), \\
\mathrm{Cov}\{N_n(t), N_n(s)\} &= n\{F(s \wedge t) - F(s)F(t)\}
\end{aligned}$$

and the variance of $W_n(t)$ is $F(t)\{1 - F(t)\}$. The covariance of $U_n(t)$ and $U_n(s)$ is

$$\mathrm{Cov}\{U_n(s), U_n(t)\} = \bar{F}_-(s \vee t) - \bar{F}(s^-)\bar{F}(t^-).$$

Furthermore, the covariance of $W_n(s)$ and $U_n(t)$, $t < s$, is

$$\begin{aligned}
\mathrm{Cov}\{W_n(s), U_n(t)\} &= n^{-1}\mathrm{E}\Big\{\sum_{i=1}^n 1_{\{t \leq X_i \leq s\}} + \sum_{i \neq j = 1}^n 1_{\{X_i \leq s\}} 1_{\{X_j \geq t\}}\Big\} \\
&\quad - nF(s)\bar{F}(t^-) \\
&= F(s) - F(t) - F(s)\bar{F}(t^-) \\
&= -\bar{F}(s)F(t^-).
\end{aligned}$$

Exercise 9.4.4. Let $\Lambda_n(t) = \int_0^t 1_{\{Y_n(s) > 0\}} Y_n^{-1}(s)\, dN_n(s)$, prove the weak convergence of the process $A_n(t) = n^{\frac{1}{2}}\{\Lambda_n(t) - \int_0^t 1_{\{Y_n(s) > 0\}}\, d\Lambda(s)\}$ to a centered Gaussian process, for every t such that $\bar{F}(t) > 0$.

Answer. For every t such that $\bar{F}(t) > 0$, the process $A_n(t)$ is written

$$\begin{aligned}
A_n(t) &= n^{\frac{1}{2}} \int_0^t 1_{\{Y_n(s) > 0\}} \Big\{\frac{dN_n(s)}{Y_n(s)} - \frac{dF(s)}{\bar{F}(s^-)}\Big\} \\
&= \int_0^t 1_{\{Y_n(s) > 0\}} \frac{dW_n(s)}{Y_n(s)} \\
&\quad + n^{\frac{1}{2}} \int_0^t 1_{\{Y_n(s) > 0\}} \Big\{\frac{1}{Y_n(s)} - \frac{1}{\bar{F}(s^-)}\Big\}\, dF(s),
\end{aligned}$$

using the previous notations and by the weak convergence of (W_n, U_n), the first term converges weakly to $A_1(t) = \int_0^t \bar{F}^{-1}(s^-)\, dW(s)$, the second term converges weakly to $A_2(t) = -\int_0^t \bar{F}^{-2}(s^-) U(s)\, dF(s)$ and the process $A_n(t)$ converges weakly to their sum.

Exercise 9.4.5. Let F be a continuous distribution function and let $\bar{F}_n(t) = \prod_{s \le t}\{1 - \Delta\Lambda_n(s)\}$, with Λ_n defined in Exercise 9.4.4, prove the convergence in probability of the process $\bar{F}_n$ on every interval $[0, \tau]$ such that $\bar{F}(\tau) > 0$.

Answer. Let $\tau_n = \sup\{t : \bar{F}_n(t) > 0\}$, for every $T_i < \tau_n$ we have

$$\bar{F}_n(T_i) = \bar{F}_n(T_{i-1})\Big\{1 - \Delta\Lambda_n(T_i)\Big\},$$
$$\Delta\Lambda_n(T_i) = 1 - \frac{\bar{F}_n(T_i)}{\bar{F}_n(T_{i-1})},$$

with $\bar{F}_n(T_0) = \bar{F}_n(0) = 1$ and $\bar{F}_n(t) = \sum_{i=1}^n 1_{\{T_i \le t\}}\{\bar{F}_n(T_i) - \bar{F}_n(T_{i-1})\} + 1$, hence

$$F_n(t) = -\sum_{i=1}^n 1_{\{T_i \le t\}}\{\bar{F}_n(T_i) - \bar{F}_n(T_{i-1})\},$$
$$\bar{F}_n(t) = \int_0^t \bar{F}_n(s^-)\, d\Lambda_n(s).$$

For a continuous distribution function F, $\bar{F}(t) = \exp\{-\Lambda(t)\}$ and $dF = \bar{F}\, d\Lambda$.

As $\sup_{t \in [0,a]} |\Delta\Lambda_n(T_i)| = O(n^{-1})$, it converges a.s. to zero and

$$\begin{aligned}
\prod_{T_i \le t}\{1 - \Delta\Lambda_n(T_i)\} &= \exp\Big\{\sum_{T_i \le t} \log\{1 - \Delta\Lambda_n(T_i)\Big\} \\
&= \exp\Big\{-\sum_{T_i \le t} \Delta\Lambda_n(T_i)\Big\} + o(1) \\
&= \exp\{-\Lambda_n(t)\} + o(1)
\end{aligned}$$

with an a.s. $o(1)$. The uniform convergence in probability of the process Λ_n on every interval $[0, a]$ strictly included in $[0, \tau_n]$ implies

$$\lim_{n\to\infty} \sup_{t \in [0,a]} \Big| \prod_{T_i \le t}\{1 - \Delta\Lambda_n(T_i)\} - \exp\{-\Lambda(t)\} \Big| = 0,$$

this proves the uniform convergence in probability of $\bar{F}_n$ to $\bar{F}$ on $[0, a]$.

9.5 Distribution of a right-censored variable

On a probability space $(\Omega, \mathcal{A}, P)$, let X be a failure time and let C be an independent right-censoring variable such that $P(X < C)$ is strictly positive, the censored failure time and its indicator variable δ are defined as

$$T = X \wedge C, \quad \delta = 1\{X \le C\}. \tag{9.3}$$

They determine the counting processes

$$N(t) = \delta 1_{\{T \le t\}}, \quad Y(t) = 1_{\{T \ge t\}}, \tag{9.4}$$

and let $\mathbb{F} = (\mathcal{F}_t)_{t \ge 0}$ be the filtration generated on $(\Omega, \mathcal{A}, P)$ by N, $\mathcal{F}_t = \sigma(\{X \le s\}, 0 \le s \le t)$.

Let F be the distribution function of the variable X and let $\bar{F}(x) = 1 - F(x)$. The censoring variable C has a distribution function G and a survival function $\bar{G}$. The end-points of the support of the variables X, and respectively C, are $\tau_F = \sup\{t : \bar{F}(t) > 0\}$, and respectively $\tau_G = \sup\{t : \bar{G}(t) > 0\}$.

Exercise 9.5.1. Let $\bar{H} = \bar{G}\,\bar{F}$ be the survival function of the variable T and let $\tau = \sup\{t : \bar{H}(t) > 0\} = \tau_F \wedge \tau_G$, write the probabilities $P(\delta = 1)$ and $P(\delta = 0)$, and prove that the cumulated hazard function for T and X and identical on $[0, \tau]$.

Answer. The probabilities are

$$P(\delta = 1) = P(T \le C) = \int_0^\tau \bar{G}(t^-)\, dF(t),$$

$$P(\delta = 0) = P(T > C) = \int_0^\tau \bar{F}(t)\, dG(t).$$

The cumulated hazard function of X is defined on $[0, \tau]$ as

$$\Lambda(t) = \int_0^t \frac{dF}{\bar{F}^-} = \int_0^t \frac{\bar{G}^- \, dF}{\bar{G}^- \, \bar{F}^-} = \int_0^t \frac{\bar{G}^- \, dF}{\bar{H}^-}$$

reversely Λ defines $\bar{F}$ by (9.1).

Exercise 9.5.2. Let F and G have densities f and respectively g, write the probabilities $H^1(t)$ of $\{T \le t, \delta = 1\}$ and $H^0(t)$ of $\{T \le t, \delta = 0\}$, prove that

$$\bar{F}(t) = \bar{H}(t) + \bar{F}(t) \int_0^t \frac{dH^0(s)}{\bar{F}(s)}.$$

Answer. The probabilities are

$$\begin{aligned}H^1(t) &= P(T \le t, \delta = 1) = P(T \le t, T \le C)\\ &= \int_0^t \bar{G}(s^-)\,dF(s),\\ H^0(t) &= P(T \le t, \delta = 0) = P(T \le t, T > C)\\ &= \int_0^t \bar{F}(s)\,dG(s).\end{aligned}$$

The equality is equivalent to

$$\bar{H}(t) = \bar{F}(t)\Big\{1 - \int_0^t \frac{dH^0(s)}{\bar{F}(s)}\Big\}$$

where $H^0(t) = \int_0^t \bar{F}(s)\,dG(s)$ and $\int_0^t \bar{F}^{-1}(s)\,dH^0(s) = G(t)$, it is therefore equivalent to $\bar{H}(t) = \bar{F}(t)\bar{G}(t)$.

Exercise 9.5.3. Let F and G have densities, write the densities of the variables T and (T, δ).

Answer. The density of T is $h(t) = f(t)\bar{G}(t) + g(t)\bar{F}(t)$ and the density of (T, δ) is

$$\begin{aligned}l(T,\delta) &= \{f(T)\}^\delta \{\bar{F}(T)\}^{1-\delta} = \{\lambda(T)\}^\delta \bar{F}(T)\\ &= \{\lambda(T)\}^\delta \exp\{-\Lambda(T)\}\end{aligned}$$

where $P(\delta = 1) = \int_0^\infty \bar{G}(t)\,dF(t)$. Using (9.4), the logarithm of the density is also written as

$$\log l(T,\delta) = \int_0^\tau \log \lambda(t)\,dN(t) - \int_0^\tau \lambda(t)Y(t)\,dt.$$

Exercise 9.5.4. Write the expectations of $N(t)$, $Y(t)$ defined by (9.4), and of $\log l(T, \delta)$, under a probability P_0 such that X has an intensity λ_0 and a distribution function F_0, C has a distribution function G_0. Define the Kullback–Leibler information of (T, δ) for probabilities P with intensity λ and P_0 with intensity λ_0.

Answer. The expectations of $N(t)$ and $Y(t)$ under P_0 are

$$E_0 N(t) = \int_0^t \bar{G}_0(s)\,dF_0(s), \quad E_0 Y(t) = \bar{H}_0(t),$$

the expectation of $l(T,\delta)$ under P_0 is deduced

$$\begin{aligned} E_0 \log l(T,\delta) &= \int_0^\tau \{\log\lambda(s)\}\bar{G}(s)\,dF_0(s) - \int_0^\tau \lambda(s)\bar{H}_0(s)\,ds \\ &= \int_0^\tau \left\{\log\lambda(s) - \frac{\lambda(s)}{\lambda_0(s)}\right\} f_0(s)\bar{G}_0(s)\,ds. \end{aligned}$$

The Kullback–Leibler information of (T,δ) is $K(P_\lambda, P_{\lambda_0}) = E_0\{\log l_0(T,\delta) - \log l(T,\delta)\}$, therefore

$$\begin{aligned} K(P_\lambda, P_{\lambda_0}) &= E_0\{\log l_0(T,\delta) - \log l(T,\delta)\} \\ &= \int_0^\tau \left\{\log\frac{\lambda_0(t)}{\lambda(t)} - \frac{\lambda_0(t)}{\lambda(t)} + 1\right\} f_0(t)\bar{G}_0(t)\,dt. \end{aligned}$$

Exercise 9.5.5. Find the Hellinger distance between the densities of (T,δ) under probabilities P_λ and P_{λ_0} under the same censoring distribution G. Give an approximation for λ in an ε-neighbourhood of λ_0.

Answer. The density of (T,δ) under a probability P is $f\bar{G}1_{\{\delta=1\}} + g\bar{F}1_{\{\delta=0\}}$, the Hellinger distance of probabilities P related to densities (f,g) and P_0 related to (f_0,g_0) is

$$\begin{aligned} h^2(P,P_0) &= \frac{1}{2}\int_0^\tau \left(f^{\frac{1}{2}} - f_0^{\frac{1}{2}}\right)^2 \bar{G} + \frac{1}{2}\int_0^\tau \left(\bar{F}^{\frac{1}{2}} - \bar{F}_0^{\frac{1}{2}}\right)^2 g \\ &= \frac{1}{2}\int_0^\tau \left\{f + f_0 - 2(ff_0)^{\frac{1}{2}}\right\}\bar{G} \\ &\quad + \frac{1}{2}\int_0^\tau \left(\bar{F} + \bar{F}_0 - 2\bar{F}^{\frac{1}{2}}\bar{F}_0^{\frac{1}{2}}\right) g \end{aligned}$$

where $\int_0^\tau (f\bar{G} + g\bar{F}) = \bar{F}(0)\bar{G}(0) = 1$ for all f and g. Let $\bar{F} = e^{-\Lambda}$ and $\bar{F}_0 = e^{-\Lambda_0}$, then

$$h^2(P,P_0) = 1 - \int_0^\tau \lambda^{\frac{1}{2}}\lambda_0^{\frac{1}{2}} e^{-\frac{\Lambda+\Lambda_0}{2}}\bar{G} - \int_0^\tau e^{-\frac{\Lambda+\Lambda_0}{2}} g.$$

By the change of variables $u = -\bar{G}$ and $v = e^{-\frac{\Lambda+\Lambda_0}{2}}$, we obtain

$$\int_0^\tau e^{-\frac{\Lambda+\Lambda_0}{2}} g = 1 - \int_0^\tau \frac{\lambda+\lambda_0}{2} e^{-\frac{\Lambda+\Lambda_0}{2}}\bar{G},$$

it follows that

$$h^2(P,P_0) = \frac{1}{2}\int_0^\tau \left\{\lambda^{\frac{1}{2}}(t) - \lambda_0^{\frac{1}{2}}(t)\right\}^2 e^{-\frac{\Lambda(t)+\Lambda_0(t)}{2}}\bar{G}(t)\,dt.$$

Let $\lambda_\varepsilon(t) = \lambda_0(t)\{1 + \varepsilon u(t)\}$ for every t, then $\Lambda_\varepsilon(t) = \Lambda_0(t) + \varepsilon \int_0^t u\, d\Lambda_0$ and $\lambda_\varepsilon^{\frac{1}{2}}(t) = \lambda_0^{\frac{1}{2}}(t)\{1 + \frac{1}{2}\varepsilon u(t) + o(\varepsilon)\}$, so

$$h^2(P_\varepsilon, P_0) = \frac{\varepsilon^2}{8} \int_0^\tau u^2(t) e^{-\varepsilon \int_0^t u\, d\Lambda_0} \lambda_0(t) \bar{G}(t)\, dt + o(\varepsilon^2)$$
$$= \frac{\varepsilon^2}{8} \int_0^\tau u^2(t) \lambda_0(t) \bar{G}(t)\, dt + o(\varepsilon^2).$$

Exercise 9.5.6. Let $N_n(t) = \sum_{1 \le i \le n} \delta_i 1\{T_i \le t\}$, write the predictable compensator of N_n according to $Y_n(t) = \sum_{1 \le i \le n} 1\{T_i \ge t\}$ and prove that the process $A_n(t)$ is a martingale.

Answer. Under P_0, the process N_n is written as

$$N_n(t) = \sum_{1 \le i \le n} 1_{\{C_i \ge X_i\}} 1_{\{T_i \le t\}}$$
$$= \int_0^t 1_{\{C_i \ge s\}}\, d1_{\{T_i \le s\}}$$

where $\{C_i \ge s\}$ is $\mathcal{F}_{s^-}$ measurable and the process Y_n is predictable, the predictable compensator of $N_n(t)$ is therefore

$$\widetilde{N}_n(t) = \int_0^t Y_n(s) \lambda(s)\, ds$$

and their difference is a local martingale $M_n(t)$. By integration of the predictable process Y_n with respect to the local martingale M_n, the difference

$$\Lambda_n(t) - \int_0^t 1_{\{Y_n(s) > 0\}}\, d\Lambda(s) = \int_0^t 1_{\{Y_n(s) > 0\}} \frac{dN_n(s) - Y_n(s) d\Lambda(s)}{Y_n(s)}$$

is a local martingale.

Exercise 9.5.7. For independent samples $(X_i)_{i=1,\dots,n}$ and $(C_i)_{i=1,\dots,n}$, let T_i and δ_i be defined by (9.3) and let $N_n(t) = \sum_{i=1}^n \delta_i 1_{\{T_i \le t\}}$ and $Y_n(t) = \sum_{i=1}^n 1_{\{T_i \ge t\}}$.

Let $W_n = n^{-\frac{1}{2}}(N_n - nH^1)$ and $U_n = n^{-\frac{1}{2}}(Y_n - n\bar{H}^-)$, prove the weak convergence of the process (W_n, U_n) to a centered Gaussian process. Deduce the weak convergence of the process $A_n(t) = n^{-\frac{1}{2}}\{\Lambda_n(t) - \int_0^t \bar{H}^{-1}(s^-)\, dH^1(s)\}$ on every interval $[0, a]$ such that $a < \tau$.

Answer. The expectation of $N_n(t)$ is $nH^1(t)$ and the expectation of the process $N_n^0(t) = \sum_{i=1}^n (1 - \delta_i) 1_{\{T_i \le t\}}$ is $nH^0(t)$, then

$$Y_n(t) = n - N_n(t^-) + N_n^0(t^-)$$

and its expectation is $n\bar{H}(t^-)$. The processes W_n and $W_n^0 = n^{-\frac{1}{2}}(N_n^0 - nH^0)$ are empirical processes and they converges weakly to $W = B \circ H^1$ and respectively $W^0 = B \circ H^0$, where B is the standard Brownian bridge. The weak convergence of (W_n, U_n) to a centered Gaussian process (W, U) follows. The covariance of $N_n(t)$ and $N_n(s)$ is

$$\mathrm{Cov}\{N_n(t), N_n(s)\} = n\{H^1(s \wedge t) - H^1(s)F(t)\},$$

the covariance of $U_n(t)$ and $U_n(s)$ is

$$\mathrm{Cov}\{U_n(s), U_n(t)\} = \bar{H}^0(s \vee t) - \bar{H}^0(s)\bar{H}^0(t),$$

the covariance of $W_n(s)$ and $U_n(t)$, $t < s$, is

$$\begin{aligned}\mathrm{Cov}\{W_n(s), U_n(t)\} &= n^{-1}P(T_i \leq s \wedge C_i, T_i \geq t) \\ &\quad + \sum_{i \neq j=1}^{n} P(X_i \leq s \wedge C_i)P(T_j \geq t) - nH^1(s)\bar{H}(t^-) \\ &= \int_{t^-}^{s} \bar{G}(u^-)\, dF(u).\end{aligned}$$

For every t such that $Y_n(t) > 0$, the process $A_n(t)$ is written

$$\begin{aligned}A_n(t) &= n^{\frac{1}{2}} \int_0^t 1_{\{Y_n(s)>0\}} \Big\{ \frac{dN_n(s)}{Y_n(s)} - \frac{\bar{G}(s^-)\, dF(s)}{\bar{H}(s^-)} \Big\} \\ &= \int_0^t 1_{\{Y_n(s)>0\}} \frac{dW_n(s)}{Y_n(s)} \\ &\quad + n^{\frac{1}{2}} \int_0^t 1_{\{Y_n(s)>0\}} \Big\{ \frac{1}{Y_n(s)} - \frac{1}{\bar{H}(s^-)} \Big\} \bar{G}(s^-)\, dF(s),\end{aligned}$$

by the weak convergence of (W_n, U_n), the first term converges weakly to $A_1(t) = \int_0^t \bar{H}^{-1}(s^-)\, dW(s)$, the second term converges weakly to

$$A_2(t) = -\int_0^t \bar{H}^{-2}(s^-)U(s)\, dH^1(s) = -\int_0^t \bar{H}^{-1}(s^-)U(s)\, d\Lambda(s)$$

and the process $A_n(t)$ converges weakly to their sum.

Exercise 9.5.8. Calculate the covariances of the processes $M_n = N_n - \int_0^t Y_n\, d\Lambda$ and A_n.

Answer. The process M_n belongs to $\mathcal{M}_0^{2,loc}$ and the covariance of $M_n(s)$ and $M_n(t)$ is

$$E\{M_n(s)M_n(t)\} = \int_0^{s \wedge t} E(Y_n)\,(1 - \Delta\Lambda)\, d\Lambda$$

and by the a.s. uniform convergence of Y_n to $\bar{H}_-$ on $[0, s \wedge t]$, it converges to $\int_0^{s\wedge t} \bar{H}_-(1 - \Delta\Lambda)\, d\Lambda$. For every t such that $Y_n(t) > 0$, we have

$$\begin{aligned} n^{\frac{1}{2}} A_n(t) &= \int_0^t Y_n^{-1}\, dN_n - \Lambda(t) \\ &= \int_0^t Y_n^{-1}\, dM_n \end{aligned}$$

and the covariance of $A_n(s)$ and $A_n(t)$ is

$$C_n(s,t) = E \int_0^{s\wedge t} Y_n^{-1}(1 - \Delta\Lambda)\, d\Lambda$$

and it converges to $C(s,t) = E \int_0^{s\wedge t} \bar{H}_-^{-1}(1 - \Delta\Lambda)\, d\Lambda$. The covariance of $M_n(s)$ and $A_n(t)$ is therefore $n^{-\frac{1}{2}} \int_0^{s\wedge t}(1 - \Delta\Lambda)\, d\Lambda$.

Exercise 9.5.9. Calculate approximations of the variance of $\log \bar{F}_n(t)$ and $\bar{F}_n(t)$ for $t < \tau$.

Answer. By definition of $\bar{F}_n$ we have

$$\bar{F}_n(t) = \int_0^t \bar{F}_n(s^-)\, d\Lambda_n(s)$$

where the variance of $\Lambda_n(t)$ converges to

$$V(t) = \int_0^t \frac{1 - \Delta\Lambda(s)}{\bar{F}(s^-)\bar{G}(s^-)}\, d\Lambda(s)$$

and the variance of $\bar{F}_n(t)$ converges to

$$\mathrm{Var}\bar{F}_n(t) = \int_0^t \bar{F}^2(s^-)\bar{G}^2(s^-)V(s).$$

By a first order expansion of the logarithm, the logarithm of $\bar{F}_n(t)$ is

$$\log \bar{F}_n(t) = \sum_{i=1}^n \log\Big\{1 - \frac{1}{\Delta\Lambda_n(T_i)}\Big\} = -\sum_{i=1}^n \frac{1}{\Delta\Lambda_n(T_i)} + o_p(1),$$

where $\Delta\Lambda_n(t) = O_p(n)$. By the a.s. convergence of Λ_n to Λ, uniformly on $[0,t]$ for every $t < \tau$, and arguing like for the sum of the empirical probabilities of binomial variables, the variable

$$Var_n(t) = \sum_{i=1}^n \frac{1}{\Delta\Lambda_n(T_i)}\Big\{1 - \frac{1}{\Delta\Lambda_n(T_i)}\Big\}$$

converges in probability to the variance of $\log \bar{F}_n(t)$. By the differentiability of the logarithm, the variance of $\bar{F}_n(t)$ has the approximation

$$\mathrm{Var}\bar{F}_n(t) = \frac{\mathrm{Var}\log \bar{F}_n(t)}{\bar{F}^2(t)} + o(1)$$

and the variable $Var_n(t)\bar{F}_n^2(t)$ converges in probability to the variance of $\bar{F}_n(t)$.

9.6 Spatial jump processes

On a measurable subspace $(E, \mathcal{E})$ of $\mathbb{R}^d$, a Poisson process with stationary intensity λ is defined by the probability

$$P(k; A) = e^{-\lambda(A)} \frac{\lambda^k(A)}{k!}$$

for every measurable set A such that the measure $\lambda(A)$ of A is strictly positive.

Exercise 9.6.1. Calculate the covariance of $N(A)$ and $N(B)$ for measurables sets A and B.

Answer. By the independence of the increments of N, the expectation of $N(A)N(B)$ is

$$\begin{aligned}
E\{N(A)N(B)\} &= E\{N(A \setminus B)N(B)\} + E\{N(A \cap B)N(B)\} \\
&= E\{N(A \setminus B)\}E\{N(B)\} + E\{N^2(A \cap B)\} \\
&\quad +E\{N(A \cap B)N(B \setminus A)\} \\
&= \lambda(A \setminus B)\lambda(B) + \lambda^2(A \cap B) + \lambda(A \cap B) \\
&\quad +\lambda(A \cap B)\lambda(B \setminus A) \\
&= \lambda(A \setminus B)\lambda(B) + \lambda(A \cap B)\lambda(B) + \lambda(A \cap B) \\
&= \lambda(A)\lambda(B) + \lambda(A \cap B)
\end{aligned}$$

and the covariance of $N(A)$ and $N(B)$ is

$$\begin{aligned}
\mathrm{Cov}\{N(A), N(B)\} &= E\{N(A)N(B)\} - \lambda(A)\lambda(B) \\
&= \lambda(A \cap B).
\end{aligned}$$

Exercise 9.6.2. Let N be a spatial Poisson process with intensity λ, calculate the probability $P(N(A) = k \mid N(B) = n)$ for integers $k \leq n$ and measurables sets $A \subset B$.

Answer. By the independence of the increments of the Poisson process, we have $P(N(A) = k, N(B \setminus A) = n-k) = P(N(A) = k)P(N(B \setminus A) = n-k)$, therefore

$$\begin{aligned}
P(N(A) = k \mid N(B) = n) &= \frac{P(N(A) = k, N(B \setminus A) = n - k)}{P(N(B) = n)} \\
&= \frac{\{\lambda(A)\}^k \{\lambda(B \setminus A)\}^{n-k}}{\{\lambda(B)\}^n} \frac{n!}{k!(n-k)!},
\end{aligned}$$

thus $P(N(A) \mid N(B) = n)$ has the binomial distribution with parameters n and the probability $\lambda(A)\lambda^{-1}(B)$.

Exercise 9.6.3. Write the Laplace transform of a Poisson process with stationary intensity λ on a measurable subspace $(E, \mathcal{E})$ of $\mathbb{R}^d$.

Answer. Let $(A_n)_{n\geq 1}$ a partition of E, the probabilities $P_i(A_n)$ are Poisson variables with intensities $\lambda_i(A_n)$ for $i = 1, 2$. The Laplace transform of a Poisson process X with stationary intensity λ is

$$\begin{aligned} L(t) = E_P e^{-tX} &= \sum_{n\geq 0} e^{-\lambda(A_n)} \sum_{k\geq 1} e^{-tk} \frac{\lambda^k(A_n)}{k!}, \\ &= \sum_{n\geq 0} e^{-\lambda(A_n)} \sum_{k\geq 1} [\exp\{-t + \log \lambda(A_n)\}]^k \frac{1}{k!} \\ &= \sum_{n\geq 0} \exp\{-\lambda(A_n)(1 - e^{-t})\} \\ &= \int_E \exp\{-\lambda(1 - e^{-t})\}. \end{aligned}$$

Exercise 9.6.4. Find the affinity and the Kullback information of the probabilities of Poisson process with intensity measures λ_1 and λ_2 on a measurable subspace $(E, \mathcal{E})$ of $\mathbb{R}^d$.

Answer. The affinity of P_1 and P_2 is

$$\begin{aligned} \rho(P_1, P_2) = \int_E (P_1 P_2)^{\frac{1}{2}} &= \sum_{n\geq 0} \{P_1(A_n) P_2(A_n)\}^{\frac{1}{2}} \\ &= \sum_{n\geq 0} e^{-\frac{1}{2}\{\lambda_1(A_n)+\lambda_2(A_n)\}} \sum_{k\geq 0} \frac{\{\lambda_1(A_n)\lambda_2(A_n)\}^{\frac{k}{2}}}{k!} \\ &= \sum_{n\geq 0} \exp[-\frac{1}{2}\{\lambda_1^{\frac{1}{2}}(A_n) - \lambda_2^{\frac{1}{2}}(A_n)\}^2] \\ &= \int_E \exp\{-\frac{1}{2}(\lambda_1^{\frac{1}{2}} - \lambda_2^{\frac{1}{2}})^2\}. \end{aligned}$$

The information $K(P_1, P_2)$ is

$$K(P_1, P_2) = \sum_{n\geq 0} P_1(A_n) \log \frac{dP_1(A_n)}{dP_2(A_n)},$$

thus

$$K(P_1,P_2) = \sum_{n\geq 0} e^{-\lambda_1(A_n)} \sum_{k\geq 0} \frac{\lambda_1^k(A_n)}{k!} \log\Big[e^{-\{\lambda_1(A_n)-\lambda_2(A_n)\}} \frac{\lambda_1^k(A_n)}{\lambda_2^k(A_n)}\Big]$$

$$= \sum_{n\geq 0}\Big[\lambda_2(A_n) - \lambda_1(A_n) + \lambda_1(A_n)\log\frac{\lambda_1(A_n)}{\lambda_2(A_n)}\Big]$$

$$= \int_E \Big(\lambda_2 - \lambda_1 + \lambda_1 \log\frac{\lambda_1}{\lambda_2}\Big)$$

where $\int_E \lambda_1 = \int_E \lambda_2 = 1$, therefore $K(P_1,P_2) = K(\lambda_1,\lambda_2)$.

Exercise 9.6.5. Let Y be a spatial Poisson process with stationary intensity λ on a measurable space $(E,\mathcal{E},d)$ of $\mathbb{R}^d$, and let X be an exponential process with parameter μY, $\mu > 0$, write the density of X.

Answer. For every A in $\mathcal{E}$, the density of $X(A)$ has an exponential distribution with parameter $\mu Y(A)$ where $Y(A)$ is a Poisson variable with intensity $\lambda(A)$, therefore

$$f(x,A) = \mu E\{Y(A)e^{-\mu Y(A)x}\}$$

$$= \mu e^{-\lambda(A)} \sum_{k\geq 0} \frac{\lambda^k(A)}{k!} k e^{-\mu Y(A)x}\}$$

$$= \mu\lambda(A)e^{-\lambda(A)} e^{-\lambda(A)\{1-e^{-\mu Y(A)x}\}}.$$

Exercise 9.6.6. Let Y be a spatial exponential process with a stationary random measure μ on a measurable space $(E,\mathcal{E})$ of $\mathbb{R}^d$ and let X be a Poisson process with intensity Y on $(E,\mathcal{E})$. Find the distribution of X, is it stationary on $(E,\mathcal{E})$?

Answer. For every A in $\mathcal{E}$, $Y(A)$ is an exponential variable with intensity $\mu(A)$ and $X(A)$ has a Poisson distribution with parameter $Y(A)$, its probabilities are

$$\pi_k(A) = P(X(A)=k) = E[\pi_k\{Y(A)\}]$$

$$= E\Big\{e^{-Y(A)}\frac{Y^k(A)}{k!}\Big\}$$

$$= \mu(A)\frac{1}{k!}\int_0^\infty e^{-y\{\mu(A)+1\}} y^k\, dy$$

$$= \frac{\mu(A)^k}{\{\mu(A)+1\}^{k+1}}.$$

By the stationarity of the measure μ on $(E, \mathcal{E})$, $\mu(A \cup B) - \mu(B) = \mu(A)$ for all disjoint sets A and B of $\mathcal{E}$, then

$$\pi_k(A \cup B) - \pi_k(B) = \frac{\mu(A \cup B)}{\{\mu(A \cup B) + 1\}^{k+1}} - \frac{\mu(B)}{\{\mu(B) + 1\}^{k+1}}$$
$$= \frac{\mu(A) + \mu(B)}{\{\mu(A) + \mu(B) + 1\}^{k+1}} - \frac{\mu(B)}{\{\mu(B) + 1\}^{k+1}}$$

and the distribution of X is not stationary on $(E, \mathcal{E})$.

Chapter 10

Continuous processes

10.1 Brownian motion

The Brownian motion $(B_t)_{t\geq 0}$ is a Gaussian process starting from zero, with continuous sample paths on $\mathbb{R}$ and independent increments, its expectation is zero, its variance is $B_t^2 = t$ and

$$E(B_s B_t) = s \wedge t. \tag{10.1}$$

For $s < t$, the property (10.1) implies its second order stationarity

$$E(B_t - B_s)^2 = (t-s)^2 = E(B_{t-s}^2)$$

and the Gaussian variables $B_t - B_s$ and B_{t-s} have the same distribution. For all θ in $\mathbb{R}$ and $t > 0$, the Laplace transform of B_t is

$$E\{e^{-\theta B_t}\} = e^{\frac{\theta^2 t}{2}}$$

and the process

$$Y_\theta(t) = \exp\{-\theta B_t - \frac{1}{2}\theta^2 t\} \tag{10.2}$$

has the expectation 1.

Let $T_a = \inf\{s : B_s \geq a\}$, the events $\{T_a < t\}$ and $\{B_t < a\}$ are equivalent and the variable T_a is a stopping time with respect to the filtration $(\mathcal{F}_t)_{t\geq 0}$ such that $\mathcal{F}_t$ is the sigma-algebra $\sigma(B_s, s \leq t)$, it follows that the Laplace transform of T_a is $L_{T_a}(x) = e^{a\sqrt{2x}}$.

Exercise 10.1.1. Prove that for the Brownian motion X on $[0,1]$, there exists δ in $]0, \frac{1}{2}[$ such that $h^{\delta-\frac{1}{2}} \sup_{|t-s|\leq h} |X_t - X_s|$ converges a.s. to zero as h tends to zero.

Answer. The L^2 norm of $X_t - X_s$ is $E|X_t - X_s|^2 = |t-s|$ and $h^{2\delta-1}\sup_{|t-s|\le h} E|X_t - X_s|^2 = h^{2\delta}$ converges to zero. Let $(t_j)_{j\le J_h}$ in $[0,1]$ such that $|t_i - t_{i+1}| \le h$ and $t_0 = 0 < t_1 < \cdots < t_i < t_{i+1} < \cdots < t_{J_h} = 1$. It follows that for every $\varepsilon > 0$

$$\sum_{t_i \le J_h} P(h^{\delta-\frac{1}{2}}|X_{t_i} - X_{t_{i+1}}| > \varepsilon) \le \frac{hJ_h}{h^{2\delta-1}\varepsilon^2}$$

where J_h has the order h^{-1}. Hence

$$\frac{hJ_h}{h^{2\delta-1}\varepsilon^2} = \frac{O(h^{1-2\delta})}{\varepsilon^2}$$

and it converges to zero as h tends to zero, it follows that $h^{\delta-\frac{1}{2}}\sup_{|t-s|\le h}|X_t - X_s|$ converges a.s. to zero.

Exercise 10.1.2. Calculate the covariance of $X_t = \int_0^t B_u\,du$ and X_s for the Brownian motion B on $\mathbb{R}_+$.

Answer. For $0 \le s < t$, we have

$$\begin{aligned}
E(X_s X_t) &= \int_0^s \int_0^t E(B_u B_v)\,du\,dv \\
&= \int_0^s \int_u^t u\,du\,dv + \int_0^s \int_0^t 1_{\{v<u\}} v\,du\,dv \\
&= \int_0^s u(t-u)\,du + \int_0^s v(s-v)\,dv \\
&= \frac{s^2 t}{2} - \frac{s^3}{3} + \frac{s^3}{6} = \frac{s^2}{2}\Big(t - \frac{s}{3}\Big)
\end{aligned}$$

and the variance of X_t is calculated by symmetry

$$\begin{aligned}
E(X_t^2) &= \int_0^t \int_0^t E(B_u B_v)\,du\,dv = 2\int_0^t \int_u^t u\,du\,dv \\
&= 2\int_0^t u(t-u)\,du = \frac{t^3}{3}.
\end{aligned}$$

Exercise 10.1.3. Calculate the covariance of the process $X_t = t^{-p}\int_0^t x^p\,dB_x$, $p > 0$.

Answer. Integrating by parts the integral defining X_t, it is written as

$$X_t = B_t - pt^{-p}\int_0^t x^{p-1} B_x\,dx.$$

The covariance of X_t and X_s, $0 < s < t$ is then

$$\begin{aligned}
C(s,t) &= s - \frac{p}{s^p}\int_0^s x^{p-1}E(B_xB_t)\,dx - \frac{p}{t^p}\int_0^t x^{p-1}E(B_xB_s)\,dx \\
&\quad + \frac{p^2}{s^pt^p}\int_0^s\int_0^t x^{p-1}y^{p-1}E(B_xB_y)\,dx\,dy \\
&= s - \frac{sp}{p+1} - \frac{p}{t^p}\int_0^t x^{p-1}(s\wedge x)\,dx \\
&\quad + \frac{p^2}{s^pt^p}\int_0^s\int_0^t x^{p-1}y^{p-1}(x\wedge y)\,dx\,dy \\
&= s - \frac{sp}{p+1} - \frac{s(t-s)^p}{t^p} - \frac{ps^{p+1}}{(p+1)t^p} + \frac{p^2}{s^pt^p}I_p
\end{aligned}$$

where the last integral is

$$\begin{aligned}
I_p &= \int_0^t y^{p-1}\int_0^{s\wedge y} x^p\,dx\,dy + \int_0^s x^{p-1}\int_0^{x\wedge t} y^p\,dy\,dx \\
&= \frac{s^{p+1}}{p+1}\int_s^t y^{p-1}\,dy + \frac{1}{p+1}\int_0^s y^{2p}\,dy + \frac{1}{p+1}\int_0^s x^{2p}\,dx \\
&= \frac{s^{p+1}(t-s)^p}{p(p+1)} + \frac{2s^{2p+1}}{(p+1)(2p+1)}.
\end{aligned}$$

Exercise 10.1.4. Prove that the covariance of the Brownian motion implies it is a martingale with respect to the filtration $(\mathcal{F}_t)_{t\geq 0}$ and the process $(B_t^2 - t)_{t\geq 0}$ is a $(\mathcal{F}_t)_{t\geq 0}$-martingale. Let F be an increasing function, on $\mathbb{R}_+$, prove that $(B\circ F_t)_{t\geq 0}$ and $(B^2\circ F_t - F_t)_{t\geq 0}$ are martingale with respect to $(\mathcal{F}_t)_{t\geq 0}$.

Answer. The conditional expectation $E(B_t \mid B_s)$, for $s < t$, is a linear function of B_s, let $E(B_t \mid B_s) = aB_s$ then $E(B_tB_s) = aE(B_s^2)$ and $a = 1$, it follows that $(B_t)_{t\geq 0}$ is a martingale.

As $B_t - B_s$ and B_{t-s} have the same distribution,

$$E(B_t - B_s \mid B_s) = EB_{t-s} = 0.$$

The martingale property of the Brownian motion implies

$$\begin{aligned}
E(B_t^2 - B_s^2 \mid \mathcal{F}_s) &= E\{(B_t - B_s)^2 + 2(B_t - B_s)B_s \mid \mathcal{F}_s\} \\
&= E\{(B_t - B_s)^2 \mid \mathcal{F}_s\},
\end{aligned}$$

hence $E(B_t^2 - B_s^2 \mid \mathcal{F}_s) = E(B_{t-s}^2) = t - s$.

By a change of scale, the process $(B \circ F_t)_{t \geq 0}$ has the expectation zero, the variance F_t and the covariance $F_t \wedge F_s$ at s and t, the Gaussian variables $B \circ F_t - B \circ F_s$ and $B \circ (F_t - F_s)$ have the same distribution.

Let $\mathcal{F}_t = \sigma(B \circ F_s, s \leq t)$, like for the Brownian motion we have

$$\begin{aligned} E(B^2 \circ F_t - B^2 \circ F_s \mid \mathcal{F}_s) &= E\{(B \circ F_t - B \circ F_s)^2 \\ &\quad +2(B \circ F_t - B \circ F_s) B \circ F_s \mid \mathcal{F}_s\} \\ &= E\{(B \circ F_t - B \circ F_s)^2 \mid \mathcal{F}_s\} \\ &= E(B^2 \circ (F_t - F_s) = F_t - F_s, \end{aligned}$$

therefore $(B^2 \circ F_t - F_t)_{t \geq 0}$ is a martingale with respect to $(\mathcal{F}_t)_{t \geq 0}$.

Exercise 10.1.5. Write the density of B_t conditionally on B_s, for $s < t$ and for $s > t$ in $\mathbb{R}_+$.

Answer. For $s < t$, B_t has the conditional expectation $E(B_t \mid B_s) = B_s$. The property

$$E(B_t^2 - B_s^2 \mid B_s) = E(B_{t-s}^2) = t - s$$

entails $E(B_t^2 \mid B_s) = B_s^2 + t - s$ and its conditional variance satisfies

$$\begin{aligned} \operatorname{Var}(B_t \mid B_s) &= t - s, \\ \operatorname{Var}(B_t) &= \operatorname{Var}(B_t \mid B_s) + \operatorname{Var} E(B_t \mid B_s) = t, \end{aligned}$$

the density of B_t conditionally on B_s is the Gaussian density with expectation B_s and with variance $t - s$.

For $s > t$, the expectation of B_t conditionally B_s is a linear function of B_s, $E(B_t \mid B_s) = aB_s$ which entails $E(B_s B_t) = t = as$ and

$$E(B_t \mid B_s) = \frac{tB_s}{s}.$$

The conditional variance satisfies

$$\begin{aligned} \operatorname{Var}(B_t \mid B_s) &= E(B_t^2 \mid B_s) - \frac{t^2 B_s^2}{s^2}, \\ \operatorname{Var}(B_t) &= E \operatorname{Var}(B_t \mid B_s) + \operatorname{Var} E(B_t \mid B_s), \\ E \operatorname{Var}(B_t \mid B_s) &= t - \frac{t^2}{s}, \end{aligned}$$

let $E(B_t^2 \mid B_s) = aB_s^2 + bB_s + c$, then $E(B_t^2) = t = as + c$, hence $a = 1$, $c = t - s$ and $E(B_t^2 \mid B_s) = B_s^2 + t - s$

$$\operatorname{Var}(B_t \mid B_s) = \frac{(s^2 - t^2) B_s^2}{s^2} + t - s.$$

The density of B_t conditionally on B_s is the Gaussian density with expectation $s^{-1}tB_s$ and with variance $\mathrm{Var}(B_t \mid B_s)$.

Exercise 10.1.6. Prove that the process Y_θ defined by (10.2) is a martingale with respect to $(\mathcal{F}_t)_{t\geq 0}$.

Answer. For all $s < t$, Y_θ satisfies

$$\begin{aligned}
E\{Y_\theta^{-1}(s)Y_\theta(t) \mid B_s\} &= E[\exp\{\theta(B_t - B_s) - \frac{1}{2}\theta^2(t-s) \mid B_s\}] \\
&= E[\exp\{\theta B_{t-s} - \frac{1}{2}\theta^2(t-s)\}] \\
&\geq e^{-\frac{1}{2}\theta^2(t-s)}
\end{aligned}$$

by convexity, therefore

$$E\{Y_\theta(t) \mid B_s\} \geq Y_\theta(s)e^{-\frac{1}{2}\theta^2(t-s)}$$

where the exponential is lower than one. As its expectation is $EY_\theta(t) = 1$, we have $EY_\theta(t) = EY_\theta(s)$ for all s and t, it follows that

$$E\{Y_\theta(t) \mid B_s\} = Y_\theta(s)$$

for $s < t$, and Y_θ is an exponential martingale with respect to $(\mathcal{F}_t)_{t\geq 0}$.

Exercise 10.1.7. Let a and b be strictly positive, prove that the stopping times T_a and $T_{a+b} - T_a$ are independent and $T_{a+b} - T_a$ has the same distribution as T_b.

Answer. For $T_a = \inf\{s : B_s \geq a\}$, the equivalence of $\{T_a > t\}$ and $\{B_t < a\}$ implies

$$\begin{aligned}
P(T_{a+b} - T_a > t \mid \mathcal{F}_{T_a}) &= P(B_{T_a+t} < a + b \mid \mathcal{F}_{T_a}) \\
&= P(B_{T_a+t} - B_{T_a} < b \mid \mathcal{F}_{T_a}) \\
&= P(B_t < b) = P(T_b > t),
\end{aligned}$$

which does not depend on T_a therefore T_a and $T_{a+b} - T_a$ are independent and T_{a+b} has the same distribution as $T_a + T_b$.

Exercise 10.1.8. Prove that for $a > 0$, the stopping time $T_a = \inf\{t : |B_t| > a\}$ has the inverse Gaussian density

$$f_a(t) = \frac{a}{\sqrt{2\pi t^3}}e^{-\frac{a^2}{2t}},\ t \geq 0.$$

Answer. The symmetry of the Gaussian distribution $\mathcal{N}(0,t)$ entails $P(B_t > a) = P(B_t < -a)$ and

$$P(T_a \le t) = 1 - \frac{2}{\sqrt{2\pi t}} \int_a^\infty e^{-\frac{x^2}{2t}}\,dx$$
$$= 1 - \frac{2}{\sqrt{2\pi}} \int_{\frac{a}{\sqrt{t}}}^\infty e^{-\frac{x^2}{2}}\,dx,$$

its derivative is the density of T_a

$$f_a(t) = \frac{2}{\sqrt{2\pi t^3}} e^{-\frac{a^2}{2t}}.$$

Exercise 10.1.9. For every stopping time τ for the filtration $(\mathcal{F}_t)_{t\ge0}$ the process $(B_{t+\tau} - B_t)_{t\ge0}$ is a Brownian motion independent of $\mathcal{F}_s$ for every $s \le \tau$.

Answer. For every $x > 0$, the variable $B_{t+(\tau\wedge x)} - B_t$ is $\mathcal{F}_{t+x}$-measurable and it has the same centered Gaussian distribution as the Brownian motion $B_{\tau\wedge x}$, with covariance

$$E(B_{\tau\wedge x} B_{\tau\wedge y}) = \tau \wedge x \wedge y.$$

The variable $B_{t+\tau} - B_t$ has the same Gaussian distribution as B_τ which depends only on τ.

Exercise 10.1.10. Let β be a right-continuous function of $L_2(\mathbb{R})$ with left-hand limits, the process

$$Y(t) = \int_0^t \beta(s)\,dB_s$$

is a transformed Brownian motion, calculate the increasing function of its quadratic variations, its Laplace transform and prove that the process

$$Y_\lambda(t) = \exp\Big\{\lambda \int_0^t \beta(s)\,dB_s - \frac{\lambda^2}{2} \int_0^t \beta^2(s)\,ds\Big\}$$

is a local martingale with mean 1.

Answer. Let $V_Y(t) = \int_0^t \beta^2(s)\,ds$ be the increasing function of the quadratic variations of the process Y and let $L_{Y_t}(\lambda) = \exp\{\frac{1}{2}\lambda^2 V_Y(t)\}$ be the Laplace transform of Y_t. For all $s < t$, Y_λ satisfies

$$E\{Y_\lambda^{-1}(s) Y_\lambda(t) \mid B_s\} = e^{-\frac{1}{2}\lambda^2(t-s)} E[\exp\{\lambda(B_t - B_s) \mid B_s\}]$$
$$= e^{-\frac{1}{2}\lambda^2(t-s)} E[\exp\{\lambda B_{t-s}\}]$$
$$\ge e^{-\frac{1}{2}\lambda^2(t-s)}$$

by convexity and the properties of the Brownian motion, therefore

$$E\{Y_\lambda(t) \mid B_s\} \geq Y_\lambda(s)e^{-\frac{1}{2}\lambda^2(t-s)}$$

where the exponential is lower than one. As its expectation is $EY_\lambda(t) = 1$, we have $EY_\lambda(t) = EY_\lambda(s)$ for all s and t, it follows that

$$E\{Y_\lambda(t) \mid B_s\} = Y_\lambda(s)$$

for $s < t$, and Y_λ is a martingale with respect to $(\mathcal{F}_t)_{t\geq 0}$.

Exercise 10.1.11. Let β be a function of $L_2(\mathbb{R})$, let $Y(t) = \int_0^t \beta(s)\,dB_s$ and for $a > 0$, let $T_{Y,a} = \inf\{s : Y(s) = a\}$, calculate a lower bound for the Laplace transform of $T_{Y,a}$.

Answer. The martingale property of the process Y_λ and a convexity argument provide a lower bound for the Laplace transform of $T_{Y,a}$

$$\begin{aligned} e^{a\lambda} &= E\exp\Big\{\frac{1}{2}\lambda^2 V_Y(T_{Y,a})\Big\} \leq E\Big[\frac{1}{T_{Y,a}}\int_0^{T_{Y,a}} \exp\Big\{\frac{T_{Y,a}}{2}\lambda^2\beta^2(s)\Big\}\,ds\Big] \\ &\leq E\Big[\exp\Big\{\frac{T_{Y,a}}{2}\lambda^2\|\beta\|_\infty^2\Big\}\Big] \end{aligned}$$

and

$$Ee^{\lambda T_{Y,a}} \geq \exp\Big\{\frac{a\sqrt{2}}{\sqrt{\lambda}\|\beta\|_\infty}\Big\}.$$

Exercise 10.1.12. Determine the distribution of $T_a = \inf\{s : Y_\theta(s) > a\}$, for the process Y_θ defined by (10.2).

Answer. We have

$$\begin{aligned} P(T_a \leq t) &= P\Big(\sup_{s\in[0,t]} Y_\theta(s) \geq \sup_{x\in[t,1]} Y_\theta(x)\Big) \\ &= P\Big(\sup_{u\in[0,t]}\{Y_\theta(t-u) - Y_\theta(t)\} \geq \sup_{v\in[0,1-t]}\{Y_\theta(t+v) - Y_\theta(t)\}\Big) \\ &= P\Big(\sup_{u\in[0,t]}\{Y_\theta(-u) - 1\} \geq \sup_{v\in[0,1-t]}\{Y_\theta(v) - 1\}\Big) \end{aligned}$$

where the processes $(Y_\theta(-u) - 1)_{u\in[0,t]}$ and $(Y_\theta(v) - 1)_{v\in[0,1-t]}$ are independent and centered with variances

$$\begin{aligned} \mathrm{Var}\{Y_\theta(-u) - 1\} &= E\{Y_\theta^2(-u) + 1\} = e^{-\theta^2 u} + 1, \\ \mathrm{Var}\{Y_\theta() - 1\} &= E\{Y_\theta^2(v) + 1\} = e^{\theta^2 v} + 1. \end{aligned}$$

The variables

$$X = \{e^{-\theta^2 t} + 1\}^{-\frac{1}{2}} \sup_{u \in [0,t]} \{Y_\theta(-u) - 1\},$$
$$Z = \{e^{\theta^2(1-t)} + 1\}^{-\frac{1}{2}} \sup_{v \in [0,1-t]} \{Y_\theta(v) - 1\}$$

have a normal distribution and the distribution function of T_a is given by

$$P(T_a \leq t) = P(\{e^{-\theta^2 t} + 1\}X^2 \geq \{e^{\theta^2(1-t)} + 1\}Z^2)$$
$$= P\Big(\frac{Z^2}{X^2} \leq \frac{e^{-\theta^2 t} + 1}{e^{\theta^2(1-t)} + 1}\Big).$$

Let

$$u = \frac{e^{-\theta^2 t} + 1}{e^{\theta^2(1-t)} + 1}$$

in $]e^{-\theta^2}, 1]$, then

$$t = -\frac{1}{\theta^2} \log \frac{1-u}{ue^{\theta^2} - 1},$$

the variables Z^2 and X^2 have a $\Gamma(\frac{1}{2}, \frac{1}{2})$ distribution and the distribution function of T_a is calculated with the change of variables $y = z^{-1}v$ in the integrals

$$P\Big(\frac{Z^2}{X^2} \leq u\Big) = \int_0^\infty P(Z^2 \leq uy)\, dP(X^2 \leq y)$$
$$= \frac{2}{\pi} \int_0^\infty y^{-\frac{1}{2}} e^{-\frac{y}{2}} \int_0^{uy} v^{-\frac{1}{2}} e^{-\frac{v}{2}}\, dv\, dy$$
$$= \frac{2}{\pi} \int_0^u \int_0^\infty e^{-\frac{v(z+1)}{2z}}\, dv z^{-\frac{3}{2}}\, dz$$
$$= \frac{2}{\pi} \int_0^u \frac{dz}{z^{\frac{1}{2}}(z+1)}.$$

The expansion of the integrand and the change of variables $z = t^2$ yield

$$\frac{1}{z^{\frac{1}{2}}(z+1)} = \frac{1}{z^{\frac{1}{2}}} - \frac{z^{\frac{1}{2}}}{1+z},$$
$$\frac{z^{\frac{1}{2}}\, dz}{1+z} = \frac{2t^2}{1+t^2}\, dt = 2\Big(1 - \frac{1}{1+t^2}\Big)\, dt,$$

hence

$$P\Big(\frac{Z^2}{X^2} \le u\Big) = \frac{4}{\pi}\Big\{\sqrt{u} - u + \arctan(u)\Big\}.$$

Exercise 10.1.13. Let B be a Brownian motion sampled at random times $(S_i)_{i \le n}$ such that $S_i - S_{i-1} = X_i S_n$ for $i \le n$, with $S_0 = 0$.

(a). Find the distribution of B as the variables X_i are independent uniformly distributed.

(b). Find the distribution of B as $(X_i)_{i \le n}$ has a Dirichlet distribution.

Answer. The density f_{S_n} of the variables $\{B(S_i)\}_{i \le n}$ is the product of the centered Gaussian densities of $\{B(S_i) - B(S_{i-1})\}$, for $i = 1, \dots, n$.

(a). Under the condition of independent uniform variables X_i, the density f_{S_n} at $y = (y_1, \dots, y_n)$ factorizes

$$f_{S_n}(y) = \frac{1}{(2\pi S_n)^{\frac{n}{2}}} \prod_{i=1}^n \int_{\mathbb{R}} \frac{e^{-\frac{y_i^2}{2S_n x_i}}}{\sqrt{x_i}}\, dx_i.$$

(b). The density f_{S_n} at $y = (y_1, \dots, y_n)$ under a Dirichlet distribution D_a is

$$\begin{aligned} f_{S_n}(y) &= \frac{\Gamma_{\bar{a}_n}}{(2\pi)^{\frac{n}{2}} \prod_{i=1}^n \Gamma_{a_i}} \int_C \prod_{i=1}^n \frac{e^{-\frac{y_i^2}{2S_n x_i}}}{\sqrt{S_n x_i}} \prod_{i=1}^{n-1} x_i^{a_i - 1} \Big(1 - \sum_{j=1}^{n-1} x_j\Big)^{a_n - 1} dx_i \\ &= \frac{\Gamma_{\bar{a}_n}}{(2\pi S_n)^{\frac{n}{2}} \prod_{i=1}^n \Gamma_{a_i}} \int_C \prod_{i=1}^n e^{-\frac{y_i^2}{2S_n x_i}} \prod_{i=1}^{n-1} x_i^{a_i - \frac{1}{2}} \Big(1 - \sum_{j=1}^{n-1} x_j\Big)^{a_n - \frac{1}{2}} dx_i \end{aligned}$$

where the integral is on the set $C = \{x \in]0,1[^n : \sum_{i=1,\dots,n} x_i = 1\}$.

Exercise 10.1.14. For the Brownian motion B and $\lambda > 0$, calculate

$$\int_0^\infty E\Big\{\Big(\int_0^t e^{\lambda B_s}\, ds\Big)^n e^{\lambda B_t}\Big\} e^{-\alpha t}\, dt.$$

Answer. Let $\varphi_\lambda = \frac{\lambda^2}{2}$ so that $E(e^{-\lambda B_t}) = e^{t\varphi_\lambda}$, the expectation is calculated using the property of independent increments for B

$$\begin{aligned}\phi_{nt}(\lambda) &= E\Big\{\Big(\int_0^t e^{-\lambda B_s}\,ds\Big)^n e^{-\lambda B_t}\Big\}\\
&= n!E\int_0^t ds_1\int_0^{s_1} ds_2\cdots\int_0^{s_{n-1}} ds_n e^{-\lambda(B_t+B_{s_1}+\cdots+B_{s_n})}\\
&= n!E\int_0^t ds_1\int_0^{s_1} ds_2\cdots\int_0^{s_{n-1}} ds_n e^{-\lambda(B_t-B_{s_1})}\\
&\quad .e^{-2\lambda(B_{s_1}-B_{s_2})-\cdots-n\lambda(B_{s_{n-1}}-B_{s_n})-(n+1)\lambda B_{s_n}}\\
&= n!\int_0^t ds_1\int_0^{s_1} ds_2\cdots\int_0^{s_{n-1}} ds_n \exp\{(t-s_1)\varphi_\lambda\}\\
&\quad .\exp\{(s_1-s_2)\varphi_{2\lambda}+\cdots+s_n(\varphi_{(n+1)\lambda}\}\\
&= n!\int_0^t ds_1\int_0^{s_1} ds_2\cdots\int_0^{s_{n-1}} ds_n \exp\{t\varphi_\lambda\}\\
&\quad .\exp\{s_1(\varphi_{2\lambda}-\varphi_\lambda)+\cdots+s_n(\varphi_{(n+1)\lambda}-\varphi_{n\lambda}\},\end{aligned}$$

then

$$\begin{aligned}\int_0^\infty \phi_{nt}(\lambda)e^{-\alpha t}\,dt &= n!\prod_{k=1}^n \frac{1}{\varphi_{k\lambda}-\varphi_{(k+1)\lambda}}\int_0^\infty e^{-(\alpha-\varphi_\lambda)t}\,dt\\
&= \frac{n!}{\alpha-\varphi_\lambda}\prod_{k=1}^n \frac{1}{\varphi_{k\lambda}-\varphi_{(k+1)\lambda}}\\
&= \frac{(-1)^n 2^n(\lambda^{-2n}n!}{\alpha-\varphi_\lambda}\prod_{k=1}^n\frac{1}{1+2k},\end{aligned}$$

as $\varphi_{k\lambda}-\varphi_{(k+1)\lambda}=\frac{\lambda^2}{2}\{k^2-(k-1)^2\}$.

10.2 Diffusion processes

The Ornstein–Uhlenbeck process $(X_t)_{t\geq 0}$ is a diffusion process with constant coefficients solution of the equation

$$dX_t = \alpha X_t\,dt + \sigma\,dB_t, \tag{10.3}$$

with initial value X_0, where $(B_t)_{t\geq 0}$ is the Brownian motion and $\sigma > 0$. Equivalently, it is written as

$$X_t = X_0 e^{\alpha t} + \sigma\int_0^t e^{\alpha(t-s)}\,dB_s,$$

its expectation is $E(X_t) = E(X_0)e^{\alpha t}$ and its variance is

$$\mathrm{Var}(X_t) = \sigma^2 \int_0^t e^{2\alpha(t-s)}\, ds.$$

A diffusion process with a functional drift $(\alpha_t)_{t\geq 0}$ and with initial value X_0 is defined by the equation

$$dX_t = \alpha_t X_t\, dt + \sigma\, dB_t, \tag{10.4}$$

or $dX_t = \alpha_t X_t\, dt + \sigma_t\, dB_t$, with a functional variance of $L^2(\mathbb{R}_+)$.

A diffusion process with an auto-regressive drift $(\alpha(x))_{x\in\mathbb{R}}$ and with initial value X_0 is solution of the equation

$$dX_t = \alpha(X_t)X_t\, dt + \sigma\, dB_t.$$

Exercise 10.2.1. Write an integral expression of X_t given by equation (10.4) and give its expectation and its variance conditionally on X_s, for $s < t$.

Answer. The equation $dX_t = \alpha_t X_t\, dt$ with initial value X_0 has the solution $X_t = X_0 e^{A_t}$ where $A_t = \int_0^t \alpha_s\, ds$ and (10.4) has the solution

$$X_t = X_0 e^{A_t} + \sigma \int_0^t e^{A_t - A_s}\, dB_s.$$

The conditional expectation of $X_t - X_s$ is

$$\begin{aligned} E(X_t - X_s \mid X_s) &= E(X_0)e^{A_t - A_s} + \sigma \int_s^t e^{A_t - A_u}\, dE(B_u \mid X_s) \\ &= E(X_0)e^{A_t - A_s} - \sigma s B_s \int_s^t e^{A_t - A_u} u^{-2}\, du \end{aligned}$$

and its conditional variance is

$$\begin{aligned} \mathrm{Var}(X_t) &= \sigma^2 \int_s^t e^{2(A_t - A_u)}\, d\mathrm{Var}(B_u \mid B_s) \\ &= \sigma^2 \int_s^t e^{2(A_t - A_u)}\, (du + 2s^2 B_s^2 u^{-3}\, du). \end{aligned}$$

The equation (10.4) with a functional variance has the solution

$$X_t = X_0 e^{A_t} + \int_0^t e^{A_t - A_s} \sigma_s\, dB_s,$$

the conditional expectation of $X_t - X_s$ is

$$\begin{aligned} E(X_t - X_s \mid X_s) &= E(X_0)e^{A_t - A_s} + \int_s^t e^{A_t - A_u} \sigma_u\, dE(B_u \mid X_s) \\ &= E(X_0)e^{A_t - A_s} + s B_s \int_s^t e^{A_t - A_u} \sigma_u u^{-2}\, du \end{aligned}$$

and its conditional variance is

$$\begin{aligned}\mathrm{Var}(X_t) &= \int_s^t e^{2(A_t-A_u)}\sigma_u^2\, d\mathrm{Var}(B_u \mid B_s)\\ &= \int_s^t e^{2(A_t-A_u)}\sigma_u^2\,(du + 2s^2B_s^2u^{-3}\,du).\end{aligned}$$

Exercise 10.2.2. For $a > 0$ and for the process $X_t = \mu t + \sigma B_t$ on $\mathbb{R}_+$, let $T_a = \inf\{t : |X_t| > a\}$, prove that the probability $P(T_a < t)$ converges to zero as t tends to zero and write the Laplace transform of T_a.

Answer. Let $\mu > 0$, we have

$$P(T_a < t) = P(|X_t| > a) \le P(\sigma|B_t| > |a - \mu t|) \le \frac{\sigma^2 t}{(a-\mu t)^2}$$

and it converges to zero as t tends to zero. If $\mu < 0$, $a - \mu t$ is replaced by $a + \mu t$.

The events $\{T_a > t\}$ and $\{B_t < a\}$ are equivalent and T_a is a stopping time with respect to the filtration $\mathbb{F} = (\mathcal{F}_t)_{t\ge 0}$ generated by X_t, and $(X_t - \mu t)_{t\ge 0}$ is a martingale time with respect to $\mathbb{F}$. For every t

$$E\{e^{\theta(X_t-\mu t)}\} = E(e^{\theta\sigma B_t}) = e^{\frac{\theta^2\sigma^2 t}{2}}$$

and the same equality

$$E\Big\{e^{\theta(X_\tau-\mu\tau)-\frac{\theta^2\sigma^2\tau}{2}}\Big\} = 1$$

is true at every $\mathbb{F}$-stopping time τ, then at T_a

$$\begin{aligned}1 &= E\Big\{e^{\theta(a-\mu T_a)}e^{-\frac{\theta^2\sigma^2 T_a}{2}}\Big\},\\ e^{-\theta a} &= E\Big(e^{-\theta\mu T_a-\frac{\theta^2\sigma^2 T_a}{2}}\Big)\end{aligned}$$

and the Laplace transform of T_a is

$$L_{T_a}(\lambda) = E(e^{\lambda T_a}) = e^{-\theta_\lambda a},$$

where $\theta_\lambda^2\sigma^2 + 2\theta_\lambda\mu + 2\lambda = 0$, under the condition $\mu^2 > 2\lambda\sigma^2$.

Exercise 10.2.3. For $a > 0$ and $\mu > 0$, prove that the process $X_t = \mu t + \sigma B_t$ is a submartingale with respect to the filtration generated by X_t.

Answer. For $s < t$, the covariance of X_s and X_t is

$$E\{(X_s - \mu s)(X_t - \mu t)\} = \sigma^2 s$$

and the conditional expectation of the Brownian motion implies

$$E(X_t - \mu t \mid X_s - \mu s) = \sigma B_s = X_s - \mu s,$$
$$E(X_t \mid X_s) = \mu t + \sigma E(B_t \mid B_s = \frac{X_s - \mu s}{\sigma})$$
$$= X_s + \mu(t-s) \geq X_s,$$

X is therefore a submartingale.

Exercise 10.2.4. Find the distribution of $T_a = \inf\{t : |X_t| \geq a\}$, $a > 0$.

Answer. The event $\{T_a < t\}$ is equivalent to $\{-a - \mu t \leq \sigma B_t \leq a - \mu t\}$, it has the probability

$$P(T_a < t) = 1 - P(B_t > a - \mu t) - P(B_t < -a - \mu t)$$
$$= 1 - \frac{1}{\sqrt{2\pi t}} \int_{a-\mu t}^{\infty} e^{-\frac{x^2}{2t}}\, dx - \frac{1}{\sqrt{2\pi t}} \int_{-\infty}^{-a-\mu t} e^{-\frac{x^2}{2t}}\, dx$$
$$= 1 - \frac{1}{\sqrt{2\pi}} \int_{\frac{a-\mu t}{\sqrt{t}}}^{\infty} e^{-\frac{x^2}{2}}\, dx - \frac{1}{\sqrt{2\pi}} \int_{-\infty}^{\frac{-a-\mu t}{\sqrt{t}}} e^{-\frac{x^2}{2}}\, dx$$

and its derivative is the density of T_a

$$f_a(t) = \frac{1}{\sqrt{2\pi}} \Big\{ e^{-\frac{(a-\mu t)^2}{2t}} - e^{-\frac{(a+\mu t)^2}{2t}} \Big\}$$
$$= \frac{\sqrt{2} e^{-\frac{a^2+\mu^2 t^2}{2t}} \sinh(a\mu)}{\sqrt{\pi}}.$$

Exercise 10.2.5. Let C be a predictable process with respect to the filtration $\mathbb{F} = (\mathcal{F}_t)_{t\geq 0}$ of the Brownian motion B, such that $v_t^2 = \int_0^t E(C_s^2)\, ds$ is finite for every y, prove that the process

$$Y_\theta(t) = \exp\Big\{\theta \int_0^t C_s\, dB_s - \frac{\theta^2 v_t^2}{2}\Big\}$$

is a martingale with expectation 1 and determine $dY_1(t)$. Find a bound for $E\{Y_\theta^a(t)\}$, a in $]0,1[$.

Answer. The variable $X_t = \int_0^t C_s\, dB_s$ has a centered Gaussian distribution with variance v_t^2 and the process $(X_t)_{t\geq 0}$ is martingale with respect to $\mathbb{F}$ such that

$$E(e^{\theta X_t}) = \frac{1}{\sqrt{2\pi} v_t} \int_0^t e^{\theta x - \frac{x^2}{2v_t^2}}\, dx = e^{\frac{\theta^2 v_t^2}{2}},$$

therefore $E\{Y_\theta(t)\} = 1$ for every t. For $s < t$, the variable $\int_s^t C_u\, dB_u$ is independent of X_s and

$$E\{Y_\theta^{-1}(s)Y_\theta(t) \mid Y_\theta(s)\} = E\Big[\exp\Big\{\theta \int_s^t C_u\, dB_u - \frac{1}{2}\theta^2(v_t^2 - v_s^2) \mid Y_\theta(s)\Big\}\Big]$$
$$\geq e^{-\frac{1}{2}\theta^2(v_t^2-v_s^2)}$$

by convexity, where the exponential is lower than 1. As $Y_\theta(s)$ and $Y_\theta(t)$ have the same expectation, we have

$$E\{Y_\theta(t) \mid Y_\theta(s)\} = Y_\theta(s)$$

and the process Y_θ is a martingale.

The process $Y_1(t)$ is solution of the diffusion equation

$$dY_t = -v_t Y_t\, dt + C_t Y_t\, dB_t.$$

For a in $]0,1[$, the process $Y_{\theta,a}(t) = \exp\Big\{a\theta \int_0^t C_s\, dB_s - \frac{a^2\theta^2 v_t^2}{2}\Big\}$ is a martingale with expectation 1, then

$$E\{Y_\theta^a(t)\} = E\Big\{Y_{\theta,a}(t) e^{-\frac{a(1-a)\theta^2 v_t^2}{2}}\Big\}$$
$$= \Big\{e^{-\frac{(1-a)\theta^2 v_t^2}{2}}\Big\}^a < 1.$$

Exercise 10.2.6. Let C be a predictable process with respect to the filtration $\mathbb{F} = (\mathcal{F}_t)_{t\geq 0}$ of the Brownian motion B, such that $E\Big\{\exp\{\frac{1}{2}\int_0^t C_s^2\, ds\Big\}$ is finite for every t, prove that the process

$$Y_\theta(t) = \exp\{\theta \int_0^t C_s\, dB_s - \frac{1}{2}\theta^2 \int_0^t C_s^2\, ds\}$$

is a martingale with expectation 1.

Answer. The variable $X_t = \int_0^t C_s\, dB_s$ has a centered Gaussian distribution with variance $V_t^2 = \int_0^t C_s^2\, ds$ conditionally on $\mathcal{F}_t^-$, and such that

$$E(e^{\theta X_t} \mid \mathcal{F}_t^-) = \frac{1}{\sqrt{2\pi} V_t} \int_0^t e^{\theta x - \frac{x^2}{2V_t^2}}\, dx$$
$$= e^{\frac{\theta^2}{2}\int_0^t C_s^2\, ds},$$

then $E(e^{\theta X_t - \frac{\theta^2}{2}\int_0^t C_s^2\, ds} \mid \mathcal{F}_t^-) = 1$ for every $t > 0$, since C_s is $\mathcal{F}_t^-$ measurable. Let $s < t$, by the same argument $X_t - X_s$ is a centered Gaussian variable with variance $V_{s,t}^2 = \int_s^t E(C_u^2 \mid \mathcal{F}_s)\,du$ conditionally on $\mathcal{F}_s$ and we obtain

$$E(e^{\theta(X_t - X_s)} \mid \mathcal{F}_s) = e^{\frac{\theta^2}{2}\int_s^t E(C_u^2|\mathcal{F}_s)\,du},$$

it follows that $E(e^{\theta(X_t-X_s)-\frac{\theta^2}{2}\int_s^t E(C_u^2|\mathcal{F}_s)\,du} \mid \mathcal{F}_s) = 1$. By convexity, the process

$$Z_\theta(t) = e^{\theta X_t - \frac{\theta^2}{2}\int_0^t C_u^2\,du}$$

is such that for $s < t$

$$E\Big\{\frac{Z_\theta(t)}{Z_\theta(s)} \mid \mathcal{F}_s\Big\} = E\Big(e^{\frac{\theta^2}{2}\int_s^t \{E(C_u^2|\mathcal{F}_s) - C_u^2\}\,du} \mid \mathcal{F}_s\Big) \le 1,$$

hence $E\{Z_\theta(t) \mid \mathcal{F}_s\} \le Z_\theta(s)$. As $E\{Z_\theta(t)\} = 1$ for every $t > 0$, it follows that $E\{Z_\theta(t)\} = E\{Z_\theta(t)\}$ for all $s < t$ and Z_θ is a martingale.

10.3 Brownian bridge

The definition of the Brownian bridge is introduced in Chapter 6.

Exercise 10.3.1. Let g in $L^1([0,1])$ and let Z be the Brownian bridge, find the variance of the variable $Y = \int_0^1 g_s Z_s\,ds$.

Answer. Let $G_t = \int_0^t g_s\,ds$, the variable Y is Gaussian and centered, its variance is

$$\begin{aligned} E(Y^2) &= 2\int_0^1 s g_s 1_{\{s<t\}} g_t\,ds\,dt - \Big(\int_0^1 s g_s\,ds\Big)^2 \\ &= 2\int_0^1 (G_1 - G_s) s\,dG_s - \Big(\int_0^1 s\,dG_s\Big)^2, \end{aligned}$$

integrating by parts implies $\int_0^1 s\,dG_s = G_1 - \int_0^1 G_s\,ds$ and

$$2\int_0^1 s\,G_s\,dG_s = G_1^2 - \int_0^1 G_s^2\,ds,$$

therefore

$$\begin{aligned} E(Y^2) &= \int_0^1 G_s^2\,ds - 2G_1\int_0^1 G_s\,ds + G_1^2 - \Big(G_1 - \int_0^1 G_s\,ds\Big)^2 \\ &= \int_0^1 G_s^2\,ds - \Big(\int_0^1 G_s\,ds\Big)^2. \end{aligned}$$

Exercise 10.3.2. Prove that for the Brownian bridge and $\alpha > 0$ and $\beta > \frac{1}{2}\alpha$

$$P(Z_t < \alpha(1-t) + \beta t, 0 \le t \le 1) \le \frac{1}{2(\alpha+\beta)}.$$

Answer. The Gaussian variable $Y_t = \{\alpha(1-t)+\beta t\}^{-1} Z_t$ has the variance

$$v_t = \frac{t(1-t)}{\{\alpha(1-t)+\beta t\}^2},$$

its first derivatives are

$$v_t' = \frac{\alpha-(\alpha+\beta)t}{\{\alpha(1-t)+\beta t\}^3},$$

$$v_t'' = \frac{2\alpha(\alpha-2\beta)}{\{\alpha(1-t)+\beta t\}^4},$$

such that $v_t'' < 0$ hence the variance is concave and maximum at $(\alpha+\beta)^{-1}\alpha$ where it is

$$v_M = \frac{1}{2(\alpha+\beta)},$$

it follows that the variance of $\sup_{t\in[0,1]} Y_t$ has the upper bound v_M and by the inequality (1.3) we obtain

$$P(Z_t > \alpha(1-t) + \beta t, 0 \le t \le 1) = P(\sup_{0\le t\le 1} Y_t > 1)$$
$$\le \frac{1}{2(\alpha+\beta)}.$$

Exercise 10.3.3. Let Z be the Brownian bridge, prove that for $a > 0$ and $\lambda > 0$

$$P\Big(\sup_{a\le t\le 1} Z_t > \lambda t\Big) \le \frac{1-a}{a\lambda^2}.$$

Answer. The variable $t^{-1}Z_t$ is Gaussian, centered with a decreasing variance $t^{-1}(1-t)$ on $[a,1]$, its maximum is $a^{-1}(1-a)$ then the variance of $\sup_{a\le t\le 1} t^{-1}Z_t$ is bounded by $a^{-1}(1-a)$. The inequality is deduced from the inequality (1.3).

10.4 Markov process

Exercise 10.4.1. Write the transition probabilities of the Brownian motion and, for real and bounded functions ψ on $[0,1]$ and ϕ on $[0,1]^2$, write the expectations of the processes $\int_0^t \psi(B_s)\,ds$ and of the variable $\int_0^1 \int_0^1 \psi(B_u, B_v)\,du\,dv$.

Answer. The Brownian motion is a stationary Markov process with transition probabilities

$$\begin{aligned}\pi_{A'}(A;t-s) &= P(B_t \in A \mid B_s \in A') \\ &= \int 1_{\{y\in A\}}1_{\{x\in A'\}}P(B_{t-s} = y-x)\,dx\,dy \\ &= \int 1_{\{y\in A\}}1_{\{x\in A'\}}f_{t-s}(y-x)\,dx\,dy,\end{aligned}$$

where f_t is the Gaussian density with mean zero and variance t

$$f_t(x) = \frac{1}{\sqrt{2\pi t}}e^{-\frac{x^2}{2t}}$$

with respect to Lebesgue's measure on $\mathbb{R}$. The process $\int_0^t \psi(B_s)\,ds$ has the expectation

$$\begin{aligned}E\int_0^t \psi(B_s)\,ds &= \int_{\mathbb{R}}\int_0^t \psi(x)P(B_s = x)\,ds\,dx \\ &= \int_0^t \int_{\mathbb{R}} \psi(x)\,dF_s(x)\,ds.\end{aligned}$$

On $\mathbb{R}^2$, let π be the measure such that

$$\pi \otimes F_B(A;A') = \int_{\mathbb{R}^2_+}\int_0^1\int_0^1 1_{\{y\in A\}}1_{\{x\in A'\}}\pi_x(dy;t-s)f_s(x)\,dx\,ds\,dt,$$

then

$$\begin{aligned}E\int_0^1\int_0^1 \psi(B_u,B_v)\,du\,dv &= \int_{\mathbb{R}^2}\int_0^1\int_0^1 \psi(x,y) \\ &\qquad \cdot f_{v-u}(y-x)f_u(x)\,dx\,dy\,du\,dv \\ &= \int_{\mathbb{R}^2} \psi(x,y)\,\pi\otimes F_B(dy;dx).\end{aligned}$$

Exercise 10.4.2. Let $(\tau_k)_{k\geq 0}$ be a homogeneous Markov jump process with conditional distribution functions

$$F_{|jj'}(x) = P(\tau_k \leq x | J_{k-1} = j, J_k = j'),\ x \geq 0,$$

for the sojourn duration variables, write the transition functions between states, the conditional distribution of the sojourn time in a state j, and the conditional distribution function of T_k.

Answer. The conditional transition function from a state j to j' is

$$F_{j'|j}(x) = p_{jj'} F_{|jj'}(x),$$
$$= P(\tau_k \leq x, J_k = j' | J_{k-1} = j), \; x \geq 0,$$

and the conditional transition probabilities are

$$p_{jj'} = P(J_k = j' | J_{k-1} = j) = \int_E dF_{j'|j}.$$

The distribution of the sojourn time in state j is

$$F_j(x) = \sum_{j' \in J(j)} F_{j'|j}(x)$$

where $J(j)$ is the set of states accessible from j. The conditional distribution function of T_k with initial state j_0 is

$$H_k(j_0, j, t) = P(T_k \leq t, J_k = j \mid J_0 = j_0)$$
$$= \sum_{j_1 \in E, \cdots, j_{k-1} \in E} \int_0^t \int_0^{y_{k-1}} \cdots \int_0^{y_2} \prod_{i=1}^k F_{j_i|j_{i-1}}(dy_i).$$

Exercise 10.4.3. For a homogeneous Markov jump process, write the Laplace transform of T_k as $J_k = j$.

Answer. The Laplace transform of τ_k as $J_k = j'$, conditionally on $J_{k-1} = j$ is determined by the distribution $F_{j'|j}$

$$L_k(\lambda, j' \mid j) = E(e^{-\lambda \tau_k} 1_{J_k = j'} \mid J_{k-1} = j) = \int_0^\infty e^{-\lambda x} \, dF_{j'|j}(x),$$

and by the independence of the variables τ_k, the Laplace transform of (T_k, J_k) is

$$\int e^{-\lambda t} H_k(j_0, j, dt) = \sum_{(j_1, \cdots, j_{k-1}) \in E^{k-1}} \int_0^{t - \sum_{i=1}^{k-1} y_i} \int_0^{y_{k-1}} \cdots \int_0^{y_2}$$
$$\prod_{i=1}^k e^{-\lambda y_i} F_{j_i|j_{i-1}}(dy_i)$$
$$= \sum_{(j_1, \cdots, j_{k-1}) \in E^{k-1}} \prod_{i=1}^k L_i(\lambda, j_i \mid j_{i-1})$$

where the kth state is $j_k = j$.

Exercise 10.4.4. Write the hazard functions associated to the conditional distribution functions $F_{|jj'}$ and the transition functions $F_{j'|j}$, prove they are constant if the distribution functions are exponential.

Answer. The conditional hazard functions are

$$\lambda_{|jj'}(x) = P\{\tau_k \leq x \mid J_{k-1} = j, J_k = j', \tau_k \geq x\} = \frac{f_{|jj'}(x)}{\bar{F}_{|jj'}(x)},$$

$$\lambda_{j'|j}(x) = P\{\tau_k \leq x, J_k = j' \leq x \mid J_{k-1} = j, \tau_k \geq x\} = \frac{f_{j'|j}(x)}{\bar{F}_j(x)},$$

where $\bar{F}_j = \sum_{j' \in E} \bar{F}_{j'|j}$.

With exponential distribution functions $F_{|jj'}(x) = 1 - e^{\mu_{|jj'} x}$ and $F_{j'|j}(x) = 1 - e^{\nu_{j'|j} x}$, we have $\lambda_{|jj'}(x) \equiv \mu_{|jj'}$ and

$$\lambda_{j'|j}(x) = \frac{\nu_{j'|j} e^{\nu_{j'|j} x}}{\sum_{j' \in E} e^{\nu_{j'|j} x}} = \nu_{j'|j} \frac{\bar{F}_{j'|j}(x)}{\bar{F}_j(x)}.$$

10.5 Stationary Gaussian processes

A process $(X_t)_{t \in T}$ is stationarity if for all s and t in T, $(X_u)_{u \in [s,t]}$ and $(X_u)_{u \in [s+h,t+h]}$ have the same distribution, it has a second order stationarity if its covariance function is stationary

$$E\{(X_t - EX_t)(X_s - EX_s)\} = \Gamma(s,t) = \gamma(t-s).$$

A stationary covariance function has a spectral representation as

$$\gamma(t) = \int_{\mathbb{R}} e^{it\lambda} \, d\mu(\lambda).$$

For every second order stationarity process $(X_t)_{t \in \mathbb{R}}$ with expectation zero, there exists a process Z with independent increments such that

$$X_t = \int_{\mathbb{R}} e^{it\lambda} \, dZ(\lambda),$$

Z is a complex isometry of $L^2(\mathbb{R}, \mu)$. If μ has a density with respect to the Lebesgue measure, this is the spectral density of X

$$f(\lambda) = \frac{1}{2\pi} \int_{\mathbb{R}} e^{-it\lambda} \, d\Gamma_t.$$

Exercise 10.5.1. Let X be centered second order stationary Gaussian process, prove that for every $x > 0$ linear combinations of X_t and X_{t-x} are stationary.

Answer. The covariance of the process $Y_t = aX_t + bX_{t-x}$ are

$$E(Y_sY_t) = (a^2+b^2)\Gamma_{t-s} + ab(\Gamma_{t-x-s} + \Gamma_{t-s+x})$$

and it depends only on $t-s$.

Exercise 10.5.2. Let X be a centered Gaussian process on $\mathbb{R}_+$ with a covariance function $R(s,t) = E(X_sX_t)$ in $L^2(\mathbb{R}_+^2)$ and let $(e_k(t))_{t\in\mathbb{R}_+,k\geq 0}$ be an orthonormal basis of $L^2(\mathbb{R}_+)$, prove that the process X and the function $R(s,t)$ have expansion on the basis $(e_k(t))_{t\in\mathbb{R}_+,k\geq 0}$ and write their quadratic norms.

Answer. The variable $Y_k = \int_{\mathbb{R}_+} X_t e_k(t)\,dt$ is a centered Gaussian variable with variance

$$\begin{aligned} E(Y_k^2) &= \int_{\mathbb{R}_+^2} E(X_sX_t)e_k(s)e_k(t)\,ds\,dt \\ &= \int_{\mathbb{R}_+^2} R(s,t)e_k(s)e_k(t)\,ds\,dt \end{aligned}$$

denoted λ_k, and $Z_k = \lambda_k^{-\frac{1}{2}} Y_k$ is a normal variable such that X has the expansion

$$X_t = \sum_{k\geq 0} Y_k e_k(t) = \sum_{k\geq 0} \lambda_k^{\frac{1}{2}} Z_k e_k(t)$$

on the orthonormal basis. It has the squared $L^2(\mathbb{R}_+)$ norm

$$\int_{\mathbb{R}_+} X_t^2\,dt = \sum_{k\geq 0} \lambda_k Z_k^2$$

where $\int_{\mathbb{R}_+} e_k^2(t)\,dt = 1$. The function $R(s,t)$ has the expansion

$$R(s,t) = \sum_{k\geq 0} \lambda_k e_k(s)e_k(t)$$

and it has the norm

$$\|R\|_{L^2}^2 = \int_{\mathbb{R}_+^2} R(s,t)\,ds\,dt = \sum_{k\geq 0} \lambda_k^2$$

where $\int_{\mathbb{R}_+} e_k(t)e_j(t)\,dt = 0$ for $k \neq j$.

10.6 Spatial processes

Let α be finite measure on a $(E, \mathcal{E})$, an exponential process X on E is defined for every set A of E by the exponential distribution function of $X(A)$, with parameter $\alpha(A)$

$$P\{X(A) \leq x\} = 1 - e^{-\alpha(A)x}, \ x \geq 0,$$

a Gamma process X on E is defined for every set A of E by a $\Gamma(\alpha(A), \lambda)$ distribution for $X(A)$

$$P\{X(A) \leq x\} = \frac{\lambda^{\alpha(A)}}{\Gamma(\alpha(A))} \int_0^x y^{\alpha-1} e^{-\lambda y}\, dy, \ x \geq 0,$$

a Dirichlet process X is defined on a partition $(B_1, \ldots, B_k)$ of $(E, \mathcal{E})$ by the Dirichlet distribution with parameters $\alpha(B_1), \ldots, \alpha(B_k)$ of the random variable $(X(B_1), \ldots, X(B_k))$ on the subset $S = \{(x_1, \ldots, x_k), \sum_{i=1}^k x_i = 1\}$ of $\mathbb{R}^k$. A Gaussian process on E is defined for every set A of E by measures α and β so that the distribution for $X(A)$ is Gaussian $\mathcal{N}(\alpha(A), \beta^2(A))$.

Exercise 10.6.1. Let X be an exponential process with measure α on E, determine the covariance of $X(A)$ and $X(B)$ and the Laplace transform $(X(A), X(B))$, for measurables sets A and B of E.

Answer. The variables $X(A)$ and $X(B)$ are independent if $A \cap B = \emptyset$, and their covariance is zero. If $A \cap B$ is not empty, the variables $X(A \setminus B)$ and $X(B)$ are independent and $A = (A \setminus B) \cup (A \cap B)$, the covariance of $X(A)$ and $X(B)$ is

$$\begin{aligned} \text{Cov}\{X(A), X(B)\} &= \text{Cov}\{X(A \cap B), X(B)\} = \text{Var}\{X(A \cap B)\} \\ &= \frac{1}{\alpha^2(A \cap B)}. \end{aligned}$$

The Laplace transform $(X(A), X(B))$ is the product of the transforms on nonintersecting sets

$$\begin{aligned} L(s,t) &= E\{e^{sX(A)+tX(B)}\} \\ &= E\{e^{sX(A\setminus B)}\}\, E\{e^{(s+t)X(A\cap B)}\}\, E\{e^{tX(B\setminus A)}\} \\ &= \frac{\alpha(A \setminus B)}{\{\alpha(A \setminus B) + s\}} \frac{\alpha(A \cap B)}{\{\alpha(A \cap B) + s + t\}} \frac{\alpha(B \setminus A)}{\{\alpha(B \setminus A) + t\}}. \end{aligned}$$

Exercise 10.6.2. Let X be a Gaussian process with measures (α, β) on $E \times E$, write the density and the Fourier transform of $(X(A_1), \ldots, X(A_k))$ for a vector of k distinct sets $(A_1, \ldots, A_k)$ of $E^{\times k}$.

Answer. The Gaussian variable $X_A = (X(A_1), \dots, X(A_k))$ has the expectation $(\alpha(A_1), \dots, \alpha(A_k))$ and the covariance matrix Σ with diagonal $(\beta(A_i))_{i=1,\dots,k})$ and with covariance

$$\mathrm{Cov}\{X(A_i), X(A_j)\} = \mathrm{Var}\{X(A_i \cap A_j)\} = \beta(A_i \cap A_j),$$

for $i \neq j$ in $\{i = 1, \dots, k\}$. With distinct sets $(A_1, \dots, A_k)$, the matrix Σ is not singular and the density of $(X(A_1), \dots, X(A_k))$ at $x = (x_1, \dots, x_k)$ is

$$f(x) = \frac{1}{|\Sigma|^{\frac{1}{2}} (2\pi)^{\frac{k}{2}}} e^{-\frac{x^T \Sigma^{-1} x}{2}}$$

and for $t = (t_1, \dots, t_k)$ its Fourier transform is

$$\widehat{\phi}(t) = E(e^{it^T X_A}) = e^{-\frac{t^T \Sigma t}{2}}.$$

Exercise 10.6.3. Let X be a Gamma process $\Gamma(\alpha, \lambda)$ with measure a α on E, determine the Laplace transform $(X(A), X(B), X(C))$, for measurables sets A, B and C of E.

Answer. The Laplace transform of $(X(A), X(B))$ is

$$\begin{aligned} L(s,t) &= E\{e^{sX(A)+tX(B)}\} \\ &= E\{e^{sX(A\setminus B)}\} E\{e^{(s+t)X(A\cap B)}\} E\{e^{tX(B\setminus A)}\} \\ &= \frac{\lambda^{\alpha(A\setminus B)}}{(\lambda+s)^{\alpha(A\setminus B)}} \frac{\lambda^{\alpha(A\cap B)}}{(\lambda+s+t)^{\alpha(A\cap B)}} \frac{\lambda^{\alpha(B\setminus A)}}{(\lambda+t)^{\alpha(B\setminus A)}} \end{aligned}$$

and $(X(A), X(B), X(C))$ has the product Laplace transform

$$\begin{aligned} L(s,t,u) &= E\{e^{sX(A)+tX(B)+uX(C)}\} \\ &= E\{e^{sX(A\setminus(B\cup C))}\} E\{e^{tX(B\setminus(A\cup C))}\} E\{e^{uX(C\setminus(A\cup B))}\} \\ &\quad E\{e^{(s+t)X(A\cap B\setminus C)}\} E\{e^{(s+u)X(A\cap C\setminus B)}\} \\ &\quad E\{e^{(u+t)X(B\cap C\setminus A)}\} E\{e^{(s+t+u)X(A\cap B\cap C)}\}, \end{aligned}$$

and each term is determined as the Laplace transform of a Gamma variable with parameter the measure of the corresponding set

$$\begin{aligned} L(s,t,u) = {} & \frac{\lambda^{\alpha(A\setminus(B\cup C))}}{(\lambda+s)^{\alpha(A\setminus(B\cup C))}} \frac{\lambda^{\alpha(B\setminus(A\cup C))}}{(\lambda+t)^{\alpha(B\setminus(A\cup C))}} \frac{\lambda^{\alpha(C\setminus(A\cup B))}}{(\lambda+u)^{\alpha(C\setminus(A\cup B))}} \\ & \frac{\lambda^{\alpha(A\cap(B\setminus C))}}{(\lambda+s+t)^{\alpha(A\cap B\setminus C)}} \frac{\lambda^{\alpha(A\cap(C\setminus B))}}{(\lambda+s+u)^{\alpha(A\cap C\setminus B)}} \\ & \frac{\lambda^{\alpha(B\cap C\setminus A)}}{(\lambda+s+u)^{\alpha(B\cap C\setminus A)}} \frac{\lambda^{\alpha(A\cap B\cap C)}}{(\lambda+s+t+u)^{\alpha(A\cap B\cap C)}}. \end{aligned}$$

Exercise 10.6.4. For a Dirichlet process X on $(E, \mathcal{E})$, calculate the expectation and the covariance of $X(A)$ and $X(B)$, for measurables sets of E.

Answer. If B in E, the variable $X(B)$ has a Beta distribution on $[0,1]$, with parameter $(\alpha(B), \alpha(B^c))$, its expectation and its variance are

$$E\{X(B)\} = \frac{\alpha(B)}{\alpha(E)},$$
$$\mathrm{Var}\{X(B)\} = \frac{\alpha(B)\{\alpha(E) - \alpha(B)\}}{\alpha^2(E)\{\alpha(E)+1\}}.$$

Let $A \cap B = \emptyset$ in E, there exists C such that (A, B, C) is a partition of E and the variable $(X(A), X(B), X(C))$ has a Dirichlet distribution with parameter $(\alpha(A), \alpha(B), \alpha(C))$, the expectation of $X(A)$ is $E\{X(A)\} = \alpha^{-1}(E)\alpha(A)$ and the independence of the variables $X(A)$ and $X(B)$ entails that the covariance of $X(A)$ and $X(B)$ is zero.

If $A \cap B$ is not empty, A and B are the unions of the non intersecting sets $A = (A \cap B) \cup (A \setminus B)$ and $B = (A \cap B) \cup (B \setminus A)$, the expectations of $X(A)$ and $X(B)$ are the sums

$$E\{X(A)\} = \frac{\alpha(A \cap B) + \alpha(A \setminus B)}{\alpha(E)} = \frac{\alpha(A)}{\alpha(E)},$$
$$E\{X(B)\} = \frac{\alpha(B)}{\alpha(E)},$$

the variables $X(A)$ and $X(B)$ are dependent and their covariance reduces to the variance of $X(A \cap B)$

$$\mathrm{Cov}\{X(A), X(B)\} = \frac{\alpha(A \cap B)\{\alpha(E) - \alpha(A \cap B)\}}{\alpha^2(E)\{\alpha(E)+1\}}.$$

Exercise 10.6.5. Write the distribution of the variable Y defined on $(\mathbb{R}^{k-1}, \mathcal{R}^k)$ as $Y(B) = E\{X(B)\}$, for a Dirichlet process X.

Answer. Every $\mathcal{R}^k$-measurable set $B = B_1 \otimes \cdots \otimes B_k$ defines the density of a Dirichlet variable

$$f_{\alpha(B)}(x) = \frac{\Gamma_{\alpha(E)}}{\prod_{i=1}^k \Gamma_{\alpha(B_i)}} \prod_{i=1}^{k-1} x_i^{\alpha(B_i)-1} \Big(1 - \sum_{j=1}^{k-1} x_j\Big)^{\alpha(B_k)-1},$$

such that $\sum_{i=1}^k x_i = 1$ and the variable $Y(B)$ has the probability measure

$$P(B) = \int_{\mathbb{R}_+^{k-1}} x_1 \cdots x_{k-1}\Big(1 - \sum_{j=1}^{k-1} x_j\Big) f_{\alpha(B)}(x)\, dx_1 \cdots dx_{k-1}$$

$$= \frac{\Gamma_{\alpha(E)}}{\prod_{i=1}^k \Gamma_{\alpha(B_i)}} \int_{\mathbb{R}_+^{k-1}} \prod_{i=1}^{k-1} x_i^{\alpha(B_i)} \Big(1 - \sum_{j=1}^{k-1} x_j\Big)^{\alpha(B_k)} dx_1 \cdots dx_{k-1}$$

$$= \frac{\prod_{i=1}^k \alpha(B_i)}{\alpha(E)}.$$

Exercise 10.6.6. For a spatial Brownian motion B with a probability measure α on a metric space $(E, \mathcal{E}, d)$, calculate $E[\{B(A_1)-B(A_2)\}^2]$ and prove that $B^2(A) - \alpha(A)$ is a martingale.

Answer. The covariance of $B(A_1)$ and $B(A_2)$ is $\alpha(A_1 \cap A_2)$ and

$$E[\{B(A_1) - B(A_2)\}^2] = \alpha(A_1) - 2\alpha(A_1 \cap A_2) + \alpha(A_2),$$

and for $A_2 \subset A_1$

$$E[\{B(A_1) - B(A_2)\}^2] = \alpha(A_1) - \alpha(A_2) = \alpha(A_1 \setminus A_2).$$

The martingale property of the Brownian motion is expressed for $A_2 \subset A_1$ as

$$\begin{aligned} E\{B(A_1) - B(A_2) \mid B(A_2)\} &= E\{B(A_1 \setminus A_2) \mid B(A_2)\} \\ &= E\{B(A_1 \setminus A_2)\} = 0, \end{aligned}$$

it implies

$$\begin{aligned} E[B^2(A_1) - B^2(A_2) \mid B(A_2)] &= E[\{B(A_1) - B(A_2)\}^2 \\ &\quad +2\{B(A_1) - B(A_2)\}B(A_2) \mid B(A_2)] \\ &= E[\{B(A_1) - B(A_2)\}^2] \end{aligned}$$

and $E[B^2(A_1) - B^2(A_2) \mid B(A_2)] = E[\{B(A_1 \setminus A_2)\}^2] = \alpha(A_1 \setminus A_2)$.

Exercise 10.6.7. For a spatial Brownian motion B with a probability measure α on $(E, \mathcal{E}, d)$, and for $x > 0$ let $A_x = \inf\{A : B(A) \geq x\}$, prove that A_x is a stopping set with respect to the filtration generated by B on $(E, \mathcal{E})$ and calculate the Laplace transform of $\alpha(A_x)$.

Answer. For a set C of $\mathcal{E}$, let $\bar{C}$ be the complementary of C. The event $\{A_x \subset \bar{C}\}$ is equivalent to $\{B(C) < x\}$ and it is measurable with respect to $\mathcal{F}_C$, A_x is therefore a stopping set. The variable B_C has a centered Gaussian distribution with variance $\alpha(C)$, its Laplace transform is

$$E\{e^{-\theta B_C}\} = e^{\frac{\theta^2 \alpha(C)}{2}}.$$

At A_x, we have $B_{A_x} = x$ and

$$e^{-\theta \alpha(A_x)} = E\{e^{x\sqrt{2\theta}}\}.$$

Exercise 10.6.8. Let x and y be strictly positive, prove that the stopping sets A_x and $A_{x+y} \setminus A_x$ are independent and $A_{x+y} \setminus A_x$ has the same distribution as A_y.

Answer. The set $A_{x+y} = \inf\{A : B(A) \geq x+y\}$ is a stopping set such that

$$\begin{aligned}
P(A_{x+y} \setminus A_x \subset \bar{C} \mid \mathcal{F}_{A_x}) &= P(B_{C \cup A_x} < x + y \mid \mathcal{F}_{A_x}) \\
&= P(B_{C \cup A_x} - B_{A_x} < y \mid \mathcal{F}_{A_x}) \\
&= P(B_C < y) = P(A_y \subset \bar{C})
\end{aligned}$$

and $B_{C \cup A_x} - B_{A_x}$ is independent of $B(A_x)$ for every C, therefore $B_{A_{x+y} \setminus A_x}$ is independent of $B(A_x)$.

Bibliography

Billingsley, P. (1968). *Convergence of probability measures* (Wiley, New York).

Doléans-Dade, C. (1970). Quelques applications de la formule de changement de variable pour les semi-martingales, *Z. Wahrsch. verw. Geb.* **16**, pp. 181–194.

Feller, W. (1971). *An Introduction to Probability Theory and its Applications, Vol. 2 (second ed.)* (Wiley, London).

Hajek, J. and Sidak, Z. (1967). *Theory of rank tests* (Academic Press, London).

Lenglart, E. (1977). Relation de domination entre deux processus, *Ann. Inst. H. Poincaré* **13**, pp. 171–179.

Meyer, P. (1976). Un cours sur les intégrales stochastiques. Sém. probab. Strasbourg, *Lecture Notes, Springer–Verlag* **511**.

Neveu, J. (1970). *Bases mathématiques du calcul des probabilités* (Masson, Paris).

Neveu, J. (1972). *Martingales à temps discret* (Masson, Paris).

Shorack, G. and Wellner, J. A. (1986). *Empirical Processes* (Wiley, New York).

Stone, M. (1974). Large deviations for empirical probability measures, *Ann. Statist.* **2**, pp. 362–366.

Index